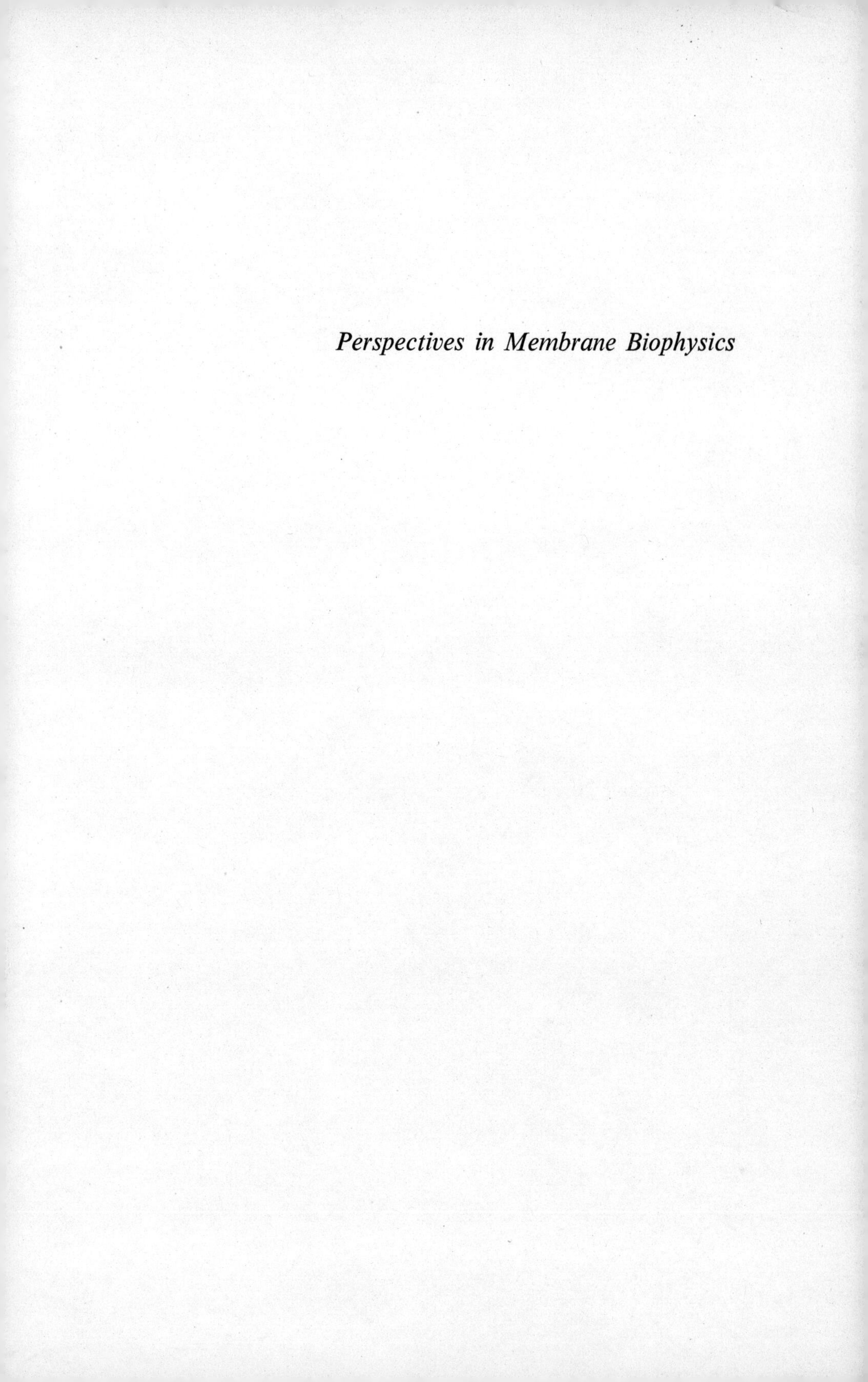

Perspectives in Membrane Biophysics

Perspectives in Membrane Biophysics

A Tribute to Kenneth S. Cole

Edited by

D. P. AGIN

Department of Physiology
University of Chicago

GORDON AND BREACH SCIENCE PUBLISHERS

New York London Paris

Gordon and Breach, Science Publishers, Inc.
440 Park Avenue South
New York, N.Y. 10016

Editorial office for the United Kingdom
Gordon and Breach, Science Publishers Ltd.
42 William IV Street
London W.C.2

Editorial office for France
Gordon & Breach
7–9 rue Emile Dubois
Paris 14e

Library of Congress catalog card number 78–178300. ISBN 0 677 15210 8. Printed in east Germany.

Dedication

Although everyone seems to be a biophysicist these days, only a small number of men have been responsible for the important increments in our knowledge. One of these men is Kenneth S. Cole. The future will know him as an archetype biophysicist of the twentieth century. Having already profoundly influenced two generations of scientists, he is now working on a third (in the eighth decade of his life).

It is a great pleasure for me to have been able to bring together what follows in tribute to Kacy Cole. This book is dedicated to him and to the young scientists of the future, the men and women who will inherit our problems and whose vision will be clearer than ours.

D. AGIN

Department of Physiology
University of Chicago

Mathematicians may flatter themselves that they possess new ideas which mere human language is as yet unable to express. Let them make the effort to express these ideas in appropriate words without the aid of symbols, and if they succeed they will not only lay us laymen under a lasting obligation, but, we venture to say, they will find themselves very much enlightened during the process, and will even be doubtful whether the ideas as expressed in symbols had ever quite found their way out of the equations into their minds.

James Clerk Maxwell

It is impossible to express a really new principle in terms of a model following old laws.

Max Planck

Contents

The Right Path

ROBERT E. TAYLOR and EDUARDO ROJAS

Laboratory of Biophysics, National Institute of Neurological Diseases and Stroke and Laboratorio de Fisiologia Celular, Facultad de Cienias, Universidad de Chile

---

"I was like a mountaineer who, not knowing his path, must climb slowly and laboriously, is forced to turn back frequently because his way is blocked but discovers, sometimes by deliberation and often by accident, new passages which lead him on for a distance. Finally, when he reaches his goal, he finds to his embarrassment a royal road which would have permitted him easy access by vehicle if he had been clever enough to find the proper start."[1]

For the understanding of excitable membranes an exceedingly important proper start was the demonstration by Curtis and Cole[2] that the impedance change during activity in the plant cell Nitella was due to an increase in the conductance of the membrane with little or no change in the membrane capacitance. A proper turning resulted from the description of the giant axon of the squid by John Z. Young at the Cold Spring Harbor Symposium in 1936.[3] Attempts to repeat the work on Nitella with whole nerve led to difficulties in interpretation and Curtis and Cole later remarked, "It is therefore necessary to further simplify matters by making transverse measurements on a single axon, but this was not considered possible until we were introduced to the squid giant axon by Dr. John Z. Young (1936). We are also very much indebted to him for his assistance in preliminary experiments which were made during the summer of 1936 at the Biological Laboratory, Cold Spring Harbor, Long Island."[4]

We shall leave to others the account of the slow and laborious climb to the Hodgkin-Huxley equations and consider a recent statement of Cole, viz., "The early inflow of sodium and the later outflow of potassium is the simple, fundamental explanation of a propagating impulse. These are calculated in detail by Hodgkin and Huxley, so an independent measure of these net flows during the passage of an impulse is certainly an important, perhaps the most important, ultimate test of the entire HH structure of analysis and synthesis."[5]

The proper path for such independent measures was opened by Richard Keynes. "The use of artificial radioactive isotopes of the principle ions involved in nervous activity provides a direct method of checking some of the findings of existing techniques, and of making certain other measurements which could not readily be made by any other means. The method is highly sensitive, so that it is well suited for studies on the isolated invertebrate nerve fibres with which much of the recent work on the mechanism of nerve conduction has been done."[6]

Measurements of the integrated ion fluxes per action potential gave satisfactory agreement with predictions.[7]

The direct measurement of the time course of the influx of sodium ions during the course of a rectangular voltage clamp pulse and during a non-propagated action potential is the subject of this communication, along with the direct measurement of the conductance changes to potassium and sodium during the action potential. Similar measurements for the time course of potassium movements during activity in nerve were not done because of the difficulties of procuring radioactive potassium.

SODIUM INFLOW IN AXONS UNDER VOLTAGE CLAMP[8]

The experimental results to be described here were done using techniques whereby a single giant axon of the squid (*Dosidicus gigas* and *Loligo forbesii*) could be internally perfused and voltage clamped. External application of ^{22}Na ions and collection of the perfusate allowed determinations of the sodium inflow. From measurements of the membrane current during, and after, the application of voltage clamp pulses the inward sodium flux as predicted by the equations of Hodgkin and Huxley could be calculated and compared to the measured flux.

Figure 1 shows the results of measurements of the total inward tracer sodium flux (open circles) resulting from voltage clamped depolarization pulses of about 3 msec. duration as a function of the absolute membrane potential during the pulse. (Potential is referred to the external solution as ground. Flux is plotted upward but inwardly directed currents would be negative quantities.)

The filled circles are the predicted fluxes as calculated from the membrane current measurements. The identification of the total flux with sodium is clear.

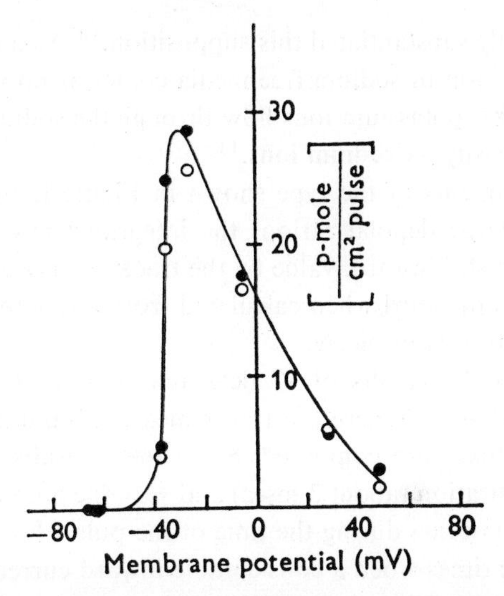

Figure 1 Comparison between the measured tracer sodium flux (0) and electrically measured ionic flux during the early transient current following a depolarizing voltage clamp pulse of about 3 msec duration. Axons from *Dosidicus gigas* internally perfused with 550 mM KF, pH 7.3 immersed in K free artificial sea water. Temp 17°C. Control experiments with tetrodotoxin were used (using the same axon) to subtract current components due to potassium. Used by permission from Atwater, I., F. Bezanilla and E. Rojas. *J. Physiol.* **201**, 657–664 (1969)

A region not shown in Figure 1 deserves a comment. For depolarizations larger than to plus some 60 odd mV the measured current is outward. One might think that this would give a calculated net negative inward sodium flux, which is unlikely, to say the least. These experiments were done with axons immersed in potassium free artificial sea water and internally perfused with sodium free potassium flouride. Many other experiments have shown quite clearly that the outward current for large depolarizations under these conditions is due to the flow of potassium ions with the same kinetics as sodium. We refer to the system which allows sodium ions to pass through the membrane in response to an electrochemical gradient as a sodium channel, without any prejudice as to the physical mechanism. This system has the property that some other ions may also pass through when the "gate" is opened in response to the appropriate potential change. Most notable of these ions are lithium, hydrazine, guanidinum and ammonium ions. That lithium may substitute for sodium is ancient history (Overton). In response to a question by Lorente de No, Hodgkin commented, "I have followed Dr. Lorente des Nó's important work on quaternary ammonium ions with great interest, but I am not clear why one should not suppose that these substances act directly as substitutes for sodium."[9] Much and later work has fully substantiated this supposition.[10] As a corrollary, action potential propagation in sodium free media comes as no surprise at all. To return to the point, potassium ions flow through the sodium channels about 1/12 to 1/25 as easily as sodium ions.[11]

In other experiments of the type shown in Figure 1, the expected result occurs, i.e., for large depolarizarions the integrated inward flux per cycle falls to a low, but still inward, value by the tracer measurement but reverses sign and becomes outward when calculated from measured membrane currents. All of the results fit nicely.

Figure 2 shows the results of 17 determinations of the type shown in Figure 1 on 6 axons;[12] Where the tracer measured and current calculated inward sodium fluxes are compared. Since these results are for pulses of relatively short duration (about 3 msec) and variable amplitude, most of the current (and flux) occurs during the time of the pulse. For longer pulses (or shorter) there are times when a considerable inward current flows following the pulse and this "tail" must be considered.

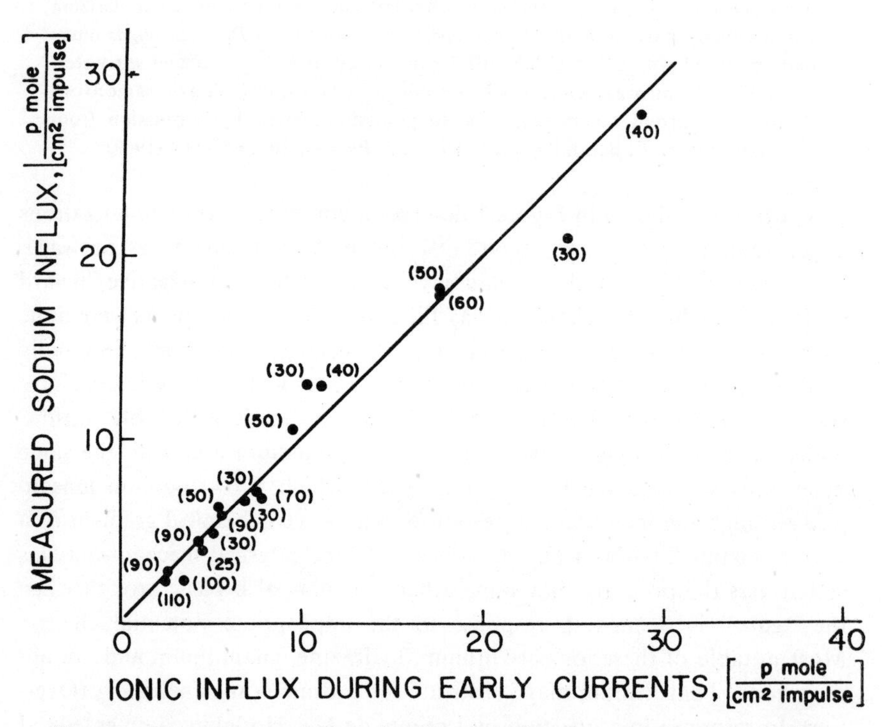

Figure 2 Measured extra influx of sodium as a function of the extra ionic flux as calculated from the measured membrane current. Pulse durations about 3 msec; pulse heights indicated. (Plotted from data of Table 1 of Atwater, I., F. Bezanilla and E. Rojas. *J. Physiol.* **201**, 657–664 (1969)

TIMING THE FLUX RECTANGULAR VOLTAGE CLAMP PULSES

The results of Figure 3 are for the total flux per cycle. In order to compare the flux which occurs *during* the pulse, for different pulse durations it is necessary to subtract that component which occurs during the tail following the pulse. Control experiments demonstrate that this current is indeed sodium and for these experiments TEA was added to the internal perfusion fluid with the result that postassium channel currents were virtually eliminated. The filled circles of Figure 4 are a plot of the measured inward sodium flux as measured with the radioactive ^{22}Na ions *minus* the tail component as computed from the measured membrane current. The dotted line is the inward flux during the long pulse shown in part A of Figure 4 as calculated from the current. The results are unequivocal and we reject any contention that the argument is circular.

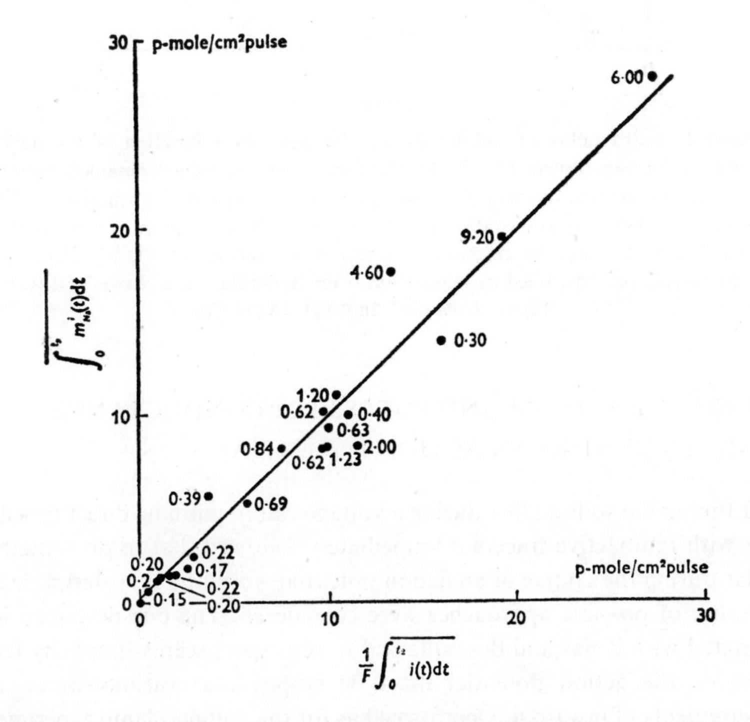

Figure 3 Measured extra influx of sodium as a function of the net extra flux obtained from the membrane current density for pulses of varying duration with the same potential during the pulse (zero absolute membrane potential). Durations are given next to corresponding points. Used by permission from Bezanilla, F., E. Rojas and R. E. Taylor. *J. Physiol.* **207**, 151–164 (1970)

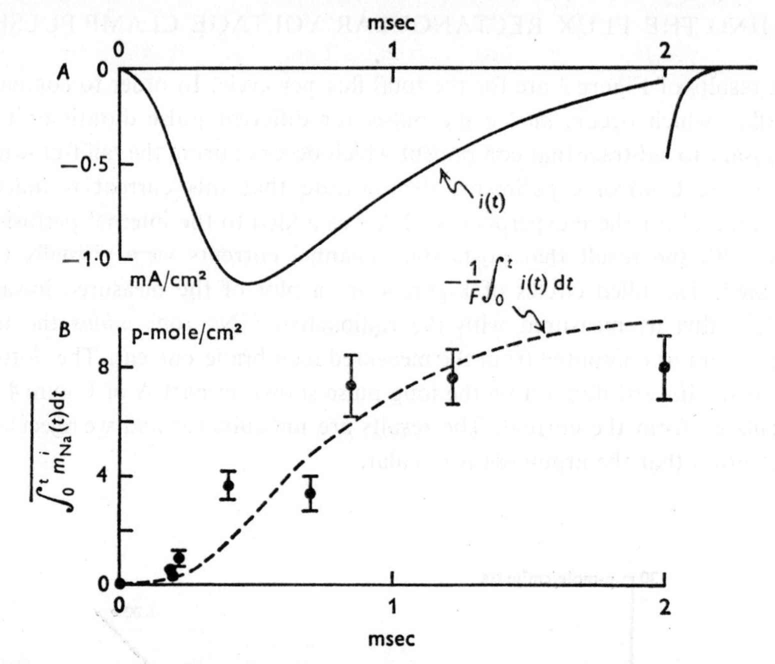

Figure 4 Extra influx of sodium during the pulse as a function of the pulse duration for experiment TN-55. A: Membrane current record obtained during the last run in experiment TN-55 corrected for the capacitative transient. This curve was integrated to give the dashed curve in B. B: Each point represents the extra influx measured by tracers minus the influx during the "tail" calculated from current records. Used by permission from Bezanilla, F., E. Rojas and R. E. Taylor *J. Physiol.* **207**, 151–164 (1970)

TIMING THE SODIUM AND POTASSIUM CONDUCTANCE CHANGES DURING AN ACTION POTENTIAL

After timing the sodium flux during a voltage clamp pulse by direct measurement with radioactive tracers it immediately followed that to do something similar during the course of an action potential would be very desirable and a number of possible approaches were considered. The one described here originated with Rojas and Bezanilla and in retrospect seems simplicity itself. Somehow the action potential must be stopped, at various times, and measurements of inward flux performed as for the voltage clamp experiments described above. The essence of the electrical part of the system is shown in Figure 5. A squid axon is prepared for simultaneous internal perfusion and voltage clamping. With the single pole-double throw switch in the open loop position the total membrane current is zero. Applying a pulse of current through the resistance connected to the internal axial wire results in the production of a non-propagated action potential.

Under these space clamped conditions the action potential is referred to as a "membrane" action potential. Uniformity of the behaviour of the membrane throughout the measuring region was checked by using external differential electrodes to measure localized membrane current densities.[13] With the switch shown in Figure 5 in the voltage clamp position, the membrane potential is under control. The problem was to switch from zero current to voltage clamp mode quickly at various times during the membrane action potential. We first tried mercury relays which worked, but there were problems associated with open time and pick-up from the relay coils. Professor Clay Armstrong was working in the adjacent laboratory in Montemar at the time and he not only suggested that we should use high speed switching transistors for this purpose but was very helpful in the process of getting something to work. The only such transistors available to us were removed from spare parts for a liquid scintillation counter and with these we succeeded in interrupting the free course of an action potential at any time we chose and switch to the voltage clamp mode within a few microseconds.

Thus, the potential at which the membrane will be clamped during the control period is preset, an action potential is initiated and then stopped suddenly at the desired moment.

Measuring the membrane current at a very short time (after the voltage clamp has charged the membrane capacitance) for various potentials during the control period yields a current voltage curve for the membrane at that time. It was found that these current voltage curves were approximately linear so that the total membrane conductance is a meaningful, and simple, quantity which we could measure as a function of time during the membrane action potential. The curve of membrane conductance vs time during the action potential was found to have a time course similar to that found by Cole and Curtis using impedance measurements at 20 KHz.[14]

If the membrane potential during the control period is close to the reversal potential for potassium channels (see above) then no net current should be flowing through these channels and the initial current just following the interruption would be proportional to the conductance of the sodium channels. These transient currents following interruption are shown just above the action potential in Figure 6A. These are unretouched superimposed photographs and the envelope of these transients is an approximate, but direct, measure of the conductance of the sodium channels as a function of time during the action potential. Convincing evidence, using radioactive tracers, that the current during these transients is indeed sodium is presented

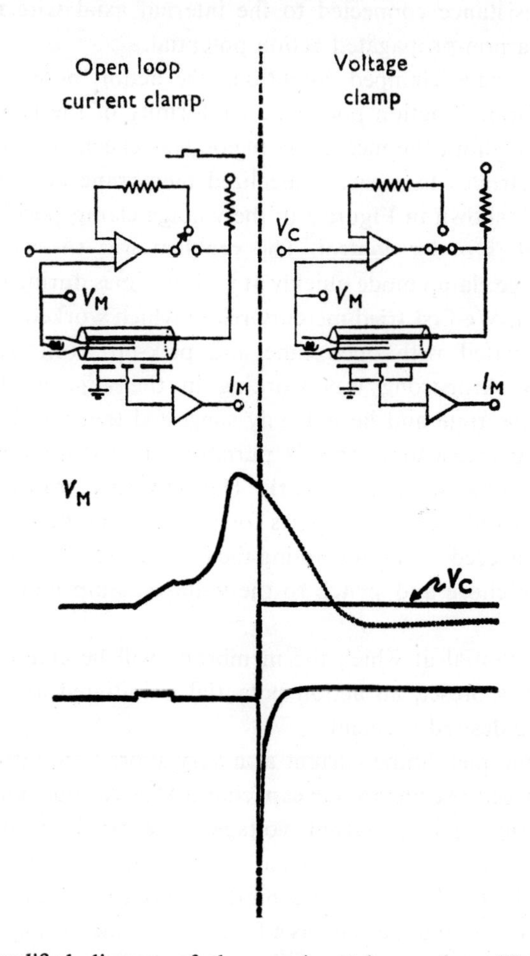

Figure 5 Simplified diagram of the experimental procedure. Upper part: a simplified diagram of the system used to switch on the membrane potential control system. Lower part: the recorded membrane potential and membrane current. On the left side of the Figure the membrane is in open loop current clamp condition because the switch at the output of the control amplifier is connected to the auxiliary feed-back loop. The action potential is excited by a short pulse of voltage through a very high resistor connected to the axial wire. On the right side of the Figure the switch connects the output of the control amplifier to the axial wire and the membrane is under voltage clamp control. The free course of the action potential is interrupted and the recorded potential is equal to the command potential. Simultaneously membrane currents are recorded as shown in the bottom of the Figure. (Used by permission from F. Bezanilla, E. Rojas and R. E. Taylor. *J. Physiol* **211**, 729–751 (1970)

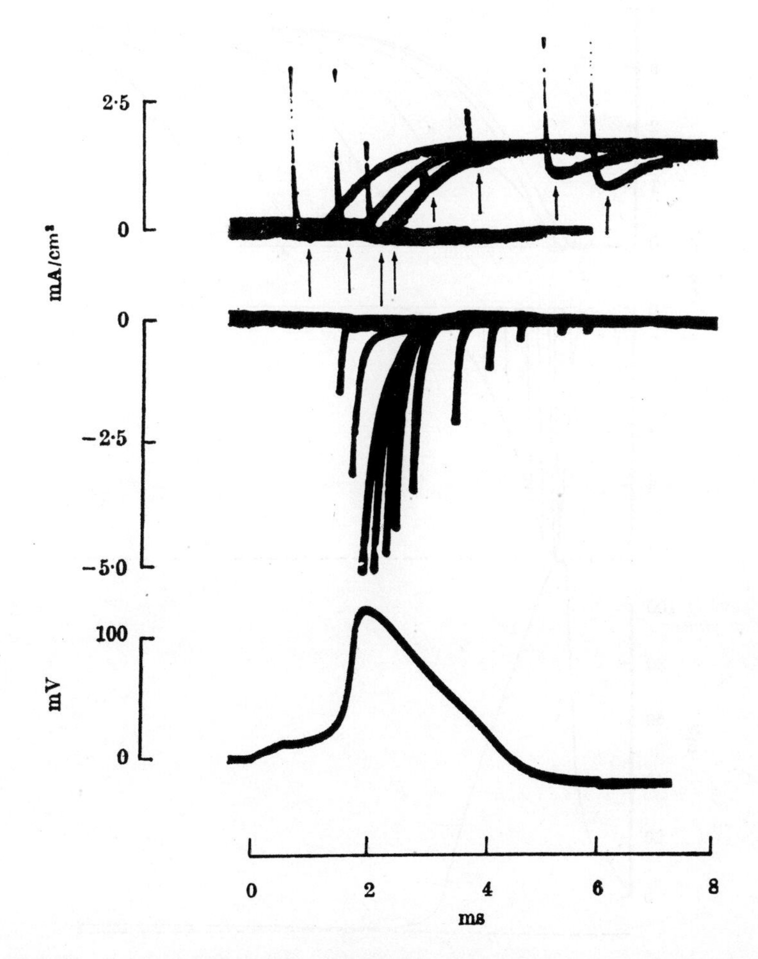

Figure 6 Current transients following interruption of action potential by switching to voltage clamp mode.

A Upper traces represent unretouched records of ionic currents when the membrane potential during the control period was 60 mV. Lower traces represent records of the ionic currents when the membrane potential during the control period was −80 MV. The capacitative current transients in the upper traces are apparent because a Z-input intensifier was used with the oscilloscope. No series resistance compensation was used. The resting potential was −60 mV. The temperature of the external seawater was 8°C. (Used by permission from Rojas, E., F. Bezanilla and R. E. Taylor, *Nature*, **225**, 747–748 (1970)

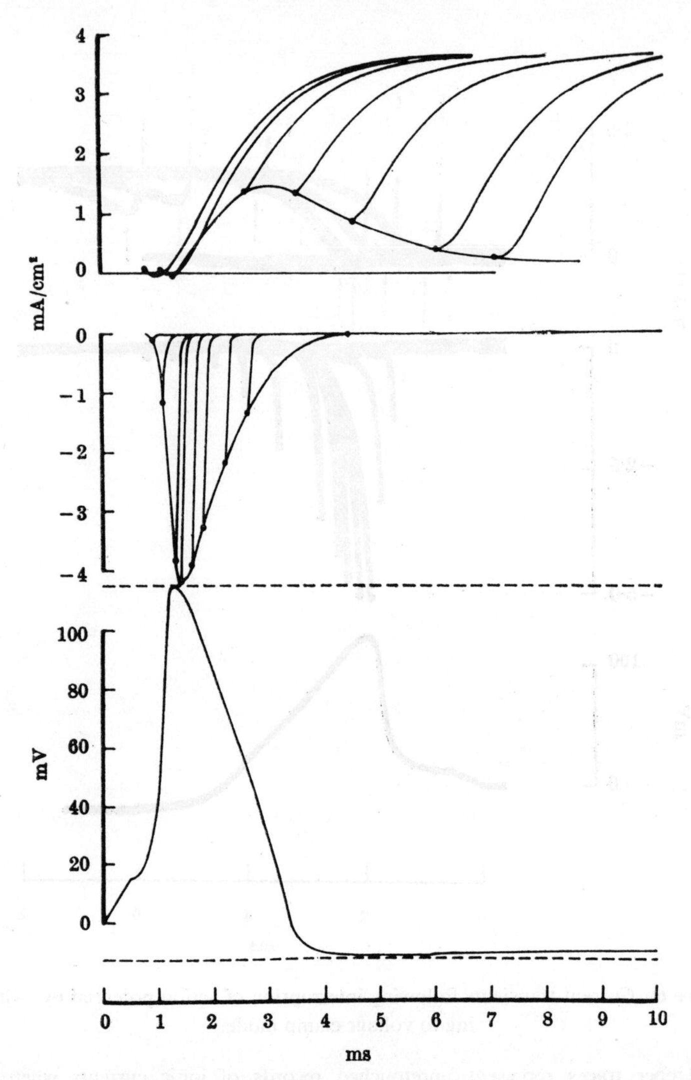

Figure 6 Current transients following interruption of action potential by switching to voltage clamp mode

B Ionic currents and membrane action potential calculated utilizing the Hodgkin and Huxley equations. The curve drawn through the points on the upper curves (initial potassium currents is proportional to the potassium conductance change. The curve drawn through the points on the center curves (initial sodium currents) is proportional to the sodium conductance change.) Resting potential equal to −60 mV. The temperature was 6.3°C. (Used by permission from Rojas, E., F. Bezanilla and R. E. Taylor, *Nature* **225**, 747–748 (1970) in the next section. The results in Figure 6A are approximate because there is no compensation for the known resistance in series with the membrane (due mostly to the Schwann cell layer) and there are no corrections for the small leakage current component. Even without these corrections the result is convincingly similar to the predictions of the Hodgkin-Huxley equations shown in Figure 6B. The variation in the time constants for the inward transients is due to the series resistance. This variation is reduced by the use of compensated feedback and is predicted by the Hodgkin-Huxley equations with added series resistance. Thus even the artifact introduced by the series resistance is instructive and confirmatory.

The upper set of curves in Figure 6A are the currents following interruption of the action potential to a membrane potential close to the reversal potential for the sodium channels. Similar considerations apply. If the instantaneous current voltage curves are reasonably linear then, with corrections for series resistance and leakage, the points indicated by the arrows give the time course of the potassium channel conductance during the action potential. Compare with the predicted curves in Figure 6B.

Figure 7 shows the uncorrected results for an axon of *Dosidicus gigas* (the Chilean giant squid) above, and below for an axon of *Loligo-forbesii* obtained in Plymouth, England. In this part, we wanted to establish the method for the tracer flux measurements and to see if the general shape of

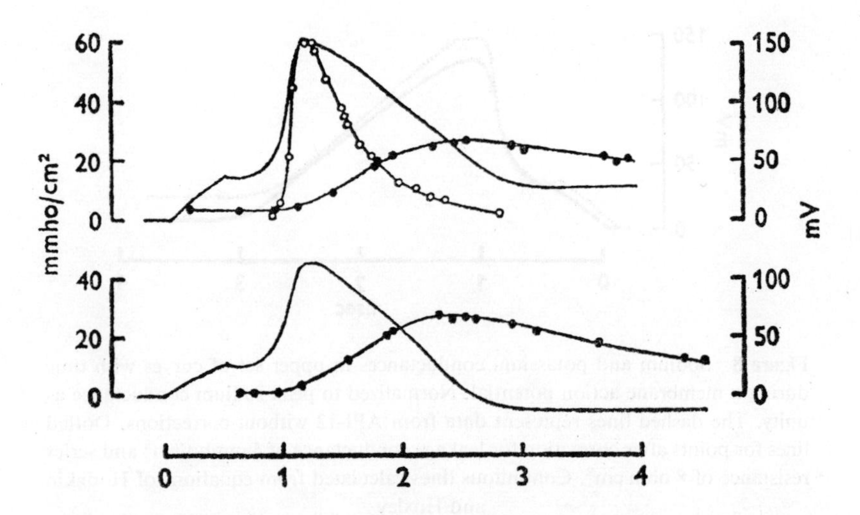

Figure 7 The temporal course of G^*_{Na} (open circles) and G^*_K (filled circles calculated with the uncorrected data from experiment API-12. Temperature 9.5°C. Resting potential −78 mV. Lower part shows G^*_K obtained on a *Loligo* axon. Resting potential −64 mV. Notice the underswing in the action potential

the curves of conductance vs time during the action potential conformed to the general ideas of the Hodgkin-Huxley-Katz proposals. Further quantitative comparisons with the predictions of the HH equations could be attempted but we feel that the main point has been demonstrated. We do not see any circular arguments here. Figure 8 shows the best quantitative comparison which we have, with corrections for leakage currents and series resistance and for the fact that the potentials during the control period following interruption were not exactly the reversal potentials for sodium and potassium channels. The various curves are identified in the legend.

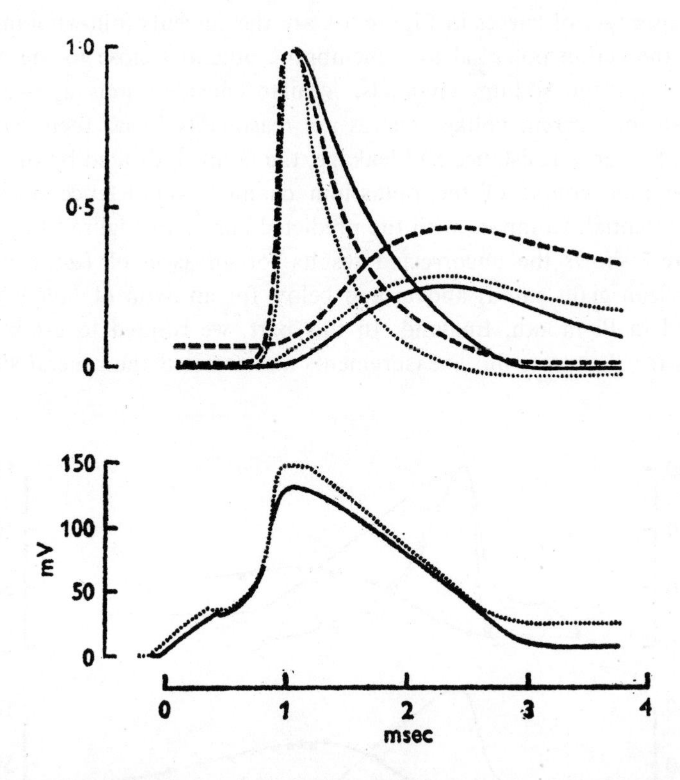

Figure 8 Sodium and potassium conductances in upper set of curves with time during a membrane action potential. Normalized to peak sodium conductance as unity. The dashed lines represent data from API-12 without corrections. Dotted lines for points after correction for leakage conductance of 5 mmho/cm^2 and series resistance of 8 ohm cm^2. Continuous lines calculated from equations of Hodgkin and Huxley

Lower curves show the measured membrane potential (dotted) compared with the calculated (continuous curve.)

Used by permission from Bezanilla, F., E. Rojas and R. E. Taylor. *Jour. Physiol.* **211**, 753–765 (1970)

TIMING THE SODIUM FLUX DURING AN ACTION POTENTIAL

All of the methods mentioned above were combined by Atwater, Bezanilla and Rojas to perform the GRAND EXPERIMENT: to get an independent measure of the net flux during the course of a "membrane" action potential.

Stimulating the perfused axon ten times a second and collecting the perfusate about every 200 seconds the action potential was interrupted at various times. The membrane current during the period of voltage control after the interruption was recorded and the component of the flux during the "tail" subtracted from the net measured flux.

For gruesome details the original papers must be consulted. In Figure 9, the results of several experiments are presented, showing the extra sodium influx as a function of the time at which the action potential was interrupted compared to the predictions as calculated from the equations of Hodgkin and Huxley. To get the time course of the sodium entry during the action potential, one may differentiate the curve in Figure 9 if he cares to.

L'envoi

There are no longer any experimental or logical reasons to doubt that under ordinary conditions the action potential of the squid axon membrane

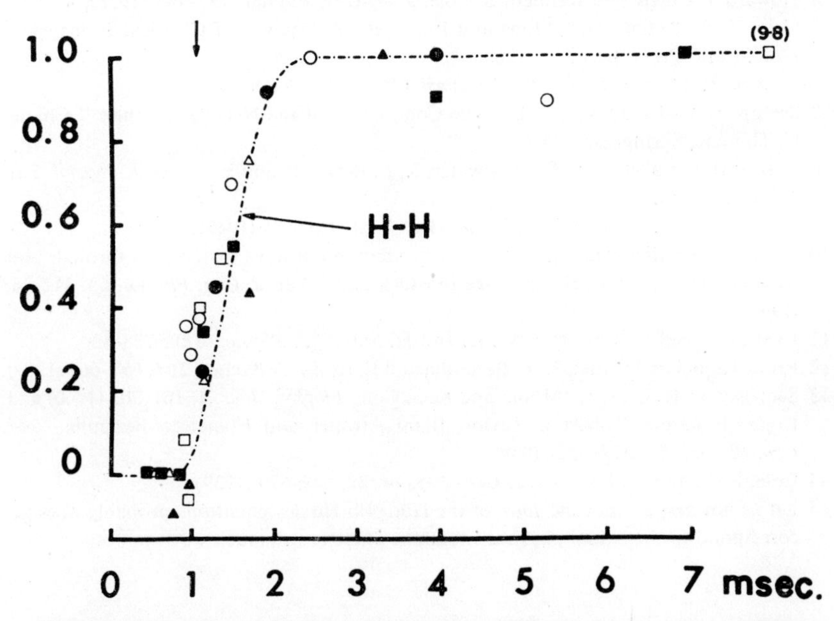

Figure 9 Normalized integrated sodium influx during the course of a membrane action potential determined using radioactive sodium and action potential interruption. The curve is not drawn through the points but was calculated using the HH equations

does indeed arise as suggested by Hodgkin, Huxley and Katz in 1949 and that the membrane conductance changes to sodium and potassium ions and the inward flow of sodium ions occur with a time course closely as calculated from the equation of Hodgkin and Huxley.[15]

We are thus confident that, although we may not have found the best or most direct path to the top, we are at least climbing the right mountain. Kenneth S. Cole was clever enough to find many of the proper starts and turnings, including the early impedance measurements with Curtis, Baker and Hodgkin, the instigation of voltage clamping the iron wire model by Bartlett, and the space, current and voltage clamping with Marmont. These were some of the most significant steps toward that most important way station we call the Hodgkin-Huxley equations.

NOTES AND REFERENCES

1 A remark of H. von Helmholtz quoted by Henry Margenau in his introduction to the second english translation of "On the Sensations of Tone as a Physiological Basis for the Theory of Music." Dover Publications, Inc. New York, 1954.

2 J. *Gen. Physiol.* **21**, 189 (1937).

3 A remarkable gathering.

4 Howard J. Curtis and Kenneth S. Cole. *Jour. Gen. Physiol.* 757–765 (1938).

5 Cole, K. S., "Membranes, Ions and Impulses—A Chapter of Classical Biophysics" Univ. Calif. Press, Berkeley, 1968.

6 Keynes, R. D. *Arc. d. Sci. Physiol.* (Paris) III, 165–176 (1949).

7 See pp. 42ff. of Hodgkin, A. L. "The Conduction of the Nervous Impulse." Charles C. Thomas, Springfield, 1964.

8 This section is abstracted from Atwater, I., F. Bezanilla and E. Rojas. *J. Physiol.* **201**, 657–664 (1969).

9 Hodgkin, A. L. *Arch. d. Sci. Physiol.* (Paris) III, 151–163 (1949).

10 For an interesting example of work with ammonium ions, which pass through both sodium and potassium channels see Binstock and Lecar. *J. Gen. Physiol.* **63**, 342–363 (1969).

11 First described by Chandler, W. K. and H. Meves, *J. Physiol.* **180**, 788–820.

12 From Table I in Atwater, I., F. Bezanilla and E. Rojas. *J. Physiol.* **201**, 657–664 (1969).

13 See: Taylor, R. E., J. W. Moore and K. S. Cole, *Biophys. Jour.* **1**, 161–202 (1960) and, Rojas, Eduardo, Robert E. Taylor, Illani Atwater and Francisco Bezanilla, *Jour. Gen. Physiol.* **54**, 532–552 (1969).

14 Cole, K. S. and H. J. Curtis. *J. Gen. Physiol.* **22**, 649–670 (1939).

15 Let us not despair that the *form* of the Hodgkin Huxley equations probably does not correspond to the actual physical situation. There is still much work to do.

PAPER 2

*The Role of the Squid Axon in Transport Studies**

L. J. MULLINS

Department of Biophysics University of Maryland School of Medicine Baltimore

---

In writing an article for a volume dedicated to Kenneth Cole, it seems appropriate to emphasize the contribution that his favorite preparation, the squid axon, has made or is likely to make to fields other than the electrical properties of the membrane. The rediscovery of the squid giant axon by J. Z. Young in 1936 and its almost immediate exploitation by Cole and his associates are too well known to require detailing here; nor do I propose to review the dramatic studies of Cole and Marmont and Hodgkin and Huxley (using voltage and current clamp methods) which led to an essentially complete experimental understanding of how the processes of excitation develop and how a nerve impulse is generated by Na and K concentration gradients.

Ion transport studies contributed in a modest way to these exciting developments by showing that nerve fibers gained the appropriate quantities of Na and lost similar quantities of K during a nerve impulse. It was also possible to show by isotope measurements that the early current during a voltage clamp was carried by Na^+. At any rate, data of this sort were mainly confirmatory and did little or nothing to provide the time resolution necessary for working out the detailed story that only electrical measurements could provide. A scheme of electrical excitation which would allow Na to run down an electrochemical gradient required that some mechanism be provided to supply energy such that Na inside the fiber could be kept at a low concentration; otherwise the Na battery of the fiber would soon run down. Thus, a study of the battery recharging mechanism was begun

* Aided by grants from the National Institute of Neurological Diseases and Stroke (NS 05846) and from the National Science Foundation (GB 8279).

under the general heading of "recovery processes in nerve". This Na extrusion mechanism or Na pump, is of general occurrence in cells and parallel studies were begun on red cells, muscle fibers, and a variety of other rather complex tissues. I should like to argue that the squid axon is superior to other cells and tissues in its ability to give clear-cut answers to the many questions that arise in connection with a study of Na extrusion mechanisms, and that as such it deserves to be considered as the model for Na pump phenomena. It may turn out that other cells have Na pumps that operate in ways significantly different from that of the squid axon. We shall not know that these differences are real, however, until experimental information on the squid axon is complete.

More experimental effort has been expended on studies of the red cell Na/K pump than on any other. As an experimental material, red cells have the advantage of being a population of single cells without any sort of barrier external to the cell membrane. These cells can be manipulated such that they lose most of their hemoglobin and substances can be incorporated into the cells in order to set new concentrations of metabolites and ions inside the cell. In addition, mammalian red cells are readily available in the laboratory and are relatively simple to manipulate. However, the cells have the disadvantage that a population of cells is bound to include some with leaky membranes. The presence of such cells can falsify ion flux measurements. While the preparation of ghosts is technically impressive, some cells do not reseal very well and all the uncertainties of leaky cells are enhanced in working with resealed cells. In addition, while it is possible to incorporate compounds such as ATP into the ghosts, the initial concentrations supplied are subject to considerable modification by metabolism during the period of resealing.

Muscle fibers have also been used for transport studies and single muscle fiber preparations would appear to offer many advantages. In the case of the 100 μ frog sartorius fibers, however, the labor of setting up such a fiber has discouraged most investigators. Although barnacle muscle fibers are ten times as big and relatively easy to use, the complex internal structure found in all muscle fibers and the worry that a compartmentalization of ions may take place in structures such as the reticulum are distinct disadvantages. It would appear, therefore, that we should understand ion transport better in simpler systems before attempting to analyze the Na pump in muscle.

The squid axon has advantages and disadvantages which must be compared with those cells discussed above. The main disadvantage of the squid axon is that the animals are expensive to catch and at present it is virtually impossible to maintain them in captivity. A second disadvantage is that the axon is covered by a Schwann cell layer which is something of a diffusion barrier although not a serious one. Advantages are that one can measure fluxes over a specified area of membrane from a single fiber, and that continuous control of the solutes both inside and outside the fiber can be maintained.

Technical improvements in the methods for measurement of ion fluxes in squid axons have been continuous. Originally, the loading of radioactive isotopes into axons depended on pretreating the axon with radioactive seawater followed by washing in inactive seawater. Subsequent studies (Caldwell and Keynes, 1960; Sjodin and Mullins, 1967) showed that such treatments, at least for K, selectively loaded the Schwann cells with isotope. The specific activity inside the axon can seldom be raised to more than 10% of that of the seawater, given the relatively few hours available for loading, while the Schwann cells can be brought to isotopic equilibrium in a very much shorter time. Microinjection, originally used by Grundfest, Kao and Altamirano (1954) for electrophysiological observations on the effects of injected substances, was developed by Caldwell *et al.* (1960) into an extremely useful technique for introducing both isotopes and substrates into axons and much of the information we have on the behavior of the Na pump has resulted from the application of this technique. With the discovery (Baker, Hodgkin and Shaw, 1962) that axons can be perfused with a simple salt solution and still retain their electrophysiological properties, the way appeared clear for a major improvement in technique for the study of active transport since in principle it ought to have been possible to introduce substrates and maintain a defined set of substrate and ion concentrations in the axon on a time-independent basis. With regard to the maintenance of a substrate-dependent Na efflux, the results from internal perfusion experiments have not been entirely satisfactory; either the fluxes are abnormally high (Shaw, 1966) or are much too high in the absence of a substrate (Canessa-Fischer *et al.*, 1969). Fluxes of Na do appear to be stable in perfused squid axons when F^- is a major constituent of the perfusion fluid but this substance irreversibly inactivates membrane ATPase, so that a sensitivity of Na efflux to ATP can hardly be expected. Another technique for the study of Na transport in squid axons is that of internal dialysis, in which a porous glass dialysis tube is inserted into the axon and dialyzable solutes in the dialysis fluid are exchanged with solutes in the axon (Brinley and Mullins, 1967). With this technique, Na efflux is stable with time, is of normal magnitude when physiological concentrations of ATP are included in the dialysis fluid, and in the absence of substrates it agrees rather closely with the value calculated from flux ratio considerations. The main purpose of the following discussion is to examine some of the results that have recently been obtained with squid axons with a view to seeing what sorts of models of active transport will accommodate these newer experimental findings.

Much experimental work carried out over the past decade has led to the following sorts of conclusions regarding active transport:

1) it is a coupled movement of Na out of and K into cells
2) it is dependent on ATP or on some substrate readily converted to ATP
3) the enzyme, membrane ATPase, is both the system that converts ATP into transport work and the pump that moves the ions involved
4) ouabain, or other cardiac glycosides are specific inhibitors of both transport and membrane ATPase.

Using techniques such as internal dialysis, it has proved possible to examine these conclusions somewhat more rigorously than was previously possible. The results obtained suggest that active transport may not be as simple as one might have thought, and that some of the conclusions listed above do not actually lend themselves readily to some of the more conventional schemes that have been proposed.

In part, this difficulty stems from the increasingly complex phenomena that have been observed in studies of Na-K transport, and in part difficulties have become apparent when well known processes have been observed in greater detail than was formerly possible. The phenomena of transport can be classified into substrate or inhibitor dependent fluxes of Na or K:

A. The "classical" Na efflux

1. Requires $[ATP]_i$
2. Proportional to $[Na]_i$
3. Stimulated by $[K]_o$ or $[Na]_o$
4. Insensitive to $[ADP]_i$ and $[Ca]_o$
5. Inhibited by $[P_i]_i$ and $[Mg]_i$
6. Glycoside inhibits but also initiates a new Na efflux

B. The glycoside dependent Na efflux (Brinley and Mullins, 1968)

1. Independent of $[ATP]_i$ or $[K]_o$
2. Dependent on $[Na]_i$
3. Requires ouabain, strophanthidin or similar compounds

C. The ATP dependent K influx

1. Requires $[ATP]_i$
2. Proportional to $[K]_o$
3. Stimulated by $[Na]_i$
4. Inhibited by $[ADP]_i$
5. Inhibited by glycoside
6. Insensitive to $[Na]_o$, $[Li]_o$, or $[Choline]_o$

D. The Ca dependent Na efflux (Baker et al., 1969 a, b)

1. Independent of glycoside or $[K]_0$
2. Dependent on $[ATP]_i$ and $[Na]_i$
3. Stimulated by $[Li]_0$
4. Dependent on $[Ca]_0$

E. The ATP dependent Na influx (Brinley and Mullins, 1968)

1. Independent of glycoside or $[K]_0$
2. Dependent on $[ATP]_i$ and $[Na]_i$

Flux (A) is that usually considered when active transport of Na is meant; point (A.1.) was not entirely settled by the injection experiments of Caldwell et al. (1960) which showed that the injection of either ATP or ArgP would restore Na efflux but that only ArgP would yield a Na efflux that was decreased when $[K]_0$ of seawater was lowered. A test of the relative efficacy of ArgP and ATP in dialyzed axons in promoting Na efflux (Mullins and Brinley, 1967) showed that ArgP was virtually inactive while ATP gave a normal Na efflux. Requirement (A.3.) is a well known one and one that is usually ascribed to a coupling between inward K transport and Na efflux. However, as I shall show later, this supposed coupling can vary over such wide limits that an obligatory coupling of Na to K becomes somewhat implausible. It may, at this point, be safer to assume that external K has a role in promoting Na transport which is independent of K transport. The finding (A.5.) is a new one (De Weer, 1970) and may give important clues as to the mechanism of transport. Although glycosides have been frequently used to define active transport, the finding (A.6.) (Brinley and Mullins, 1968) shows that it is impossible for such a definition of transport to be an accurate one. This fact is emphasized by point (B.2.) which categorizes a separate Na efflux that is glycoside dependent. The difference in substrate dependence for Na fluxes (A) and (B) is shown as Figure 1 (data of Brinley and Mullins, 1968). This flux has not been studied extensively enough to characterize it further; it differs from (D) in that the Na efflux is ATP independent. The fluxes in (C) describe what is usually called the K pump; (C.6.) lists the effect of Na-free solutions—these have a large effect on Na efflux but none on K influx.

What evidence there is suggests that fluxes (A), (B), and (C) may be brought about by mechanisms that are closely related, while (D) and (E) may be brought about by another mechanism. Further discussion must depend on the way in which one defines the mechanisms responsible for metabolism-dependent ion fluxes.

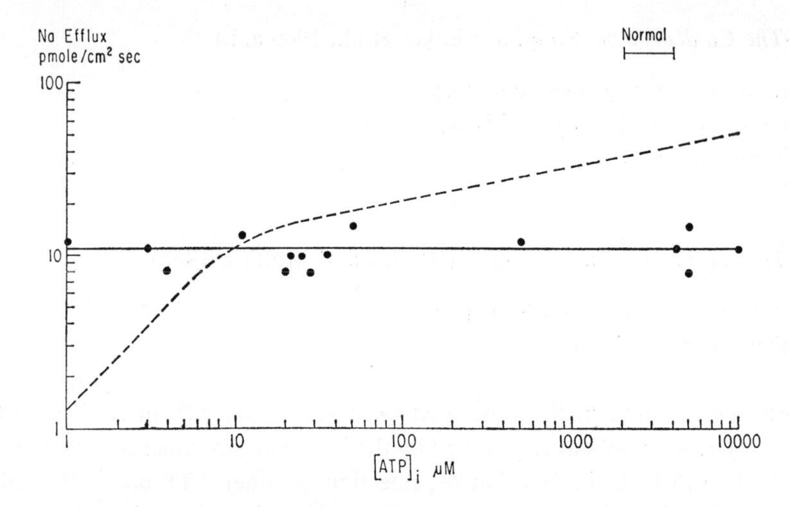

Figure 1 Na efflux is plotted as a function of [ATP] supplied to the axoplasm by dialysis. The broken line is Na efflux into seawater and the solid line and experimental points refer to seawater + 10 mM strophanthidin. Na efflux is apparently stimulated by this glycoside at low $[ATP]_i$ and inhibited at high $[ATP]_i$. The measurements show that Na efflux in glycoside is independent of [ATP]. The bracketed horizontal line labelled "normal", represents the experimentally measured range of [ATP] in fresh axoplasm. Plot uses double logarithmic scales

DEFINITIONS OF ACTIVE TRANSPORT

Active transport has been defined as that which is involved in a net movement of material from a region of lower to one of higher electrochemical potential. That is, the definition is concerned entirely with the uphill movement of solute. It is easy to imagine, however, that transport processes involving the expenditure of energy might also move material in a downhill direction. Such processes might play a physiological role in vastly accelerating the movement of solutes into the cell even though active transport in the sense of the uphill movement was not involved. It would seem therefore that a more appropriate definition of active transport might relate the amount of metabolic substrate used to the quantity of solute transported across the membrane without regard for whether the work performed was with or against an electrochemical potential gradient.

Another difficulty with a simple definition of active transport becomes apparent when coupled ion transport processes are considered. For example, there are circumstances in which the movement of sodium ion via active transport may be uphill, while the allegedly coupled movement of potassium ion in the opposite direction may involve transporting the ion against no electrochemical potential gradient or even moving it downhill. Under these circumstances, it is difficult to see whether the work involved in moving sodium against a gradient can be partially compensated by the coupled ion movement in the opposite direction or whether this is an unrealistic assumption to make. Clearly the answer to such a problem lies in a complete understanding of the mechanism of pumping and entirely consistent definitions of active transport can only come when the actual mechanism is much better understood.

Transport studies must usually rely on an operational definition of active transport and a variety of inhibitors has been used in order to furnish a means of testing whether a particular ion flux is or is not sensitive to the substrate for active transport processes, ATP. For tissues which do not have a prominent glycolytic system, cyanide is a useful inhibitor because it will rather promptly reduce the cellular ATP level to tens of micromolar and in such tissues sodium efflux is observed to fall to low levels. Efflux does not fall to the value predicted by flux ratio considerations but then given the small residual ATP concentration which is observed in the tissue it is to be expected that some sodium transport will still be energized. More specific inhibitors of sodium transport such as ouabain and related cardiac glycosides might be expected to be even more useful because in vitro studies have shown that these glycosides inhibit membrane ATPase while physiological studies show that cardiac glycosides reduce sodium efflux. Another method of operationally defining active transport is to say that when the ATP concentration inside a cell is increased from zero to any particular value, the change in ionic fluxes which follows this step-change in ATP concentration is a measure of the transport that is taking place. A difficulty with this definition is that, in principle, we can recognize two sorts of effects which ATP might have; it might induce a permeability change in the membrane by virtue of its binding to some membrane component, or it might bring about the active transport which we wish to define. It may prove exceedingly difficult to separate these two hypothetical effects of ATP so that a definition of active transport based on a change in ATP concentration may be a somewhat imperfect one. In turn, a further improvement in the definition of active transport would be to have it specify the quantity of substrate (ATP) that is split per quantity of ion transported; this would allow the definition to relate the chemical free energy utilized in ATP hydrolysis to a definite quantity of ion transported and if one found, for example, that an ion flux occurred under conditions where no ATP was hydrolyzed but where ATP was necessary, then it would be justifiable to assign another role to the effect of ATP on the system.

The difficulties with inhibitors used to define active transport can be summarized by a detailed consideration of the findings with cardiac glycosides. The following results (Brinley and Mullins, 1968) show how difficult it is to work out an entirely consistent scheme for transport based on the action of cardiac glycosides.

Effect of glycoside on Na fluxes

	ATP Absent		ATP Present	
	Control	Glycoside	Control	Glycoside
Na Influx	40 *f**	No effect	Increased	No effect
Na Efflux	1.2 *f*	12 *f*	Increased	12 *f*

* pmole/cm^2 sec.

The fact that glycoside has no effect on Na influx when ATP is absent must mean that this substance does not affect Na permeability, while the fact that Na efflux is the same in the presence or absence of ATP plus glycoside must mean that this efflux is ATP independent. Sodium influx is increased by ATP but this change is glycoside insensitive. We might consider two separate explanations for this effect. Either the increase in Na influx is a permeability change induced by ATP, or the Na influx is a carrier movement that happens to be glycoside insensitive. We cannot rule out a permeability change mechanism on the basis of looking at Na efflux since this is known to be highly sensitive to ATP; we can, however, examine the Na efflux in the presence of strophanthidin and find that this movement is independent of internal [ATP]. If ATP were to produce a permeability change leading to an increased Na influx, the simplest explanation would be that this would be independent of glycoside and that efflux as well as influx would be affected. Measurements of Na efflux in the presence of ATP and glycoside show that Na efflux is the same whether or not ATP is present (although admittedly the increment in efflux to be observed would be only a few pmole/cm^{2}sec) so that the best provisional conclusion is that the increment in Na influx with ATP present is a carrier mediated process that is glycoside insensitive.

The examples cited above were chosen to illustrate the difficulty inherent in using a glycoside to define active transport fluxes. Further details relating to mechanism will be given later in this paper, but at present it seems that the distinction between purely passive and active fluxes may be a somewhat blurred one. This is inferred because the large sodium efflux induced by ouabain in an axon that has a very low $[ATP]_i$ must have some energy source since it is clearly a movement against a large electrochemical gradient. The only energy source available under the experimental circumstances cited would appear to be the sodium concentration gradient itself so that the allegedly purely passive sodium entry into the cell could, in principle, supply the necessary energy to allow for the ouabain induced sodium efflux. In turn one infers that the inward sodium movement must be through some kind of carrier system because otherwise it is difficult to see how this purely dissipative ion movement could contribute energy for the outward movement of sodium.

Given the present capability for flux measurement, it appears that an operational definition of active transport may be: the change in flux that is produced by a change in the $[ATP]_i$ from zero to a particular value. This definition does not distinguish between downhill and uphill movements of an ion, nor does it separate possible permeability effects of ATP from carrier mediated ion movement. More importantly, it rather arbitrarily separates ion movement that may be coupled into separate ion pumps. I shall argue later that there is no evidence that ATP induces permeability changes in the membrane for either Na or K, and that evidence for the coupling of ionic fluxes is equivocal. It appears, therefore, that the provisional definition of a pump in terms of the ATP dependence of its ionic flux may be a useful one.

THE COUPLING OF ION TRANSPORT FLUXES

While the original idea of Dean (1941) in suggesting a Na pump was that this was a device that would maintain a non-equilibrium distribution of Na^+ across the cell membrane, it was noticed rather early that $[K]_0$ stimulated the loss of Na from muscle (Steinbach, 1940) and that removing $[K]_0$ from the solution bathing a nerve fiber reduced Na efflux. It seemed reasonable, therefore, to suppose that K entry was coupled to Na efflux and an early suggestion was that this coupling was 1 : 1; that is, that 1 K^+ was transported inward for each Na^+ transported outward. This assumption was convenient for studies then taking place on the control of the resting potential of muscle fibers because such a pumping mechanism was itself electroneutral and could not contribute membrane current. Subsequent work on ion fluxes has shown rather convincingly, however, that in general Na efflux is larger than K influx so that the original transport mechanism became more complicated than it might first have seemed. In the squid axon, the number of Na extruded per K taken in can be varied from greater than 3 : 1 to 1 : 1 depending on $[Na]_i$ (Sjodin and Beaugé, 1968; Mullins and Brinley, 1969).

These experimental findings make it at least worthwhile questioning whether a coupled ion pump (even if the coupling ratio is allowed to vary) correctly describes the active transport system. One might, for example, consider separate Na and K pumps, operating from a common energy source. The conventional carrier scheme is shown below:

(1) $nNa_i + X_i \leftrightarrow (Na_nX)_i \rightarrow (Na_nX)_o \leftrightarrow nNa_o + X_o$

(2) $X_o \rightarrow Y_o$

(3) $Y_o + mK \leftrightarrow (K_mY)_o \rightarrow (K_mY)_i \leftrightarrow mK_i + Y_i$

(4) $Y_i \rightarrow X_i$

Separate pumps would merely require the deletion of reactions (2) and (4) so that X is solely a Na carrier and Y is a K carrier; interconversion would not be possible, and X_o, Y_i would return to X_i, Y_o unloaded.

At first sight, such a scheme might seem highly unpromising in view of the evidence that $[K]_o$ affects Na efflux, that both K influx and Na efflux are reduced by glycosides, and that membrane ATPases require Na and K as activators for ATP hydrolysis. There are, however, some difficulties in reconciling the conventional carrier scheme with experimental findings. In particular, we might consider the K-free effect in squid axons. When an axon is transferred to K-free seawater, Na efflux falls to between 1/2 to 1/3 of control values. If, on the other hand, $[ATP]_i$ is reduced to levels of 1 μM, Na efflux will fall to 1/30 of control efflux (Brinley and Mullins, 1968). Clearly K-free seawater does not stop an ATP-dependent Na extrusion. It has been suggested (Baker *et al.*, 1969b) that the [K] outside the membrane is not zero under K-free conditions because K leakage from the axon and Schwann cells makes $[K]_o$ of the order of 1 mM. This is a reasonable suggestion but it does not help much in explaining the Na efflux under K-free conditions because K influx is linearly proportional to $[K]_o$ over the concentration range 1–20 mM so that if Na efflux under K-free conditions is 1/2 of control, then for a K_o of 1 mM, instead of the normal 10 mM of seawater, K influx will be 1/10 of control or the coupling ratio will have increased 5-fold. In an axon with normal $[Na]_i$ and $[ATP]_i$ the ratio would have been 3, so that under these altered conditions it would now be 15. This is stretching the carrier concept beyond its elastic limit!

It might be suggested that, under K-free conditions, Na efflux becomes coupled not to K influx but to Na influx. Unfortunately, the effect of Na-free conditions outside the squid axon is to increase Na efflux and this effect has been explained rather satisfactorily by Baker *et al.* (1969b) as a competition between Na_o and K_o for sites outside the axon. Under this scheme, an axon is less K-free when choline$^+$ is the external cation because the K^+ leaking from axon and Schwann cells are more able to activate external K sites. Such an arrangement makes it impossible to apply Na-free solutions as a test for the extent to which Na : Na exchange takes place. Such a hypothetical Na : Na exchange would require that Na influx increase under K-free conditions but such evidence as there is shows that Na influx is not altered by K-free seawater, nor is there a glycoside sensitive component to Na influx in axons with a normal $[ATP]_i$ and phosphoarginine. These experimental findings show that it is difficult to test for Na : Na exchange; it is possible that there is a carrier mediated Na influx which is normally uncoupled and that K-free conditions allow the coupling of this flux to Na efflux. The resulting ion movement must be postulated to be glycoside insensitive and hardly helps support the idea of a Na : K coupled pump. A glycoside insensitive, Na_0-dependent Na efflux has been observed in muscle (Horowicz and Gerber, 1965) so that the idea of such a form of transport is not unique.

To summarize the effect of nominally K-free solutions on Na efflux from squid axons, one can only say that the residual Na efflux appears uncoupled to the movement of any appreciable quantity of Na or K inward. External [K] has a large effect on Na efflux but one whose concentration dependence is different from that on K transport.

Much the same set of experimental findings is obtained if $[K]_0$ is held constant and $[Na]_i$ varied. The findings with respect to the ATP-dependent fluxes are (Mullins and Brinley, 1969):

1) Na efflux is linearly dependent on $[Na]_i$; K influx is linearly dependent on $[K]_0$.
2) K influx is a non-linear function of $[Na]_i$ and appears half saturated at $[Na]_i \sim 10$ mM.
3) K influx is unaffected by the replacement of $[Na]_0$ by $[\text{choline}]_0$.

These results suggest that K^+ and Na^+ on the outside do not compete for the K pump as otherwise one would expect an altered rate of K pumping under Na-free conditions. Both $[Na]_i$ and $[K]_0$ affect the rate of pumping of K^+ and Na^+ but they do so in a manner that is not consistent with a simple coupling of Na efflux with K influx. Allowing the coupling ratio to be set by both $[K]_0$ and $[Na]_i$ is a device for retaining the notion of coupling, but it is hardly helpful in understanding the mechanism of pumping.

At very low $[K]_0$, an increment in $[K]_0$ promotes far more Na extrusion than K uptake, so that the apparent coupling ratio is very high. On the other hand, at $[K]_0$ 10–20 mM, an increment in $[K]_0$ promotes only a small increment in Na efflux but yields the same increment in K influx as at low

$[K]_o$. This leads to a lower coupling ratio, although the limiting ratio for saturating $[K]_o$ would appear to be greater than 3 Na/K.

The rather complicated relationships between the influence of $[K]_o$ and $[Na]_i$ on the Na and K fluxes at an [ATP] of 3–4 mM are summarized in Figure 2. This shows that the pumped ion is linearly related to its own concentration, but that the ion on the opposite side of the membrane has a nonlinear effect on the rate of pumping. One of the simplest interpretations of such experimental results is that K_o is an activator of Na efflux and Na_i is an activator of K influx but that ion transport and ion activation of transport are separate effects.

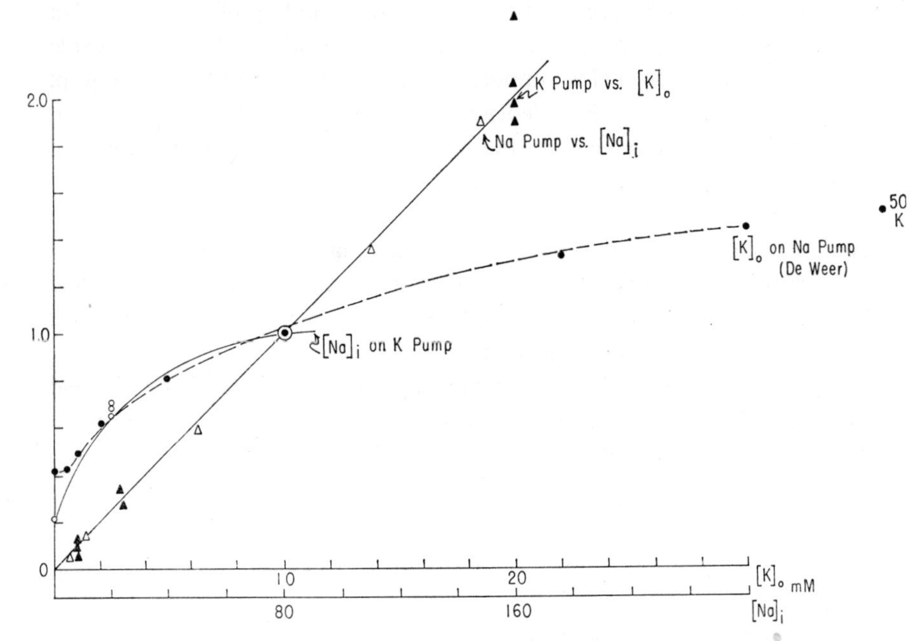

Figure 2 Ordinate is Na efflux or K influx relative to that at 10 mM $[K]_o$ seawater with $[Na]_i$ 80 mM, $[ATP]_i$ 3 mM. The abscissa is either $[K]_o$ in Na seawater or $[Na]_i$. Open circles show the effect of $[Na]_i$ on K influx from 10 K (Na) seawater, solid circles the effect of $[K]_o$ in Na seawater on Na efflux (at constant $[Na]_i$). Solid triangles give the effect of $[K]_o$ on the ATP dependent K influx while open circles show the effect of $[Na]_i$ on Na efflux (at constant $[K]_o$)

As Na and K are known to be activators of membrane ATPase, these observations suggest that "activation" of transport by Na_i and K_o relates to an improved energy supply to the ion pumps but that it is possible to distinguish between pumping and its energy supply.

Another mode of coupled ion movement which has received a great deal of attention is Na : Na exchange. The concept of this sort of interchange arose from a suggestion of Ussing that it might explain a Na efflux from muscle which appeared too large to be compatible with the energy sources available. The efflux estimate on which Ussing based this calculation proved to be erroneous—muscle has a Na efflux that makes only modest demands on the available energy sources. Meanwhile, the term "exchange diffusion" has remained as a concept and is frequently invoked to explain various sorts of measured fluxes.

There is the usual sort of difficulty in defining this coupled ion movement. The original idea was that a 1 : 1 carrier-mediated Na interchange took place which did not require ATP. A definition of exchange diffusion which has often been used is the reduction of Na efflux when $[Na]_0$ is replaced with another cation such as Li^+ or $Choline^+$. This test assumes that $[Na]_0$ has no effect on Na efflux other than providing Na for Na : Na interchange and it assumes further that the ion used to replace Na (e.g. Li) has no effect of its own.

That neither of these assumptions is likely to be generally true became apparent first in studies on frog sartorius muscle (Keynes and Swan, 1959) in which the effect of replacing Na in Ringer with Li in some cases led to a decrease in Na efflux, while in others produced no change in Na efflux. It proved possible to show that replacing Na with Li outside could, in Na-loaded muscles, lead to an *increase* in Na efflux. Other studies (Beaugé and Sjodin, 1968) have shown that Li, in addition to whatever other effects it may have, has a K-like action that stimulates Na efflux. The effects of Li in increasing or decreasing Na efflux proved to be related to $[Na]_i$; a low $[Na]_i$ favored a decrease in Na efflux, while a high $[Na]_i$ had the effect of producing an increased Na efflux into Li Ringer solution. A further complication was that the decrease in Na efflux observed in Li Ringer was not sensitive to ouabain (Horowicz and Gerber, 1965) suggesting possibly that the effect was not related to the Na carrier involved in Na : K pumping. A difficulty with the muscle fiber, however, is that one may be dealing with a complex compartment situation and the fluxes observed into Li solutions may be a reflection of this complexity rather than of a true Na : Na exchange.

For the squid axon, there can be little doubt that compartmentalization of internal Na is nonexistent; the experimental findings with Na-free solutions are, however, quite different from those of muscle. The usual effect of Li seawater on a squid axon is to increase Na efflux although the effect may not be marked in all axons. In any case, one seldom observes a decrease in Na efflux in Na-free solutions. The effect of such solutions on Na efflux has been analyzed in great detail (Baker *et al.*, 1969b) and a convincing case has been made for the idea that Na and K compete for sites on the outside of the membrane which activate Na pumping. Because K is much more potent in activating Na efflux than is Na, the effect of Na-free solutions is to make the effective [K] in seawater much greater. If there were a component of the Na efflux that were cut off by removing Na_0, one could argue that this situation is present but undetected by the method used. The maximum Na extrusion at very high $[K]_0$ appears, however, to be the same in Na-containing or Na-free seawater so that it appears unlikely that there is any Na : Na exchange in normal axons. Further confirmation of this view is found in the absence of an effect of strophanthidin on Na influx into such axons (Mullins and Brinley, 1969).

A recent finding is that the effects of Na-free and K-free solutions on the Na efflux from squid axons are reciprocal phenomena in the sense that if an axon shows a large decrease in Na efflux on applying K-free solutions, it will be relatively insensitive to external Na, while axons relatively insensitive to external K will require Na to support maximum Na extrusion (Sjodin and Beaugé, 1969; De Weer, 1970). One might think that this is a demonstration that some Na efflux is coupled Na/K and some Na/Na. Were this true, however, in normal axons there should then be a glycoside sensitive Na influx, and this remains undetected. One is, instead, led to the view that some of the pump elements in the membrane are K activated while others are Na activated. One can, of course, define a good pumping system as one in which K activation is the sole mode of operation but this merely avoids consideration of the problem of how the specificity requirement for external cation activation of the Na pump is set.

A variety of experimental treatments will abolish the K sensitivity of the squid axon; these include treatment with CN (and the reinjection of ATP) or with alkaline DNP (Baker et al., 1969b) and the injection of Arg or ADP into normal axons (De Weer, 1970). Treatment with alkaline DNP (which reduced ArgP to low levels but left ATP unaltered) is also known to induce a glycoside sensitive Na influx in the axon and it has been suggested that all of these experimental conditions which lead to an interference with metabolism also lead to a loss of K sensitivity and a glycoside sensitive Na influx which is Na : Na exchange. ATP dependent K influx is inhibited about 50% by an equimolar mixture of ADP and ATP (Mullins and Brinley, 1969) while injected P_i is without effect on the K sensitivity of the Na efflux (De Weer, 1970), so that a summary of what might be happening when the [ADP] of axoplasm is increased, is that the K influx is inhibited and an ATP dependent Na influx replaces it, while K sensitivity may be lost and replaced by an $[Na]_0$ sensitivity. This is equivalent to saying that Na/K pumping changes to Na/Na exchange.

Before accepting this scheme, however, it is prudent to ask whether the experimental findings might be explained in quite another way. In view of the suspicion that external ions may activate a mechanism without necessarily being transported by it, one may ask whether in fact the loss of the K-free effect produced by CN or other treatments is really different from the low K-free effect shown spontaneously by some fresh axons. A known effect of CN on axons which has recently been described (Blaustein and Hodgkin, 1969) is a great enhancement of the Ca efflux. Presumably this observation reflects the discharge of Ca accumulated in the mitochondria into the axoplasm and its subsequent removal from the axoplasm via a membrane pump. Therefore, at least in principle, it is possible that agents which interfere with metabolism may cause the release of material from stores in the axoplasm and that the release of this material affects the Na pump to the extent of reducing its K sensitivity. If this is so, then ADP may be a substance which changes the specificity of the K pump such that it pumps Na, and we do not have an example of Na : Na exchange under the experimental conditions described above. Such a scheme would have the advantage of making the loss of the K-free effect the same for normal and CN treated axons; i.e. it would be dependent on the release of material from the mitochondria. It would also separate from the K-free effect the effects of ADP on the K pump. Fortunately, an experimental test of this proposal is possible; if with ADP, the K pump becomes a Na pump, but the rest of the system remains unchanged, then ATP hydrolysis will take place in connection with Na/Na exchange. If, however, ATP is required in only a catalytic role, then ATP hydrolysis will not take place.

A SODIUM INFLUX PUMP

While attention has focussed mainly on active transport of Na out of and K into the cell, evidence of several sorts is accumulating that there may be a carrier mediated inward movement of Na which is not connected with the Na/K transport system. Evidence for the independence of this Na transport from Na/K transport comes from measurements of Na influx in the presence of strophanthidin; these show that Na influx is unaffected by this inhibition both when $[ATP]_i$ is normal (Mullins and Brinley, 1969) and when $[ATP]_i$ is of the order of 2 μM (Brinley and Mullins, unpublished). Sodium influx is, however, sensitive to ATP and an increased influx of about 25 *f* results from adding physiological concentrations of ATP to axons first dialyzed free of ATP.

Lithium, when used as a replacement for Na in seawater, induces a large increase in Na efflux which is ouabain insensitive (Baker *et al.*, 1969a). Since ouabain, in Na seawater, induces a Na efflux of 10–12 *f*, the observations with Li seawater suggest that this ion induces a Na efflux of 20–24 *f* and that half of this increased efflux is $[Ca]_0$ and $[ATP]_i$ dependent. Although the effect of Na substitutes on Na efflux in the presence of glycosides has not been systematically studied, Sjodin and Beaugé (1969) have shown that in the presence of strophanthidin, Na efflux in $Tris^+$ seawater is lower than in Na seawater while Baker et al. (1969a) show that $Choline^+$ seawater gives about the same Na efflux as Na seawater in the presence of ouabain.

Other evidence that makes a Na influx pump seem more plausible has been provided by Baker *et al.* (1969a) who described a ouabain insensitive, $[Ca]_0$ dependent Na efflux. This system is an exceedingly complex one so that any attempt to summarize it will be necessarily inaccurate. It does appear however, that Na and Ca compete for entry and that in seawater with a normal $[Na]_0$, Ca entry is negligible. The sensitivity of the Na efflux to [Ca] is also very small in seawater but a large $[Ca]_0$ sensitive Na efflux is apparent in Li seawater.

An interesting example of a glycoside insensitive Na efflux is the finding (Brinley and Mullins, 1968) that when $[ATP]_i$ in dialyzed axons is $\sim 1\ \mu M$ and Na efflux is $\sim 1\ f$, the application of strophanthidin increases Na efflux 10 fold; as Na influx is not affected, this cannot be ascribed to a permeability change and it seems necessary to suppose that the movement of Na down its electrochemical gradient becomes coupled to Na efflux in the presence of glycoside. Since, under the conditions of the experiment, ATP was present inside at negligible concentrations, it appears that some of the Na influx measured in the absence of ATP is carrier mediated and is capable of being coupled to Na efflux by glycoside. The efflux of Na in the presence of strophanthidin is independent of $[ATP]_i$ so that it is clear that glycoside abolishes the Na/K pump and adds a new component to the Na efflux.

The effect of strophanthidin on ATP-free axons in promoting a Na efflux differs from the effect of $[Li]_0$ in promoting a Na efflux mainly in that the Li-stimulated Na efflux requires ATP and a high $[Na]_i$. It may be that a high internal Na is necessary for the activation of a Na dependent ATPase. The observations clearly suggest that the ouabain insensitive fraction of membrane ATPase preparations may be of more than trivial interest.

The influx of Na in a carrier mediated mode is also suggested by the observations of Blaustein and Hodgkin (1969) that Ca efflux from squid axons is very likely coupled to Na entry; they suggest that a carrier mediated Na entry could provide the energy for the extrusion of Ca^{++}. Again, the Ca^{++} efflux is insensitive to glycoside and the scheme proposed is very similar to that suggested above for the effect of glycoside on Na efflux in the absence of ATP. While Li_0 was most effective in promoting Ca influx and Na efflux, Na_0 was most effective in promoting Ca efflux suggesting that if the same system is involved for both movements, then the nature of the external cation partly determines whether Na or Ca will be extruded. Other factors are $[Na]_i$ and $[ATP]_i$.

To summarize the status of the Na influx pump, it is clear that it is at present poorly defined but may have the following properties with respect to influx.

1) glycoside independent
2) $[ATP]_i$ dependent
3) $[Na]_i$ dependent
4) a competition between Ca and Na for entry
5) relative activating effect of external cations for Na efflux: Li > Na, Choline > Tris.

MEMBRANE ATPASE

There is a great deal of recent experimental evidence to persuade one that there are two forms of enzyme membrane ATPase, i.e. E_1 and E_2. Such evidence is summarized by Post et al. (1969) and consists of showing that $E_1 \sim P$ is a form which allows terminal P exchange between ATP and ADP. It is also clear that $E_1 \sim P$ and $E_2 \sim P$ show differing sensitivities to ADP and K (with respect to the loss of P by the enzyme). It seems likely, therefore, that the change from E_1 to E_2 is a conformational change in the enzyme which might be associated with ion pumping (presumably of Na), while the reverse transition $E_2 \rightarrow E_1$ might be associated with K pumping. What is actually known, however, is that Na and K are activators of the ATP hydrolysis reaction sequence and an alternate view of membrane ATPase is that it supplies energy to pump ions but that the pumps are separate molecular entities.

A choice between these alternatives is not possible at present, but as was shown earlier in this paper, the requirements for activation of pumping and the requirements for the actual transport of ions often appear to be different with respect to both [Na] and [K]. Another method of analyzing the action of the enzyme in regulating ion transport is to consider the effects of the products of ATP hydrolysis on transport. Since the energy available from ATP hydrolysis is dependent on the term $[ATP]/[ADP][P_i]$, one might expect that varying [ADP] or $[P_i]$ inside the membrane would be equivalent operations. Experiments show, however, that Na efflux is largely independent of [ADP] (Brinley and Mullins, 1968) but that it is reduced by P_i (De Weer, 1970). That fraction of K influx which is dependent on ATP is inhibited by [ADP] (Mullins and Brinley, 1969) while the K-free effect in squid axons is not influenced by large increases in $[P_i]$ (De Weer, 1970).

An increase in internal [ADP] can be expected to promote the formation of E · ATP so that one hypothesis for explaining why ADP has an inhibitory effect on K influx but not on Na efflux would be that the Na efflux is a different one from that existing before the addition of ADP, and that this new Na efflux is Na : Na exchange. This scheme rather plausibly explains why Na : Na exchange requires ATP, and why K transport is inhibited—it suggests that ATP hydrolysis is not required for Na : Na exchange. However, this is a point on which we have no data. There is one difficulty with this proposal. ADP ought to be a powerful inhibitor of ATP hydrolysis if an appreciable fraction of the enzyme is to be stabilized as E · ATP; this does not seem to be the case.

A second scheme for the action of ADP is to suppose that it allows a reaction such as the following to proceed:

$$E_1 + ATP \leftrightarrow E_1 \cdot ATP \leftrightarrow E_1 \sim P + ADP$$

$$E_1 \sim P \rightarrow E_1 + P_i$$

This would be a Na-activated ATPase, and would in effect be a short circuit on the usual cyclic scheme involving E_1 and E_2. This arrangement would avoid the necessity for supposing that ADP is highly inhibitory to ATP hydrolysis but would require that a different pathway be utilized.

Finally, it should be recognized that if enzyme and ion pumps are separate molecular entities (or functionally separate parts of the same molecule), it is possible that ADP may have separate effects on the pump and on its energy source.

ION SPECIFICITY IN TRANSPORT

The Na pump appears to have a high degree of specificity for Na; there is no evidence of transport of other ions (in the squid axon system). The inclusion of Li at a concentration 8 times that of Na has shown no effect on Na extrusion (Brinley and Mullins, 1968) so that competition for either transport or activation would appear minimal. Much the same sort of specificity is exhibited by membrane ATPase, which has an absolute requirement for Na. Such a finding argues somewhat in favor of an identity between ATPase activating and transport sites. The site on the outside of the membrane has been called the "not Na" site by Post *et al.* (1969) since its specificity for a particular monovalent cation is not great. All cations from Li to Cs show some activation of this site in the ATPase and K, Rb, and Cs are clearly transported inward by a pump that is usually called the K pump. The activation of Na extrusion by Rb and especially Cs is considerably less than for K and transport is usually less, as well. Unfortunately, experimental information is insufficient to show whether the apparent coupling ratios of Na/Cs differ greatly from those for Na/K over a range of external concentrations.

What is significant about the ion specificity of both the membrane ATPase and transport is the extreme specificity of the Na site and the distinctly lesser specificity of the K site. The K site resembles the sort of specificity found with certain cyclic polypeptides which have been introduced into artificial lipid bilayers to confer upon them K-Rb specificity, that is, a specificity that depends on a steric property of the ion. The Na specificity is different; it does not confuse Na-like ions (such as Li) with Na and suggests very strongly that it depends on the electronic structure of Na rather than on the steric properties of the ion. Na binding might be by association with a protein-metal compound because simple organic compounds such as amino acid residues or lipids seem utterly unable to provide the requisite specificity. Such a Na binding compound is, therefore, very different from a macrocyclic polypeptide ring. Thus, one infers that the chemical nature of the pumps and activating sites for Na and K are intrinsically different. Such a conclusion makes it somewhat implausible that a Na binding site can be transformed (by some conformational rearrangement of a protein) into a K binding site if the mechanisms for ion selectivity at the two sites are intrinsically different.

THE ACTIVATION OF ION TRANSPORT

Recent developments in ion transport studies using squid axons have shown that various cations applied to the outside of the membrane can have a complicated set of effects on transport. Many of these effects can be described as the activating effect of external cations on transport.

1) *K activation* This is the most familiar effect. The effect of $[K]_0$ on Na efflux is half maximal at a concentration less than 1 mM in Na-free solutions. Na competes for sites and at seawater [Na], the required [K] for half maximal Na efflux is about 10 mM.
2) *Na activation* Even when the competition of Na for K sites is allowed for, Na has an activating effect on Na efflux which varies from axon to axon such that the sum of Na efflux activated by Na and by K is roughly constant.
3) *Ca activation* In Na-free seawater, there is a Na efflux which depends on $[Ca]_0$. Na competes at the Ca activating site so that in Na seawater there is very little Ca dependent Na efflux. $[Li]_0$ appears to promote the Ca-dependent Na efflux to a greater extent than simple Na-free conditions would suggest.

4) *Ion Independent Na Efflux* There is an appreciable Na efflux into isotonic dextrose solutions. Part of this efflux is undoubtedly K-activated Na efflux even though the solution is nominally K-free but it seems unlikely that all the Na efflux is so activated. A difficulty with experiments of this sort lies in deciding on a suitable reference substance from which activating effects are to be measured. Both choline and Mg-Dextrose have been used but not enough work has been done to be sure that choline has no activating properties.

The activation of K transport inward appears simpler, although this may only reflect the fact that fewer experiments have been done on K influx than on Na efflux. At any rate, the Na concentration inside an axon has an important activating effect on K transport and one that is half saturated at about 10 mM Na. Since the Na concentration of squid axons as usually studied is often around 100 mM, the K pumping system is fully activated in most experiments. Changes in external [Na], which have large effects on the K activation of Na efflux, have no effect on K transport so that the Na/K competition previously noted does not affect the K transport system.

Given the complexity of the observations of ion activation of transport and the competition among ions for activating sites, it is clear that ion transport cannot be explained in the rather simplistic terms of a pump mechanism that moves 3 Na outward and 1 K inward per transport cycle. The best that can be done at present is to describe the observed transport as unambiguously as possible and to consider criteria that seem presently valid for deciding whether a particular transport system is related to another system or is, in fact, a separate entity.

SUMMARY

The contribution of the squid axon to active transport studies has been mainly that of showing that the energized movements of Na and K across its membrane are rather more complicated than present schemes of active transport envisage. Much more experimental work will be required before the phenomena are adequately understood and the following scheme can only be considered a more provisional one for representing transport phenomena. Figure 3 shows membrane ATPase as supplying energy to Na and K pumps; there are separate activating sites for the enzyme and transport sites for the pumps in order to accommodate the differing concentration dependencies of these functions. The Na/Ca pump is shown with an enzyme for ATP utilization. The enzyme differs from membrane ATPase in that it has no glycoside site, and it is Li activated on the outside. While information on this system is still meager, it is intended to accommodate the following experimental information: a glycoside insensitive, ATP dependent Na influx that is Na_i activated, a competition between Na and Ca for pumped entry, and a glycoside insensitive, ATP sensitive Na efflux.

The ATP independent Na efflux observed in the presence of glycoside is, rather clearly, not a property of the Na/Ca pump but is the result of transforming membrane ATPase E into E-gly, where this is a form of the enzyme

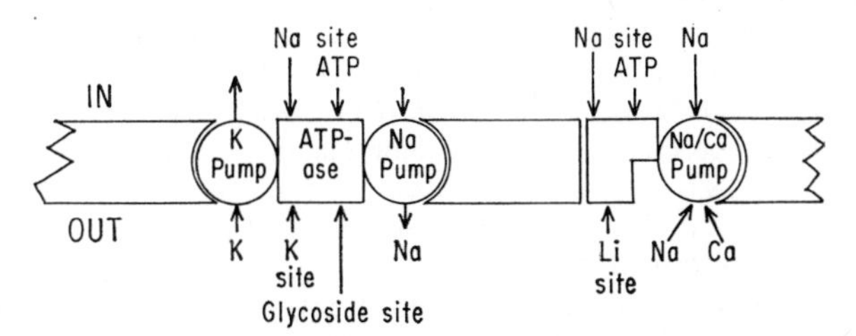

Figure 3 Two separate pumping systems for Na are shown in this diagram. On the left, membrane ATP ase, activated by Na and K, feeds energy to Na and K pumps. Glycoside applied to this system stops energy delivery from the enzyme, stops the K pump, and allows the Na pump to carry cations inward in exchange for Na_i. On the right is a glycoside insensitive enzyme that produces an ATP sensitive Na influx and efflux that is dependent on Na_i and Li_0 and Ca_0. Activation is shown by Li_0 and possibly Ca_0 is also an activator as well as a species of ion that is transported by the system

that occurs in the presence of ouabain and related substances. This change can be most easily visualized by supposing that the K pump is stopped while the Na pump is allowed to cycle freely and to carry cations inward and Na outward. Details of the activation and competition mechanisms involved in ion pumping remain to be worked out but it is quite possible that in the end the system will become even more complicated than it now appears.

REFERENCES

Baker, P. F., M. P. Blaustein, A. L. Hodgkin, and R. A. Steinhardt (1969a). The influence of calcium on sodium efflux in squid axons. *J. Physiol.* (London). **200**, 431.

Baker, P. F., M. P. Blaustein, R. D. Keynes, J. Manil, T. I. Shaw, and R. A. Steinhardt (1969b). The ouabain-sensitive fluxes of sodium and potassium in squid giant axons. *J. Physiol.* (London). **200**, 459.

Baker, P. F., A. L. Hodgkin, and T. I. Shaw (1962). Replacement of the axoplasm of giant nerve fibres with artificial solutions. *J. Physiol.* (London). **164**, 330.

Baker, P. F., and T. I. Shaw (1965). A comparison of the phosphorous metabolism of intact squid nerve with that of the isolated axoplasm sheath. *J. Physiol.* (London). **180**, 424.

Beaugé, L. A., and R. A. Sjodin (1968). The dual effect of lithium ions on sodium efflux in skeletal muscle. *J. Gen. Physiol.* **52**, 408.

Blaustein, M. P., and A. L. Hodgkin (1969). The effect of cyanide on the efflux of calcium from squid axons. *J. Physiol.* (London). **200**, 497.

Brinley, F. J., and L. J. Mullins (1967). Sodium extrusion by externally dialyzed squid axons. *J. Gen. Physiol.* **50**, 2303.

Brinley, F. J., Jr., and L. J. Mullins (1968). Sodium fluxes in internally dialyzed squid axons. *J. Gen. Physiol.* **52**, 181.

Caldwell, P. C., A. L. Hodgkin, R. D. Keynes, and T. I. Shaw (1960). Partial inhibition of the active transport of cations in the giant axons of *Loligo. J. Physiol.* (London). **152**, 591.

Caldwell, P. C., and R. D. Keynes (1960). The permeability of the squid giant axon to radioactive potassium and chloride ions. *J. Physiol.* (London). **154**, 177.

Canessa-Fischer, M., F. Zambrano, and E. Rojas (1968). The loss and recovery of the sodium pump in perfused giant axons. *J. Gen. Physiol.* **51** (5, Pt. 2), 162s.

Dean, R. B. (1941). Theories of electrolyte equilibrium in muscle. *Biol. Symp.* **3**, 331.

De Weer, P. (1970). Effects of Intracellular Adenosine-5′-diphosphate and Orthophosphate on the Sensitivity of Sodium Efflux from Squid Axons to External Sodium and Potassium. *J. Gen. Physiol.* **56**: 583.

Grundfest, H., C. Y. Kao, and M. Altamirano (1954). Bioelectric effects of ions micro-injected into the giant axon of *Loligo. J. Gen. Physiol.* **38**, 245.

Horowicz, P., and C. Gerber (1965). Effect of external potassium and strophanthidin on sodium fluxes in frog striated muscle. *J. Gen. Physiol.* **48**, 489.

Keynes, R. D., and R. C. Swan (1959). The effect of external sodium concentration on the sodium fluxes in frog skeletal muscle. *J. Physiol.* (London). **147**, 591.

Mullins, L. J., and F. J. Brinley, Jr. (1967). Some factors influencing sodium extrusion by internally dialyzed squid axons. *J. Gen. Physiol.* **50**: 2333.

Mullins, L. J., and F. J. Brinley, Jr. (1969). Potassium fluxes in dialyzed squid axons. *J. Gen. Physiol.* **53**, 704.

Post, R. L., S. Kume, T. Tobin, B. Orcutt, and A. K. Sen (1969) Flexibility of an active center in sodium-plus-potassium adenosine triphosphatase. *J. Gen. Physiol.* **54**, 306S.

Shaw, T. I. (1966) Cation movements in perfused giant axons. *J. Physiol.* (London). **182**, 209.

Sjodin, R. A., and L. A. Beaugé (1968). Coupling and selectivity of sodium and potassium transport in squid giant axons. *J. Gen. Physiol.* **51**, 152S.

Sjodin, R. A., and L. A. Beaugé (1969). The influence of potassium- and sodium-free solutions on sodium efflux from squid giant axons. *J. Gen. Physiol.* **54**, 664.

Sjodin, R. A., and L. J. Mullins (1967). Tracer and non-tracer potassium fluxes in squid giant axons and the effects of changes in external potassium concentration. *J. Gen. Physiol.* **50**, 533.

Steinbach, H. B. (1940). Na and K in frog muscle. *J. Biol. Chem.* **133**: 695.

PAPER 3

N-Shaped Characteristics in Living Membranes

HARRY GRUNDFEST

Laboratory of Neurophysiology, Department of Neurology, College of Physicians & Surgeons, Columbia University, New York

---

Kacy, it's a long time since we caulked and patched those leaky sloops. Though we have since dry sailed our separate courses, they were not too far out of each other's sight. I'm sorry that Elizabeth could not stay aboard a while longer.

The "n-shaped" current-voltage relation of certain electrogenic processes of electrically excitable membranes makes it possible for cells to generate all-or-none spikes (Hodgkin and Huxley, 1952). A negative slope conductance characteristic also occurs in various electronic systems (e.g. the tunnel diode) and is the basis for their metastable or oscillatory all-or-none transitions. Accordingly, numerous proposals to "explain" spike bioelectrogenesis of living cells have based themselves on electrical analogs with "negative resistance" characteristics. However, this formalism tends to divert attention from the fundamental aspect of bioelectric phenomena, the capacity of the very thin, relatively high resistance, ion permselective membrane to change its degree of permselectivity for one or another ion species. The nature of the changes in macromolecular configurations that cause the changes in electrical conductance is as yet unknown, but the variety of the changes and some of the characteristics of their effects are clearly evidenced. Changes in conductance on changing the membrane potential can also be induced in some varieties of lipid bilayer membranes (Mueller and Rudin, 1968).

K. S. Cole pointed out (1965, 1968) all too briefly that the n-shaped characteristic of spike generators is a consequence of two associated changes in membrane properties: (i) depolarization causes an increase in conductance and (ii) the conductance change results from increased permeability for an ion that provides an emf which is more inside positive than the resting potential. The latter condition is readily derived from the conductance form of Ohm's Law.

$$I = G(E - E_0) \quad (1)$$

$$dI/dE = dG/dE\,(E - E_0) + G \quad (2)$$

I is the current
E is the membrane potential
E_0 is the emf of the ionic battery
G is the chord conductance, I/E or R^{-1}

If E_0 is sufficiently positive the slope conductance (dI/dE) becomes negative, within certain limits. If the change in conductance (dG/dE) is very small, e.g. before Na activation takes place in the squid axon (Hodgkin and Huxley, 1952) the slope conductance will be dominated by G, which is always positive. Furthermore, as E approaches E_0 during voltage clamp measurements $E - E_0$ is reduced to the point where the slope conductance again becomes positive. Thus, the resulting n-shaped characteristic is but a consequence, albeit a very important one functionally, of the underlying fact that the electrically excitable spike generating membrane responds to depolarization with an increase in conductance toward an ion that provides an inside-positive electrochemical battery.

The recognition that the cell membrane is a heterogeneous electrochemical system (Grundfest, 1957*a*, *b*, 1959, 1961) has led to a broader analysis of its negative slope characteristics and of their bioelectric manifestations. The excitable membrane can be described in a generalized (though still relatively simplified) manner by the equivalent circuit of Figure 1. Different ionic species form electrochemical batteries with different emf's, the latter depending chiefly upon the concentration gradients for the respective ions across the membrane. Under normal conditions of animal cells the K and Cl batteries have emfs that are inside-negative, at or close to the resting potential, E_M. Other batteries that occur normally in living systems, such as those for Na and the divalent cations Ca or Mg, are normally inside-positive with respect to zero reference.

The relative contributions of the different emf's to the membrane potential are in some proportion to the mobilities of the various ionic species across the membrane. In the equivalent circuit mobilities are represented by resistances. Activation processes may increase the mobility of any given species ($\leftarrow$). Inactivation processes decrease it ($\rightarrow$). Either change results in a change in conductance and these changes may be elicited by depolarization (D)or hyperpolarization (H) of the membrane. The generalized electrogenic cell also possesses electrically inexcitable, as well as electrogencially unreactive components.

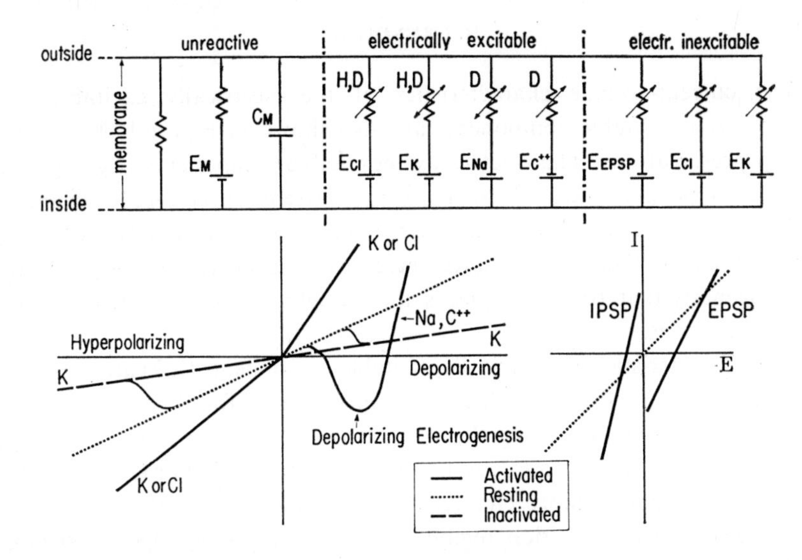

Figure 1 Generalized equivalent circuit and the voltage-current characteristics of a heterogeneous excitable membrane with a variety of ionic batteries. Above: Equivalent circuit diagram. Emf's contributed by electrogenic ion pumps are omitted. Below: Current-voltage (I-E) characteristics of electrically excitable components (left) and of electrically inexcitable (right), in voltage-clamp presentations. Origins are set at resting potential (E_M).

Membrane capacity (C_M) and invariant conductive component represent major, unreactive portion of membrane. Conductive component is subdivided into ion permselective element with E_M as average emf and nonselective element symbolized by resistance without emf. Reactive components are represented by variable resistances in series with different ionic batteries depending on different permselectivities. E_K and E_{Cl} in general are close to E_M, but E_{Na} and $E_C{}^{2+}$ (Ca or Mg in various cells) are shown as inside positive. Permselective electrically excitable channels respond to depolarizing (*D*) and/or hyperpolarizing (*H*) stimuli with activation (↑) or inactivation (↓). Electrically inexcitable depolarizing electrogenesis of receptive and synaptic membrane is indicated by inside-positive battery (E_{EPSP}). Inhibitory synaptic electrogenesis involves increased conductance for either Cl (E_{Cl}) or K (E_K).

Unreactive electrically inexcitable electrogenic components have linear (ohmic) I-E characteristics, but activation of reactive electrically inexcitable components by specific stimuli increases slope (indicating higher conductance in voltage clamp presentation). Depolarizing electrogenesis translates characteristic to right. Diagram shows inhibitory electrogenesis (IPSP) as hyperpolarizing and characteristic translated to the left. As membrane is polarized by applied currents, resting and active characteristics approach crossing beyond which the sign of recorded electrically inexcitable response is reversed relative to steady membrane potential. Reversal potential approximates equilibrium (Nernst) potential of ionic batteries that cause electrogenesis.

I-E characteristic of electrically excitable components exhibits nonlinearities which result from transition of resting membrane conductance to higher values. Only the conductance increase caused by Na or C^{2+} activation shifts the characteristic significantly along the voltage axis. Three nonlinear regions with negative slope characteristics are shown. They mark transition from E_M to E_{Na} or $E_C{}^{2+}$ by activation processes and from resting conductance to lower conductance by depolarizing and hyperpolarizing K inactivation respectively. (Grundfest, 1967)

The current-voltage characteristics of the electrically excitable and electrically inexcitable components are also shown in Figure 1, in voltage-clamp presentations. The electrically inexcitable components (right graph) behave as passive linear elements as the current or voltage is changed. However, specific stimuli activate the two varieties of electrically inexcitable membrane and their conductance increases. The degree and sign of the electrogenesis that results displaces the straight line characteristics along the voltage axis.

The characteristics of electrically excitable components (left) are markedly non-linear, on the other hand. Activation processes for K or Cl increase the slope, but the shift along the voltage axis is small or absent, since E_K and E_{Cl} tend to be close to E_M. Activation processes for Na or C^{++}, however, shift the high conductance characteristic, as called for by the appropriate ionic battery. The transition region then has a negative slope. Two more n-shaped characteristics are also present, which are generated by depolarizing and hyperpolarizing inactivation processes. In these cases dI/dE is negative because dG/dE is negative and this is independent of the nature of the ionic battery. In fact, the transition from the high conductance (resting) to the low conductance (inactivated) state entails little or no change in emf, because usually only a conductance decrease for the most mobile ion, K, is detectable.

The voltage clamp measurements of Figure 2 show that at least 3 different electrically excitable components are present in the membrane of frog atrial fibers. There is an inward current even in the presence of tetrodotoxin (TTX) which blocks Na activation. On the other hand, there is also an inward current in the presence of Mn which blocks Ca activation. When both TTX and Mn are present the inward currents are abolished, but the characteristic exhibits a region where there is little change in the outward current while the depolarization of the membrane is increased by some 50 mV. Thus, the conductance is decreasing, an indication of depolarizing K-inactivation. Accordingly, this cardiac tissue exhibits two activation processes (for Na and Ca) and a region of inactivation, for K. The Na channels are rather specific for this cation, while the Ca channels also permit entry of Na (Rougier *et al.*, 1969).

Other cells that generate spikes by activation for both Na and Ca are also known (Geduldig and Junge, 1968; Hagiwara and Nakajima, 1966). The membrane of squid giant axons can become permeable to Ca and to various "foreign" cations (Tasaki *et al.*, 1969; Yamagishi, 1970). Since the spikes are blocked by TTX it appears likely that changes in the channel structures which produce an appreciable permeability to Ca, etc., did not affect the pharmacological Na inactivation produced by TTX. Of further interest is the fact that although the spike may be of large amplitude the current associated with the electrogenesis is very small. It is therefore unlikely that these spikes can perform the function of the normal Na spikes—the propagation of the impulse in all-or-none fashion.*

The various conductance changes and their consequences are further diagrammed in Figure 3. The left ordinate is $G(V)$, with E_M (the resting potential) as the origin on the abscissa. In the center column are the voltage

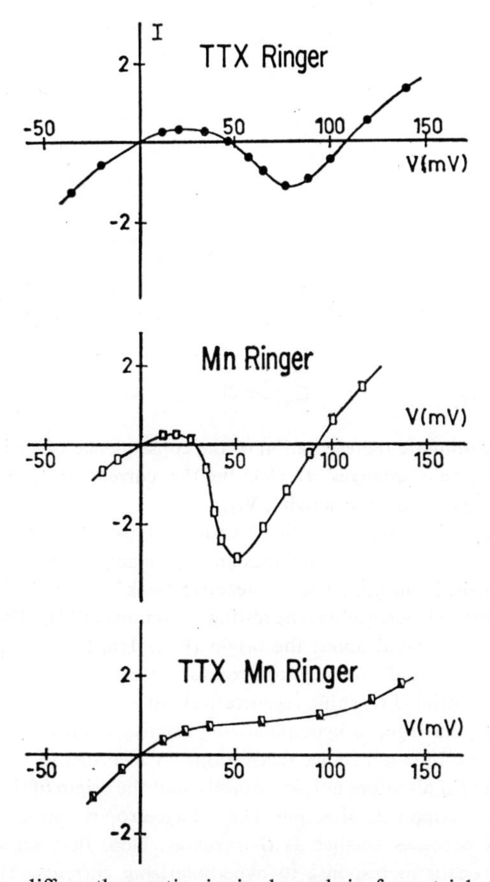

Figure 2 Three differently reactive ionic channels in frog atrial muscle. Voltage clamp measurements. An inward current develops even after Na channels are blocked by TTX (upper graph), indicating the influx of another ion with an inside positive emf. Blockage of Ca influx with Mn still leaves an inward current, but the characteristic for this ion, presumably Na, is somewhat different. When both TTX and Mn are present no inward current develops. The characteristic then exhibits a flattened region in which dg/dE is negative. This is an indication of depolarizing K inactivation and is also found in other cardiac tissue. The K channels reopen with further depolarization. (Modified from Rougier et al., 1969)

* see footnote on p. 43.

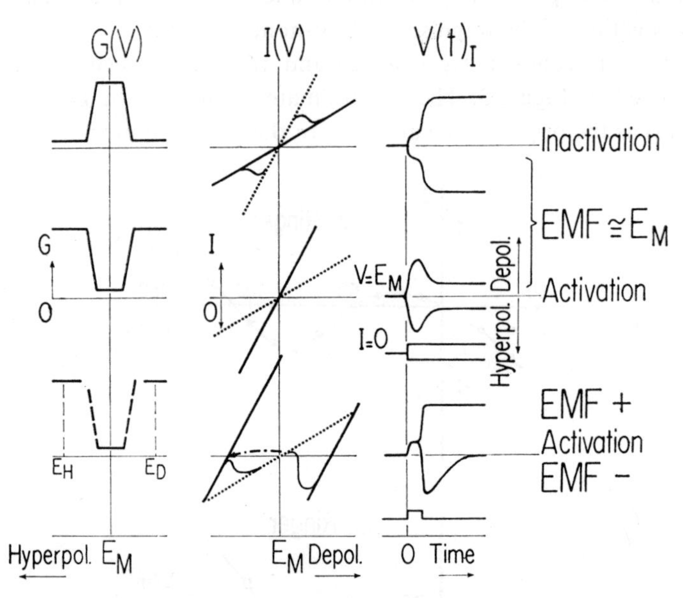

Figure 3 Diagrammatic representation of the consequence of various electrically excitable conductance changes, *G* (V), on the current voltage characteristic, I (V) and on the electrogenetic activity, $V_{(t)_I}$.

Top. Decrease of conductance (inactivation) is usually seen only when the resting conductance is high and includes an appreciable fraction of the available reactive K channels, in addition to the unreactive "leak" channels. The inactivation usually does not seriously affect the resting membranes EMF. The characteristic (middle) is merely rotated about the origin (E_M). Transition from high to low conductance is a region of negative slope conductance and when a constant current is applied the potential (I/G) shifts regeneratively to a higher value as *G* decreases. This shift is a depolarizing or hyperpolarizing inactivation response.

Middle. Increase in conductance (activation) for a species of ionic battery with an emf at or near E_M also does not significantly shift the origin of the characteristic, but rotates it in the opposite direction. The voltage drop resulting from a constant applied current becomes smaller as *G* increases. Note that activation, like inactivation, can occur in response to hyperpolarizing currents. The foregoing 4 types of responses are generated only while the current is applied.

Bottom. Increased conductance for species of ionic batteries that have emf's differing from E_M introduces possibilities for negative slope characteristics and responses which are autogenetic. They are triggered by a brief stimulus and follow a time course which is determined by the kinetics of the activation process itself. Further description in text clamp presentations of the respective characteristics. The right column shows the time course of the changes in membrane potential resulting from stimulation with applied currents. The kinetics of the development and/or subsidence of the various changes are ignored.

Inactivation is diagrammed in the upper row. As described above, the slope conductance is negative because dG/dE (equation (2) is negative and the characteristic is n-shaped in both the depolarizing and hyperpolarizing quadrants. When a sufficient constant current is applied the negative slope characteristic carries the membrane potential (E) from a low value (low $I \cdot R$ product) to a new step (high $I \cdot R$ product) as the conductance decreases. While these all-or-none changes in the membrane potential resemble those of spike electrogenesis the characteristic feature is that the change persists only as long as the current is applied. The change in $I \cdot R$ is purely dissipative and not *autogenetic* as is the spike.

Activation without a large change in the emf is diagrammed in the middle row of Figure 3. The increased conductance for K or Cl (Figure 1) tends to decrease the $I \cdot R$ product. However, the full development of the activated state is usually slow. Thus, early during the applied current the change in membrane potential is larger than during the later steady state. Since these activation responses are not regenerative they persist only while the current is applied and for a time thereafter while any small changes in E_M are dissipated.

Regenerative activation responses (lowest row in Figure 3) include a change toward a large inside-negative emf (E_H), as well as that to E_D, positive relative to E_M or, more frequently, relative to reference zero. The change to E_H might be produced by hyperpolarization or by a depolarizing stimulus. The depolarizing electrogenesis due to the shift of the membrane potential toward E_D might last indefinitely after a brief triggering pulse, but in living cells the membrane is restored to the resting potential after some time by auxiliary processes. In the squid axon Na inactivation and K activation shape the response into a spike.

When the activation process involves an inside-negative emf the membrane potential must shift rapidly toward the new value. The hyperpolarization, however, tends to quench the activation process which is induced in the first place by depolarization. The change in potential would therefore have

* Spike electrogenesis in the soma need not be an important functional requirement for many neurons, whereas it is the *sine qua non* for propagation in the axons. Thus, it is relatively unimportant whether or by what ionic mechanism a spike is generated in the soma. All axons known thus far, on the other hand, utilize the gradient of the Na battery. Presumably, Na activation causes the largest possible increase in conductance and provides the highest inward current to alter the charge on the membrane capacity.

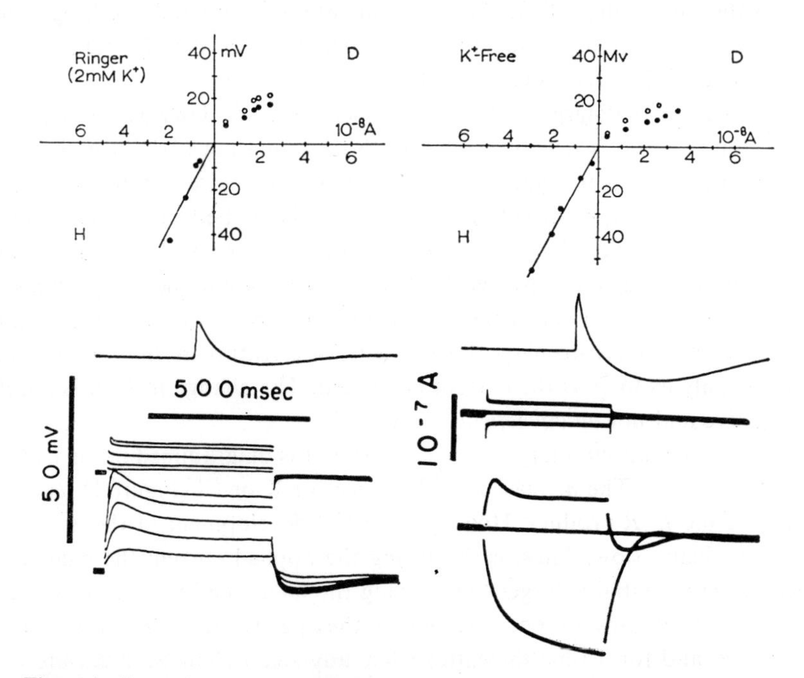

Figure 4 Example of autogenetic hyperpolarizing electrogenesis. Frog slow muscle fiber in control Ringer's solution (left) and in a K free saline (right).

Top. In both cases the steady state characteristic exhibits a curvature in the depolarizing quadrant which denotes an increase in conductance. Note that this cell does not generate spikes. The conductance change is graded, increasing with greater depolarization and is time variant, increasing with the duration of the applied current.

Middle. Neurally evoked depolarizing postsynaptic electrogenesis is succeeded by a prolonged after hyperpolarization. The latter is enhanced when E_K is made more negative by removal of K_0. Note that the peak of the hyperpolarization which therefore is due to depolarizing K activation is followed by a slower return to the resting potential.

Bottom. Responses to applied currents. The changes in membrane potential during the applied current exhibit development of a plateau as in the middle row of Figure 3. The K activation persisted after the current was terminated and caused hyperpolarization. An inward current (*right*) caused a large hyperpolarization, indicating that for an outward current the depolarizing K activation ("delayed rectification") occurred rapidly. Note also that the hyperpolarizing electrogenesis after the depolarizing current subsided more slowly than did the much larger hyperpolarization induced by the inward current. This is indicative of an autogenetic response when depolarizing K activation produces a hyperpolarizing electrogenesis. (Modified from Grundfest 1961, Figure 15; after unpublished data by Belton and Grundfest, 1961)

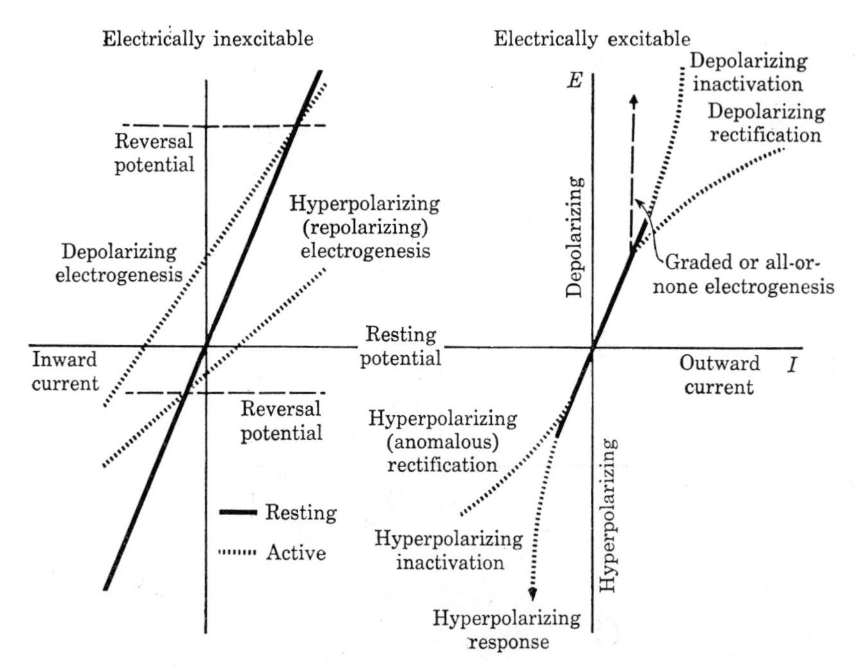

Figure 5 *I-E* relations of electrogenic membranes. *Electrically inexcitable* membranes behave as ohmic resistances, *E* changing linearly with *I*. However, the slope of the relation changes during activation of the membrane by an appropriate stimulus. Broken lines represent active membrane of (depolarizing) EPSP's and of (hyperpolarizing) IPSP's respectively. Recorded amplitudes of the responses (given by differences between resting *I-E* line and that during activity) change with change in membrane potential. Thus a characteristic feature of electrically inexcitable electrogenesis is its change in sign when *E* exceeds a reversal potential specified by intersection of resistance lines for active and passive membrane.

Electrically excitable membranes exhibit nonlinear behavior characterized by one or by several varieties of conductance changes. On stimulation with depolarizing currents many but not all cells respond with conductance increase (depolarizing activation) for Na, Ca, or Mg—which introduces a new inside-positive emf and causes graded or all-or-none depolarizing electrogenesis. Increased conductance for K or Cl (inappropriately termed "rectification") that is evoked by depolarizing stimuli or for any ions by hyperpolarizing currents is usually not regenerative. However, it can become regenerative when electrochemical conditions permit. Decreased conductance or inactivation may be evoked by depolarizing or hyperpolarizing currents and it can also be regenerative. Various ionic processes which cause nonlinear relations exhibit different degrees of time variance. (Modified from Grundfest, 1961)

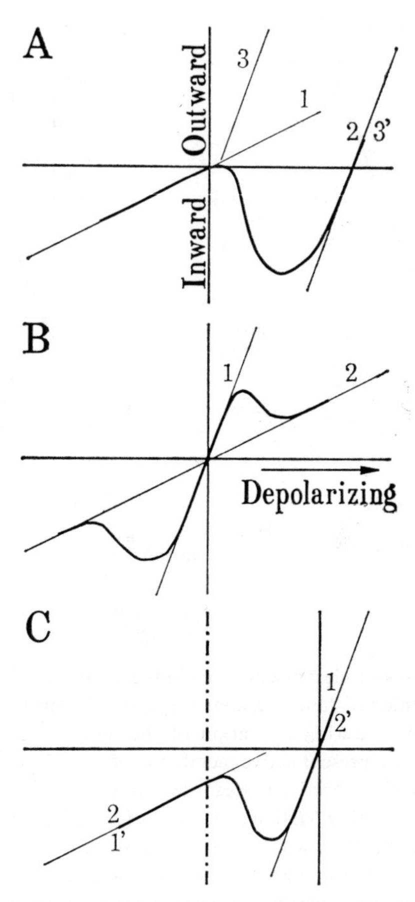

Figure 6 Comparison of various negative slope characteristics involved in different types of electrogenesis. *A*. The shift from the resting state (1) to higher conductance for a more positive ionic battery (2) results in a negative slope characteristic that causes all-or-none response of normal spike electrogenesis. K or Cl activation (3) causes little change in membrane potential. If E_K or E_{Cl} is made positive to E_M (3′) "anomalous" spikes are evoked. *B*. Inactivation processes shift the characteristic from (1) to (2), but with little or no change in E_M. In either quadrant, the transition is a negative-slope region giving rise to inactivation responses. *C*. Current-voltage characteristic of a cell depolarized by immersion in K-rich media. The original membrane potential is shown by broken line. Conductance (1) is higher than in the original state (2), but applying inward current causes a transition to state 2 and a hyperpolarizing response. If the cell is kept hyperpolarized (1′), K activation is initiated by a depolarizing stimulus and the characteristic shifts temporarily to high conductance state (2′). The result is a K spike. (Modified from Grundfest, 1966*b*)

a steep onset and a slower return to the resting potential as the charge on the membrane capacity is dissipated across the higher resistance of the hyperpolarized membrane. Purely inside-negative responses are relatively rare but may be recognized in the hyperpolarization of frog slow muscle fibers after the end of a depolarizing stimulus (Figure 4). The terminal hyperpolarization of the squid axon is also a manifestation of the shift of the membrane potential toward E_K, which is inside-negative by about 10 to 15 mV (Hodgkin and Huxley, 1952). As the K activation is terminated the resistance of the hyperpolarized membrane rises above the resting value until the resting potential is again attained (Shanes *et al.*, 1953).

The hyperpolarizing autogenetic electrogenesis conceivably might play a role in cells which generate repetitive spikes, as for example pacemaker cells. A transient hyperpolarization following a spike would tend to accelerate recovery from Na inactivation and thus might improve performance of repetitive spike discharges. Hyperpolarization by vagal stimulation or by chemical agents frequently improves spike electrogenesis in cardiac tissue.

The usefulness of inactivation processes is probably quite restricted, although their occurrence is even more widespread than was initially supposed. Depolarizing K inactivation occurs during spike electrogenesis of eel electroplaques (Nakamura *et al.*, 1965; Morlock *et al.*, 1968; Ruiz-Manresa et al., 1970) and of other weakly electric gymnotids (Bennett and Grundfest, 1959, 1966; Morlock and Janiszewski, 1970), as well as in some cardiac muscle (Figure 2), and presumably, it aids the functional performances of the cells. However, the occurrence of K inactivation as well as that for Na has several theoretical implications, not the least of which is the as yet unknown significance of the symmetry in the non-linearities of the current voltage characteristics (Figure 5). Another aspect is the interplay of the different initial states of the excitable components. The different responses diagrammed in Figure 3 result merely from different starting conditions; G high to low, or low to high; emf equal to or different from E_M. It is therefore possible sometimes to manipulate the initial conditions so as to alter the character of the response. In the diagram of Figure 6*A* the electrochemical conditions are altered to that the slope change from 1 to 3 (with emf unchanged) becomes a change from 1 to 3′ (with the emf ot the ionic battery inside positive). Such changes have been induced and have resulted in Cl spikes of skate electroplaques (Cohen *et al.*, 1961; Aljure et al., 1962, and unpublished), K spikes of *Tenebrio* muscle fibers (Figure 7) and other cells.

These generalizations on the origin and nature of the n-shaped characteristics of excitable cells permit a classification and explanation of numerous seemingly bizarre electrogenic phenomena in terms of the expanded form

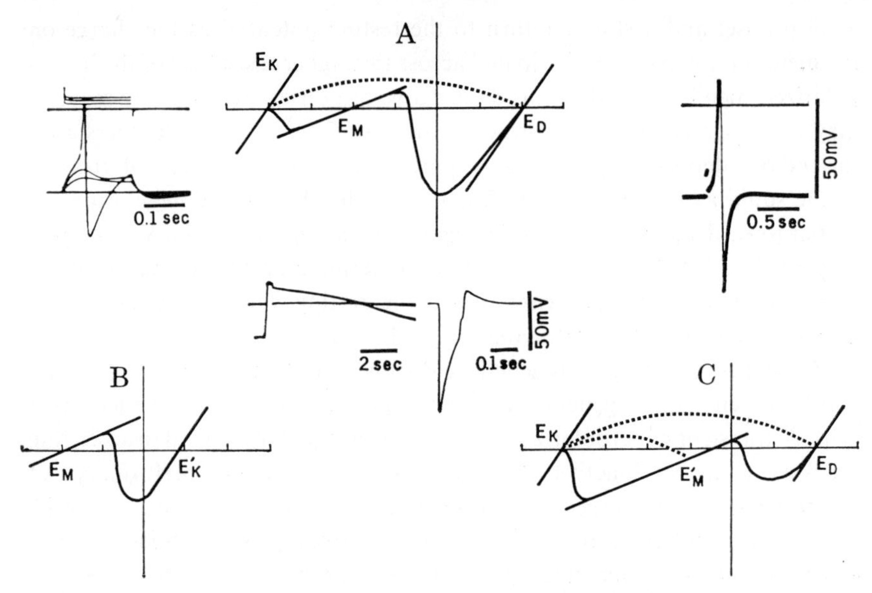

Figure 7 Intracellularly recorded "anomalous" spikes in muscle fibers of *Tenebrio molitor* and in esophageal cells of *Ascaris lumbricoides*, and explanation of the electrogenesis in diagrammatic voltage clamp presentations. *A*. Under approximately normal ionic conditions $E_D > E_M > E_K$, where E_D is the inside-positive ionic battery for depolarizing electrogenesis (Mg in *Tenebrio*). A depolarizing stimulus shifts the characteristic from E_M toward E_D, giving rise to depolarizing electrogenesis of spikes. Depolarization induces K activation and a shift of the characteristic toward E_K (dotted line), giving rise to the second phase of the spike. The latter is particularly marked in recording from *Ascaris* cell (right). Hyperpolarization quenches K activation and the potential slowly returns to E_M. *B*. *Tenebrio* muscle fibers can tolerate presence of as much as 340 mM K without appreciable change in E_M. However, $E_K > E_M$ and the response to a brief stimulus (left center record) is a long-lasting K spike. *C*. *Ascaris* esophageal cells depolarize when the bathing solution contains an increased concentration of Cl, but E_K is relatively unchanged. Thus $E_M > E_K$ and most or all of the electrogenesis is hyperpolarizing, occurring as the characteristic shifts from E_D to E_K or from E_M to E_K. The hyperpolarizing spike is brief (as in *A*) because K activation is quenched by hyperpolarization. (Grundfest, 1967, *Ascaris* data from del Castillo and Morales, 1967)

(Figure 1) of the ionic theory of Hodgkin and Huxley (1952). Reasonable explanations become available also for some phenomena that have not yet been completely analyzed by appropriate experiments. Among them, for example, are prolonged oscillatory spikes (Grundfest, 1961, Figure 26 and 27), "upside-down" spikes (*ibid*, Figure 28) and diphasic or even negative spikes (Figure 7).

Inactivation responses, generated as schematized in Figure 6*B* have been obtained in many cells. Monotonic responses to both depolarizing and

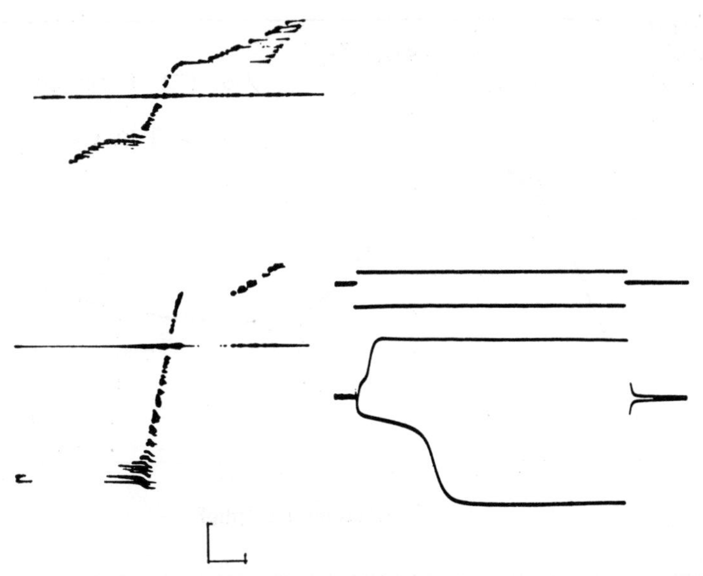

Figure 8 Depolarizing and hyperpolarizing K inactivation in an eel electroplaque. *Left*: Steady state characteristics as measured with currents that cause different degrees of change in the membrane potential. The calibrations are 10 mA/cm^2 for the ordinate and 50 mv for the abscissa. Upper graph is for cell in the standard saline, lower after replacing all Na with K. At rest the conductance is the sum of two components, contributed by the unreactive leak channels, G_L and the reactive channels G_K, which close when the membrane potential is displaced beyond certain values. Elimination of G_K causes the membrane potential to shift regeneratively to larger I · R values. In the control saline strong depolarizing currents evoked spikes the peaks of which are registered as the points on the extreme right. The G_K component of the resting cell is increased markedly in high K, as denoted by the steeper slope of the characteristic. The transition from the high conductance ($G_L + G_K$) to low conductance (G_L) in either direction is accentuated as described in the text. *Right*: The records show the inactivation responses evoked in high K by depolarizing and hyperpolarizing currents. The vertical calibration represents 10 mA/cm^2 for the current monitoring traces (upper) and 50 mv for the voltage traces (lower set). The horizontal calibration represents 2 msec. Note that the threshold for depolarizing inactivation was lower than that for hyperpolarizing inactivation. This is also seen in the characteristic (From Ruiz-Manresa, 1970)

hyperpolarizing currents are shown in Figure 8. Voltage clamp registration of the transition from a low resistance at rest to a high resistance on depolarization is shown in Figure 9. The high conductance initial steady state may be represented as the sum of the reactive conductance channels for K (G_K) and the passive leak channels (G_L). Complete K inactivation reduces the conductance to G_L alone. However, often in some cells and usually in many others, G_K may be so low in the resting state that the elimination of any open channels by inactivation is without significance on the steady state, just as eli-

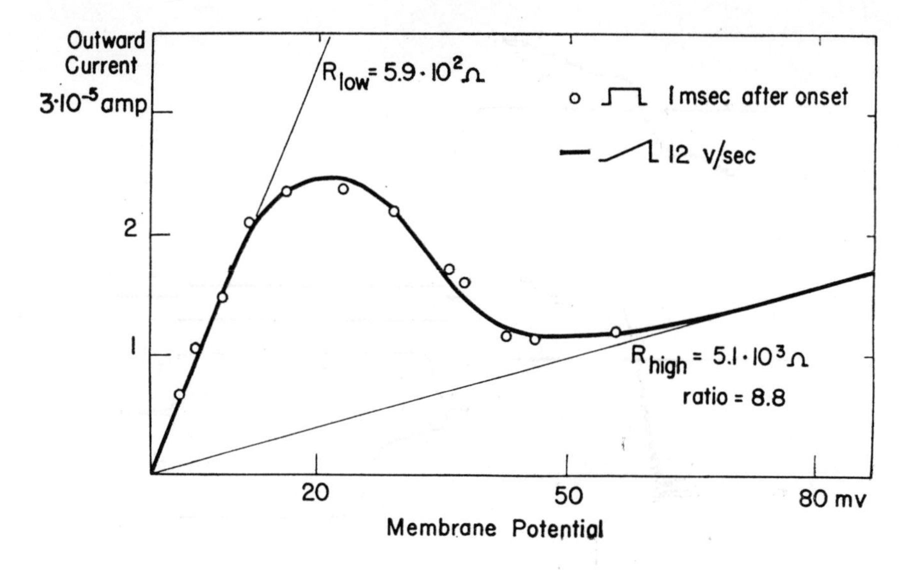

Figure 9 Voltage-current characteristic of an electroplaque from *Hypopomus sp.* showing depolarizing inactivation. The preparation was bathed in isosmotic K acetate saline. The origin is the initial membrane potential of the depolarized cell. The abscissa represents increments toward inside positivity. The heavy line shows the observed change in outward current as the cell was subjected to a "ramp" command voltage that increased at a rate of 12 v/sec. Step command voltages were also applied and the open circles show the currents measured 1 msec after the beginning of the respective pulses. The thin lines are the limiting slopes which represent the high and low states of the membrane conductance. The coincidence of the ramp and step voltage measurements indicates that the conductance change was relatively independent of the duration of the depolarization, as in the responses shown in Figure 8. In this cell, too, a large change in membrane potential occurred under current clamp conditions. A current of about 25×10^{-6} amp carried the depolarization from about 20 mv through the unstable region of negative slope, to a new steady state level of about 125 mv as the resistance changed from low to high. (From Bennett and Grundfest, 1966)

mination of G_{Na} by Na inactivation usually has no significant effect on the resting conductance. Inactivation responses may then be absent, or so small as to be poorly defined and partly or completely graded. All or most of the available G_K channels can be activated by increasing K_o, while the G_L channels are not greatly affected (Nakamura et al., 1965; Ruiz-Manresa, 1970). In such a condition $-dG/dE$ will be increased and the inactivation response becomes very marked (Figure 8).

In fact, this is the only condition in which inactivation responses occur in some cells; e.g., the hyperpolarizing responses of squid axons (Segal, 1958; Tasaki, 1959*a*; Moore, 1959) or frog nodes (Stampfli, 1958, 1959).

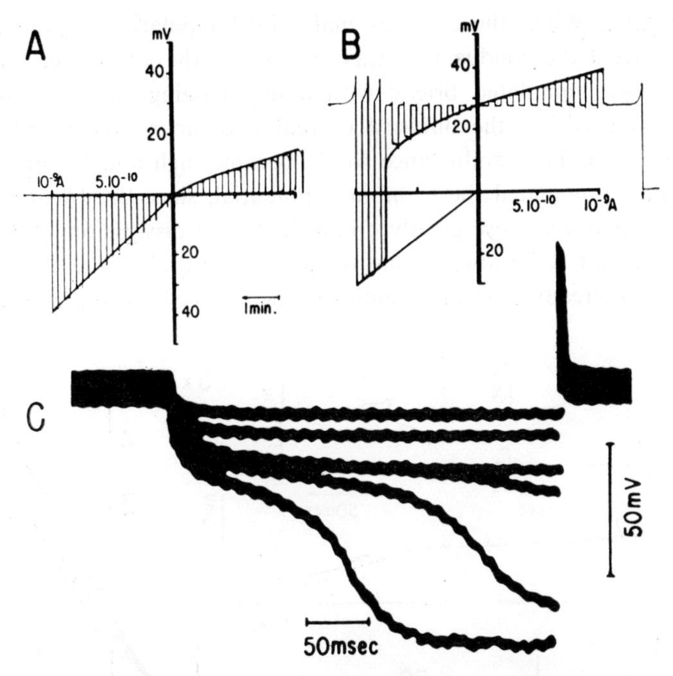

Figure 10 Hyperpolarizing inactivation responses in frog axons. *A*, *B*: I-E characteristics plotted out by periodic applications of current pulses whose amplitudes are shown on abscissa (depolarizing to the right). Ordinates: membrane potential in relative units, resting potential as the origin. During measurements in *A*, axon was in ordinary frog saline solution. *E* changed linearly with *I* for inward currents, but for outward currents it developed a curvature that betokens K activation. For measurements in *B*, the axon was depolarized by about 28 units following immersion in KCl-enriched saline. The I-E characteristic shows a high conductance even for small hyperpolarizing currents, but the characteristic became increasingly nonlinear for inward currents $>5 \times 10^{-10}$ A and returned to low conductance state with currents $>7.5 \times 10^{-10}$ A. *C*: The change in membrane potential during sufficiently large applied current is a hyperpolarizing response. Note that form of hyperpolarizing response of the frog axon resembles that of the eel electroplaque in Figure 8. (Grundfest, 1969. Modified from Stämpfli 1958, 1959)

In the resting state (Figure 10) the G_K channels contribute little to the steady state conductance. Depolarization either by applied currents or by increasing K_0 opens these K channels, increasing G_K, but they can close again when the membrane is repolarized by inward current.

The reciprocal relation (Figure 6*C*) between K spikes and hyperpolarizing inactivation responses is readily deduced from the voltage clamp data of Figure 11, on a puffer neuron bathed in high K_0. The cell, which originally had a low conductance at the resting potential of ca. -70 mV (arrow), became highly conductive and depolarized to about -15 mV in high K_0

(broken line). When the cell was maintained repolarized by an applied inward current the conductance was restored to the low value. When the cell was now stimulated briefly with a depolarizing current a K spike occurred (records) if the outward current depolarized the membrane to −50 mV or so. The conductance for K became high and the emf of this battery now dominated the membrane potential, since the *IR* drop of the applied current was now greatly diminished. The dissipation of the high conductance for K is slow, so that the spikes are prolonged. On the other hand, and as already shown in Figure 10, when the cell is maintained initially

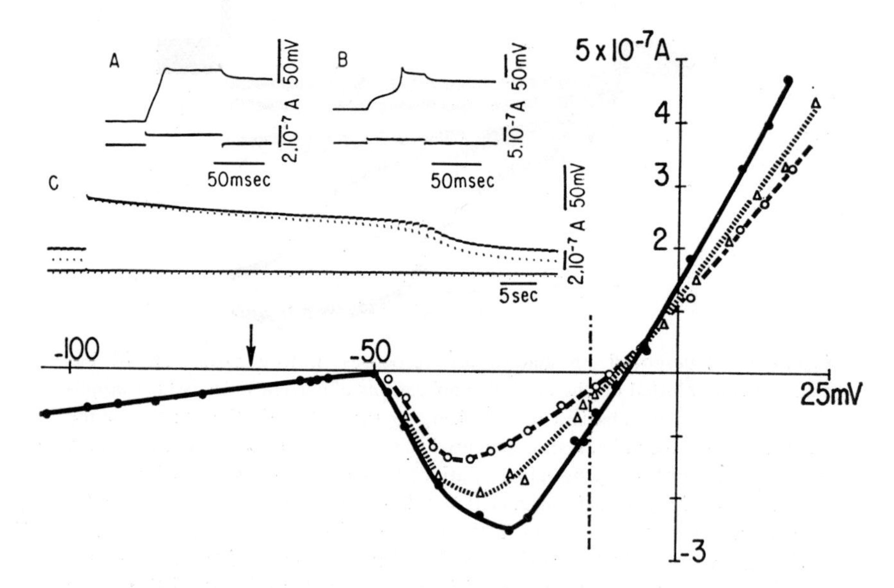

Figure 11 Regenerative electrogenesis owing to K activation in puffer neurons bathed in isosmotic KCl solution and repolarized by applied inward currents, *A–C*: Prolonged K spikes evoked by depolarizing pulses. Holding potentials were −68 mV *A*, −71 mV *B*, and −60 mV *C*. In the latter recording, small, constant repetitive inward pulses were applied. The diminished amplitudes of the hyperpolarizations show that the conductance increased during the K spike. *Graph*: Voltage clamp data from another cell which was depolarized to −14 mV (broken vertical line) by isosmotic substitution of KCl for NaCl. The holding potential was −70 mV (arrow). The conductance increase due to K activation by depolarizing stimuli and represented by a large change in slope of the I-E characteristic, was long lasting. Filled circles show peak currents shortly after the step changes in voltage. Triangles, measurements 80 msec after onset of depolarization. Open circles, 860 msec after onset. The K equilibrium potential, the point at which the 3 curves cross, is about −5 mV. The very gradual subsidence of K activation is denoted by decrease in the high slope of characteristic. End of the K activation and of spike electrogenesis will restore the characteristic to the slope at the holding potential. (Modified from Nakajima, and Kusano, 1966 and from Nakajima, 1966)

at its depolarized state repolarization to about −30 mV would bring the characteristic into the negative slope region and a hyperpolarizing inactivation response would result.

The inactivation responses shown in Figures 8 and 10 are monotonic, denoting that they probably result from only one change in conductance; the new high resistance state is also a steady state. In some cells, however, the large change in membrane potential that is induced during the depolarizing or hyperpolarizing inactivation response may initiate a secondary increase in conductance. This appears to be the case for certain cardiac tissues (e.g. Figure 2). After inactivation the conductance increases again with stronger depolarization (cf. also Figure 40 in Grundfest, 1966*a*). This phenomenon is also seen in some electroplaques (Morlock and Janiszewski, 1970).

An interplay of several conductance elements is particularly evident in certain hyperpolarizing inactivation responses, such as that in lobster muscle fiber (Reuben et al., 1961). During a constant inward current the membrane at first becomes strongly hyperpolarized in a regenerative manner, but the potential then returns spontaneously to lesser negativity. The inactivation response then has the appearance of a transient "negative spike" followed by a sustained plateau. Oscillations may also occur.

Particularly instructive in this respect, though as yet not completely analyzed, is the hyperpolarizing response of crayfish muscle fibers (Figure 12). Unlike lobster fibers, those of crayfish do not produce a hyperpolarizing response when they are bathed in a saline containing Cl. This is because the crayfish fibers develop hyperpolarizing Cl activation and the conductance increase of this activation process masks the K inactivation. However, when the fibers are depleted of intracellular Cl, by maintaining them in a medium with an impermeant anion (e.g. propionate) the conductance decrease is unmasked and they can generate hyperpolarizing responses (Figure 12*A–F*), similar to those of lobster fibers. These responses can also occur on replacing the propionate with nitrate (B') but not when Cl or Br is the anion (A') and the two graphs).

The regenerative character of the large "negative spike" is also seen in the graphs. When the fiber is hyperpolarized to a potential of about −120 mV there is first a small downward curvature of the points on the graphs, i.e., a graded inactivation occurs at first. Slightly larger hyperpolarizations trigger an abrupt change to about −300 mV. However, still larger currents do not induce proportionally high *IR* drops. Apparently another conductance comes into play, an increase this time, which develops rather more slowly. However, it finally induces a return of the potential back toward the resting value, and this new level is somewhat positive to the threshold for hyper-

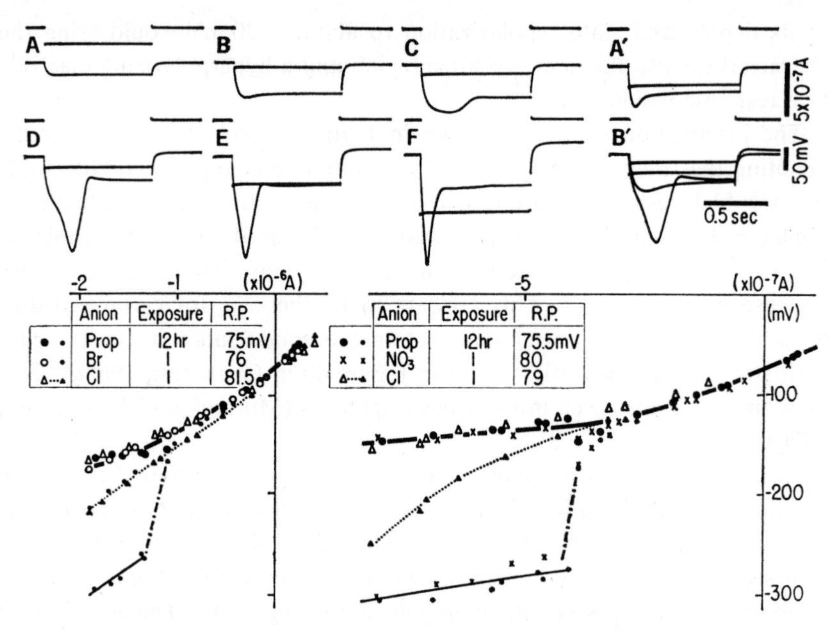

Figure 12 The results of a complex sequence of conductance changes induced in crayfish muscle fibers by hyperpolarizing currents. Records *A–F* and filled circles on right graph, from a preparation that was bathed in Cl-free (propionate) saline for 12 hours. Inward currents that drove the potential to about −120 mV caused a large further hyperpolarization (*D–F*) which subsided while the current was still applied. The return toward the resting potential had an overshoot and the steady state of the subsequent plateau was less hyperpolarizing than the threshold for the initial response. Neither the peak of the "negative spike" nor the steady state plateau were changed markedly by further increase in current, but the duration of the former decreased. *B′*: One hour after replacing propionate with NO_3. The hyperpolarizing response was still present. *A′*: After one hour in Cl the response disappeared. The graph shows the full experimental data. The large symbols denote measurements during the steady state. *Left graph*: Another experiment. The hyperpolarizing inactivation response was abolished by Br as well as Cl. Note that in both experiments the steady state characteristic was the same, no matter whether the extracellular anion was propionate, nitrate, Br, or Cl. (Unpublished data by Ozeki, Girardier, Reuben and Grundfest)

polarizing inactivation. Thus, the inactivation response is abolished and the membrane assumes potentials that reflect both the larger *IR* drop from the applied current and a shift to include a new emf. While the occurrence of these different processes is clear, the nature of the ionic battery involved in the high threshold hyperpolarizing activation remains to be clarified.

Anode break spikes may also be caused as a consequence of the conductance increase of hyperpolarizing activation. The hyperpolarizing responses of skate electroplaques (Figure 13*C*, *D*) resemble those of lobster or crayfish muscle fibers (Figure 12). However, when the hyperpolarizing responses are induced in conditions where the skate electroplaque can generate a spike (Figure 13*A*, *B*) the termination of the applied inward current also elicits anode break spikes (*C*, *D*). Both types of spikes are eliminated by reintroducing some Cl into the bathing medium (*F* to *H* and *I* to *P*), but the hyperpolarizing response itself remains.

Thus, both types of spikes are due to Cl activation. It is likely that the plateau phase of the hyperpolarizing responses involves hyperpolarizing Cl activation. If the high conductance for the anions after the current is withdrawn spike electrogenesis would occur.

The last variety of ionic interplays which I will present here manifest themselves particularly in a.c. impedance measurements, similar to the now classical findings of Cole and Curtis (1939) on the squid giant axon.

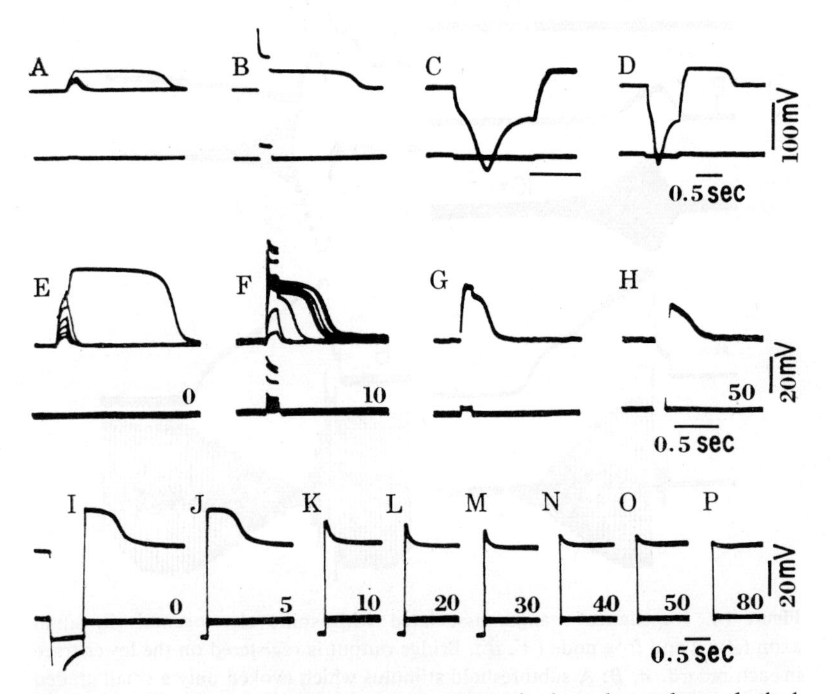

Figure 13 Cl spikes and hyperpolarizing responses in skate electroplaques bathed in Cl-free saline. *A*, *B*, *E*: Long-lasting all-or-none responses triggered by depolarizing pulses. *C*, *D*, *I*: Inward currents evoked hyperpolarizing responses and, when the current was terminated, the cells produced "anode break" spikes. Spike triggered by depolarization (*E*) was diminished and finally abolished when Cl was reintroducted to concentration of 10 mM (*F*) and 50 mM (*G* and *H*). A very strong stimulus (outward current) was applied in the latter recording. Anode break spike (*I*) was also abolished by introducing Cl (*J*–*P*). (Unpublished data and Grundfest, *et al.*, 1962)

In the latter, as in the frog node (Figure 14), the impedance decrease during the spike is a monotonic change signifying the increased conductances for both Na and K. Depolarizing K activation is normally absent in eel electroplaques, being replaced with K inactivation (Nakamura et al., 1965; Morlock *et al.*, 1968; Ruiz-Manresa *et al.*, 1970; Figure 8). This is also shown in the upper graph (*C*) of Figure 15, which presents (filled circles) the steady-state characteristic during a series of 10 msec pulses of increasing inward and outward currents. The peaks of the spikes that were evoked by the stronger depolarizing pulses are shown as open circles. The slope of the line they

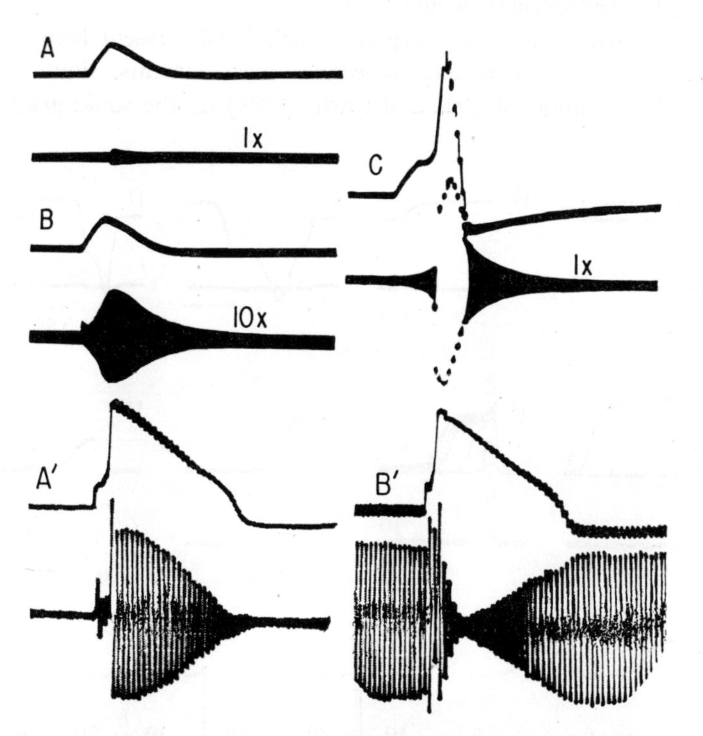

Figure 14 Impedance changes associated with spike electrogenesis in squid axon (*A*–*C*) and frog node (*A*′, *B*′). Bridge output is registered on the lower trace in each record. *A*, *B*: A subthreshold stimulus which evoked only a small graded response caused a small imbalance in the bridge. Impedance registration at lx in *A* and 10 x in *B*. *C*: A stronger stimulus evoked a spike and the characteristic impedance change. The bridge imbalance, denoting an increase in conductance, persisted during most of the hyperpolarization that terminates the squid axon spike. *A*′: In the frog axon the spike has a slower falling phase, probably because K-activation is less well developed than in the squid. The impedance change subsides in parallel with the repolarizatton. *B*′: The bridge was unbalanced initially. The decrease in impedance during the spike decreased the output. The bridge was driven with 20 K Hz (*A*′, to *C*) and 4 K Hz (*A*′, *B*′). (Modified from Tasaki, 1959 *b*).

form is steeper because Na activation increases the conductance by the value G_{Na}. In the hyperpolarizing quadrant the steady state characteristic is given by two components G_L and G_K. On depolarization the steady state conductance is reduced to G_L and the progressive closure of the G_K channels creates a negative slope region (i.e. dG_K/dE is negative), albeit not as marked as in the cell of Figure 8, when it was bathed in high K_0.

Pharmacological K inactivation is induced by Cs, particularly in the hyperpolarizing quadrant (Nakamura et al., 1965). Thus, in the middle graph of Figure 15 (Cs) the conductance for inward currents is decreased to the same value as for high outward current and is represented by the G_L channels alone. However, the K channels are open for a small range of

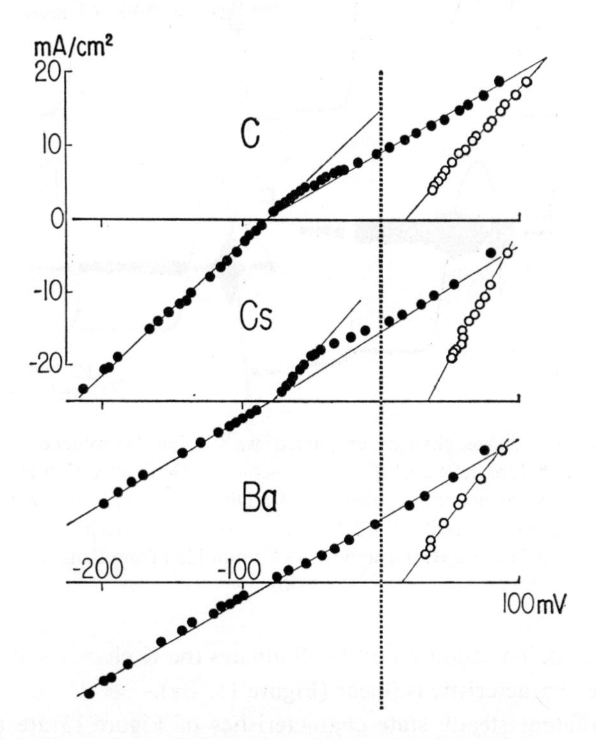

Figure 15 Changes in the steady state characteristic induced in an eel electroplaque by pharmacological K inactivation. C, measurements in control saline; Cs, after adding 5 mM CsCl; Ba, after addition of 2.5 mM $BaCl_2$. A sequence of 10 msec currents was applied for the measurements. The stronger depolarizations elicited spikes and their peaks are shown as open circles. The slope of this line indicates that the conductance at the peaks of the spikes is higher than in the steady state (filled circles). The lines of lowest slope represent G_L which is not affected by Cs or Ba. Depolarizing K activation is seen in the presence of Cs and the slope ($G_L + G_K$) is the same as the intermediate slope in the control. Further description in text. (From Ruiz-Manresa *et al.*, 1970)

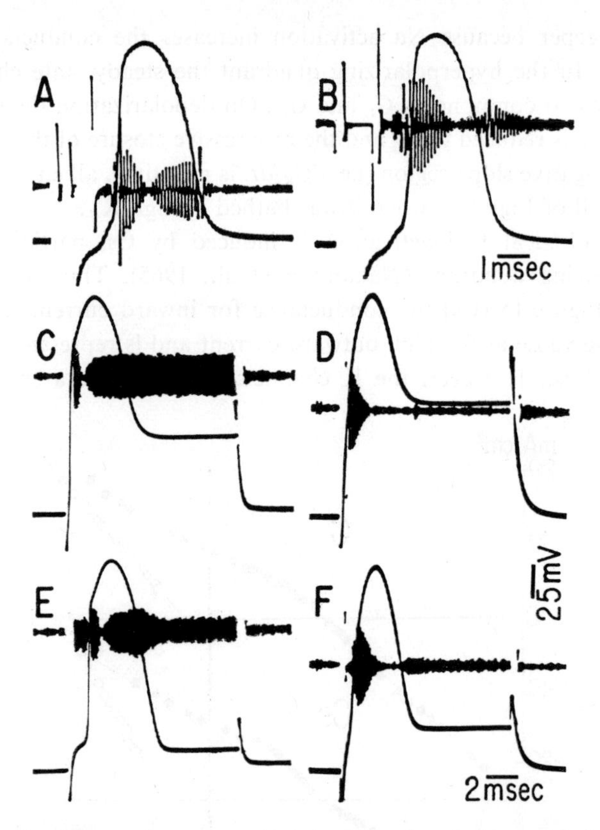

Figure 16 Impedance changes correlated with spike electrogenesis in an eel electroplaque before (left) and after pharmacological K inactivation by Cs (right). *A*, *B*: Spikes were neurally evoked. *C*, *D*: Directly evoked spikes initiated by strong, long-lasting currents. *E*, *F*: The pulses were near threshold for the spikes. Further description in text. (Grundfest, 1970; modified from Ruiz-Manresa *et al.*, 1970)

depolarizations. The addition of Ba eliminates the K channels and the entire steady state characteristic is linear (Figure 15, Ba).

The 3 different steady state characteristics of Figure 15 are reflected in different varieties of impedance changes (Figure 16 and 17). Spike electrogenesis in the control is accompanied by two periods of bridge imbalance separated by a return of the output to null. The first elevation of the neurally evoked spike is due in part to the EPSP and in part to the rise in conductance due to Na activation. The initial phase of the impedance change subsides, however, before the spike has reached its peak. The second elevation begins to increase slowly but it falls abruptly when the falling phase of the spike has declined to about 40 mV depolarization. This elevation represents an

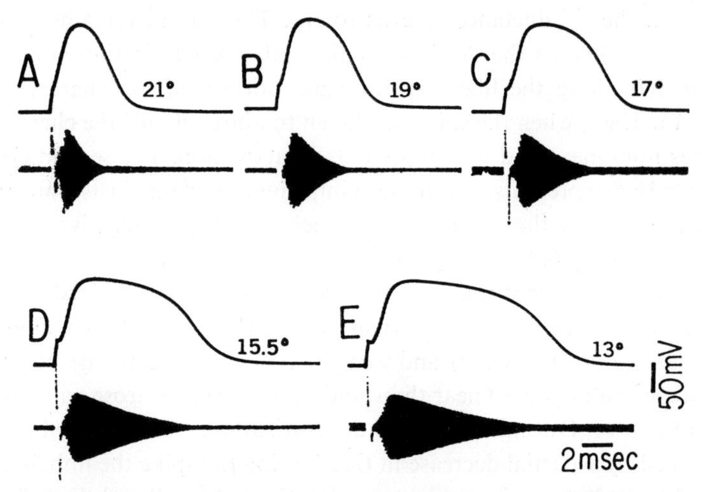

Figure 17 Spikes and impedance changes after Ba had produced complete K inactivation in an eel electroplaque. Only the impedance change due to Na conductance now developed. The onset of Na activation, as denoted by the rise in the bridge output, was not much affected by temperature changes between 21° and 13°C. However, the duration of the Na activation was greatly prolonged by lowering the temperature, indicating that Na inactivation became slower. The Q_{10} was ca 9. (From Ruiz-Manresa, 1970)

increased impedance (Morlock et al., 1968). The sequence can be deduced readily from the corresponding characteristic (*C*) in Figure 15. The initial rise in bridge output represents mainly the increase in G_{Na}. Since G_K is rapidly inactivated the total conductance during the spike, represented by the line of open circles in Figure 15, is $G_L + G_{Na}$. Since the rise in G_{Na} is transient this then will decrease with time and the return to null in the record of Figure 16*A* indicates that G_{Na} remaining at this time is equal to to G_K in the resting state, for which the bridge had been balanced initially. As G_{Na} disappears the bridge becomes unbalanced by an increase of impedance. When G_{Na} becomes zero the conductance is only that of G_L. However, as the spike falls back toward the resting potential the characteristic in Figure 15*C* will be traversed from right to left. At about 40 mV depolarization the K channels open and the negative slope of the characteristic predicts a rapid transition to the resting high conductance. The bridge imbalance then subsides very rapidly.

The change in the steady state characteristic that is induced by Cs (Figure 15) also changes the sequence of impedance changes (Figure 16*B*). The bridge is initially balanced for the resting conductance which is now G_L alone (Figure 15). The initial elevation, again due primarily to G_{Na}, now represents the full time course of Na activation, since null output recurs only when the conductance returns to G_L. The output remains null for a considerable time as the decrease of the spike depolarization sweeps from right to left along the linear steady state portion of the characteristic in Figure 15. Then, when the spike has fallen to about 40 mV the characteristic becomes non-linear and now represents K activation. The second elevation in Figure 16*B* represents a decrease in impedance and this lasts until the potential approaches the resting value, when G_K is again inactivated and the bridge returns to its balance for G_L^*.

The analysis is confirmed in the impedance data obtained when the electroplaque, before and after treatment with Cs, is excited directly by longer lasting strong (*C*, *D*) and weak (*E*, *F*) currents. In the control (*E*) the weak stimulus was so near threshold that the spikes arose with a latency of about 2 msec. During the applied depolarization, however, the bridge was unbalanced by a partial decrease in G_K. During the spike the imbalance was reversed to an increase in conductance for G_{Na}. The null and the subsequent rise in impedance are similar to those observed in *A*. However, since the current was still applied after the spike had terminated the membrane remained depolarized by about 15 mV, as at the beginning of the stimulus. The bridge output also fell to the same level as that before the spike. Thus, the impedance has increased to the maximum for strong depolarization (conductance $= G_L$) and decreased somewhat when a small depolarization had inactivated the G_K channels incompletely.

The application of Cs caused the change seen in Figure 16*F*. After the initial elevation terminated the bridge remained balanced during most of the falling phase, while the membrane potential maintained the conductance at G_L. As the spike terminated, however, the conductance increased and remained higher than in the resting state as long as the membrane was depolarized by the weak applied current. The bridge imbalance reflects the higher conductance.

For the records of Figure 16*C* and *D* the current were increased so that the plateau of depolarization after the spike carried the steady state conductance into its lowest level (G_L). In the control the bridge had been balanced for $G_L + G_K$ and the plateau output represents an increased impedance. In the Cs treated case the bridge was balanced for G_L. Thus, after G_{Na} had subsided the impedance output remained null during the persistent strong depolarization.

---

* The onset of the second elevation in Fig. 16*B* is rapid, while the decline is slow, quite in contrast with the time course of the second elevation in *A*. In the presence of Cs the negative slope of the steady state characteristic shifts the conductance rapidly from low to high. Thereafter the conductance remains high until the membrane returns to the resting potential.

When the steady state characteristic is completely linearized by Ba (Figure 15) only the initial elevation occurs during the spike (Figure 17). Thus, the imbalance is entirely monotonic, but it still differs from the registrations of squid axons or frog nodes (Figure 14). In the squid axon the high conductance outlasts the depolarizing phase of the spike because of the persistence of K activation. In the frog node the high conductance is nearly coterminous with the spike, probably because K-activation is relatively small in amphibian axons. In the eel electroplaques, however, G_{Na} terminates soon after the beginning of the falling phase of the spike.* In spikes evoked at low temperatures the peak depolarization is broad because it is sustained by the inward Na current and the fall begins when G_{Na} is terminated. The repolarization results from dissipation of the charge on the membrane capacity across the passive resistance component, the leak channels. The rise of the impedance change is practically independent of temperature in Figure 17. Thus, the impedance measurements disclose that in eel electroplaques Na activation has a Q_{10} close to unity (ca 1.1) while Na inactivation, the only other process present under the experimental conditions of Figure 17, has a Q_{10} of about 9 (Ruiz-Manresa, 1970).

REFERENCES

Belton, P., and H. Grundfest (1961). The ionic factors in the electrogenesis of the electrically inexcitable and electrically excitable membrane components of frog slow muscle fibers. *Biol. Bull.*, **121**, 382.

Bennett, M. V. L., and H. Grundfest (1959). Electrophysiology of *Sternopygus* electric organ. *Proc.21st Int. Cong. Physiol. Sci.*, p. 35.

Bennett, M. V. L., and H. Grundfest (1966) Analysis of depolarizing and hyperpolarizing inactivation responses im gymnotid electroplaques. *J. Gen. Physiol.*, **50**: 141–169.

Cohen, B., M. V. L. Bennett, and H. Grundfest (1961). Electrically excitable responses in *Raia erinacea* electroplaques. Fed. Proc., **20**, 339.

Cole, K. S. (1965). Theory, experiment and the nerve impulse. In, *Theoretical and Mathematical Biology* (T. H. Waterman and H. J. Morowitz, eds.) New York, Blaisdell, 136–171.

Cole, K. S. (1968) *Membranes, Ions and Impulses*. University of California Press.

Cole, K. S., and H. J. Curtis (1939) Electric impedance of the squid giant axon during activity. *J. Gen. Physiol.*, **22**, 649–670.

Del Castillo, J., and T. Morales (1967). The electrical and mechanical activity of the esophageal cell of *Ascaris lumbricoides*. *J. Gen. Physiol.*, **50**, 603–629.

---

* The various changes in the impedance measurements of Figures 16 and 17 were produced merely by changing the G_K channels (Figure 15) and did not affect spike electrogenesis or G_{Na}. These findings are therefore incompatible with the "interdiffusion" theory of bioelectrogenesis (Tasaki, 1968). They also indicate strongly that the G_K and G_{Na} channels are independent entities (Grundfest, 1957a, 1961), rather than a single time and state variant system. For further discussion see Grundfest (1970) and Ruiz-Manresa et al. (1970).

Geduldig, D., and D. Junge (1968). Sodium and calcium components of action potentials in one *Aplysia* giant neurone. *J. Physiol.*, **199**, 347–365.

Grundfest, H. (1957*a*). Excitation triggers in post-junctional cells. In, *Physiological Triggers* (T. H. Bullock, ed.) Washington, D. C., American Physiological Society, 119-151.

Grundfest, H. (1957*b*). The mechanisms of discharge of the electric organs in relation to general and comparative electrophysiology. *Prog. Biophys.*, **7**: 1–85.

Grundfest, H. (1957*c*). Electrical inexcitability of synapses and some of its consequences in the central nervous system. *Physiol. Revs.*, **37**, 337–361.

Grundfest, H. (1959). Evolution of conduction in the nervous system. In, *Evolution of Nervous Control* (A. D. Bass, editor). Washington, D. C., American Association for the Advancement of Science, 43–86.

Grundfest, H. (1961). Ionic mechanisms in electrogenesis. *Ann. N. Y. Acad. Sci.*, **94**, 405–457.

Grundfest, H. (1966*a*). Heterogeneity of excitable membrane: Electrophysiological and pharmacological evidence and some consequences. *Ann. N. Y. Acad. Sci.*, **137**, 901–949.

Grundfest, H. (1966*b*). Comparative electrobiology of excitable membranes. In, *Advances in Comparative Physiology and Biochemistry*, Vol. 2 (O. E. Lowenstein, editor). New York, Academic Press, 1–116.

Grundfest, H. (1967). The "anomalous" spikes of *Ascaris* esophageal cells. *J. Gen. Physiol.*, **50**, 1955–1959.

Grundfest, H. (1969). Dynamics of the cell membrane as an electrochemical system. In, *Glass Microelectrodes*, M. Lavallee, O. F. Schanne, and N. C. Hebert, editors). New York, John Wiley & Sons, 188–216.

Grundfest, H. (1970) The excitable electrogenic membrane as a heterogeneous electrochemical system: New evidence. *Israel J. Med. Sci.*, **6**, 185–194

Grundfest, H., E. Aljure, and L. Janiszewski (1962) The ionic nature of conductance increases induced in Rajid electroplaques by depolarizing and hyperpolarizing currents. *J. Gen. Physiol.*, **45**, 598 A.

Hagiwara, S., and S. Nakajima (1966). Differences in Na and Ca spikes as examined by application of tetrodotoxin, procaine and manganese ions. *J. Gen. Physiol.*, **49**: 793–806.

Hodgkin, A. L., and A. F. Huxley (1952). A quantitative description of membrane current and its application to conduction and excitation in nerve. *J. Physiol.*, **117**, 500–544.

Moore, J. W. (1959). Excitation of the squid axon membrane in isosmotic potassium chloride. *Nature*, **183**, 265–266.

Morlock, N. L., D. A. Benamy, and H. Grundfest (1968). Analysis of spike electrogenesis of eel electroplaques with phase plane and impedance measurements. *J. Gen. Physiol.*, **52**, 22–45.

Morlock, N. L., and L. Janiszewski (1970). Pacemaker and oscillatory behavior of K-conductance in electroplaques of *Eigenmannia*. *Fed. Proc.*, **31**.

Mueller, P., and O. Rudin (1968). Resting and action potentials in experimental bimolecular lipid membranes. *J. Theoret. Biol.*, **18**, 222–258.

Nakajima, S. (1966). Analysis of K-inactivation and TEA action in the supramedullary cells of puffer. *J. Gen. Physiol.*, **49**, 629–640.

Nakajima, S., and K. Kusano (1966). Behavior of delayed current under voltage-clamp in the supramedullary neurons of puffer. *J. Gen. Physiol.*, **49**, 613–628.

Nakamura, Y., S. Nakajima, and H. Grundfest (1965). Analysis of spike electrogenesis and depolarizing K-inactivation in electroplaques of *Electrophorus electricus*, L. *J. Gen. Physiol.*, **49**, 321–349.

Reuben, J. P., R. Werman, and H. Grundfest (1961). The ionic mechanisms of hyperpolarizing responses in lobster muscle fibers. *J. Gen. Physiol.*, **45**, 243–265.

Rougier, O., G. Vassort., D. Garnier, Y. M. Gargouil, and E. Coraboeuf (1969), Existence and role of a slow inward current during frog atrial action potential. *Pflug. Arch.*, **308**, 91–110.

Ruiz-Manresa, F. (1970). Electrogenesis of eel electroplaques. Conductance components and impedance changes during activity. *Ph. D. Thesis*, Columbia University, New York.

Ruiz-Manresa, F., A. C. Ruarte, T. L. Schwartz, and H. Grundfest (1970). K-inactivation and impedance changes during spike electrogenesis in eel electroplaques. *J. Gen. Physiol.*, **55**, 33–47.

Segal. J. (1958). An anodal threshold phenomenon in the squid giant axon. *Nature*, **182** 1370–1372.

Shanes, A. M., H. Grundfest, and W. H. Freygang, Jr. (1953) Low level impedance changes following the spike in the squid giant axon before and after treatment with "Veratrine" alkaloids. *J. Gen. Physiol.*, **37**, 39–51.

Stämpfli, R. (1958). Die Strom-Spannungs-Charakteristik der erregbaren Membran eines einzelnen Schnürrings und ihre Abhängigkeit von der Ionenkonzentration. *Helv. Physiol. Acta*, **16**, 127–145.

Stämpfli, R. (1959). Is the resting potential of Ranvier nodes a potassium potential? *Ann. N. Y. Acad. Sci.*, **81**, 265–284.

Tasaki, I. (1959*a*). Demonstration of two stable states of the nerve membrane in potassium-rich media. *J. Physiol.*, **148**, 306–331.

Tasaki, I. (1959*b*). Conduction of the nerve impulse. In, Neurophysiology, Sect. 1, *Handbook of Physiology*, Vol. 1 (John Field, editor). Washington, D. C., American Physiological Society, 75–121.

Tasaki, I. (1968). *Nerve Excitation. A macromolecular Approach.* Springfield, Ill., Charles C. Thomas Publ.

Tasaki, I., L. Lerman, and A. Watanabe (1969). Analysis of excitation process in squid giant axons under bi-ionic conditions. *Am. J. Physiol.*, **216**, 130–138.

Yamagishi, S. (1970). Ionic mechanism of the prolonged spike of squid axons perfused with protease. *J. Gen. Physiol.*, **55**, 145.

Intracellular Perfusion of Squid Giant Axons: A Study of Bi-ionic Action Potentials

A. WATANABE[1] AND I. TASAKI

Laboratory of Neurobiology, National Institute of Mental Health, Bethesda, Maryland

I. INTRODUCTION

Intact excitable tissues are, from a chemical point of view, very complex. This complexity is the major source of difficulty in applying physico-chemical concepts to excitation phenomena. To overcome this difficulty, the intracellular perfusion technique was invented by Baker, Hodgkin and Shaw (1961) and Oikawa, Spyropoulos, Tasaki and Teorell (1961). By this technique, it is now possible to vary the internal medium of squid giant axons and to conduct various electrophysiological measurements under well-defined experimental conditions. The method developed by Baker *et al.* differs from that invented in this laboratory by Oikawa *et al.* in the way axoplasm is removed; but the results obtained by the two different methods are reasonably consistent.

By using the technique of intracellular perfusion, a continuous effort has been made to simplify the composition of the internal and external millieu. We have established in recent years the experimental conditions under which axon excitability could be maintained with the salt of a single divalent cation species externally and the salt of a single univalent cation species internally. At present, it does not appear possible to simplify the ion requirement of the squid axon any further. The presence of divalent cation in the external medium appears to be absolutely essential. It is not

[1] Permanent address: Tokyo Medical and Dental University. Tokyo, Japan.

possible to use a divalent cation salt internally because of its strong detrimental effects on the axon membrane.

From a physico-chemical point of view, analysis of the excitation process under these simple and well-defined experimental conditions is relatively easy. Since fluxes of anions across the squid axon membrane are known to be very small, the membrane can be treated as a layer of cation-exchanger through which the internal and external cations are continuously inter-diffusing. The electric responses of the axon observed under these conditions are called "bi-ionic action potentials".

In this article, various electrophysiological properties of the squid giant axon under the bi-ionic conditions are described in great detail. Although our objective in these investigations was to elucidate the molecular mechanism of nerve excitation, very little comment is made on the implication of individual findings in relation to the theory of excitation. By separating the experimental facts from conjectures and theories in this manner, it is hoped that the reader will arrive at his own conclusion as to the significance of the findings described in this article. The major experimental findings described in this article are published in the following original papers: Tasaki, Watanabe and Singer, 1966; Watanabe, Tasaki and Lerman, 1967; Tasaki, Lerman and Watanabe, 1969.

II. METHOD

The experimental results described in this article are obtained exclusively by the method developed in this laboratory. The main advantage of this method of perfusion is that axons as small as 350 μ in diameter can be used without an increase of internal pressure. Furthermore, with our method we can constantly monitor the action potential during the removal of protoplasm, and possible injury can be detected easily.

Materials

Giant axons were obtained from *Loligo pealii* available in Woods Hole. The axon diameter ranged between 600 and 350 μ. Extensive cleaning of axons was performed when necessary (e.g., for measurements of isotope fluxes). Ordinarily, cleaning was partial; one-third to one-half of the surface of the axon swas covered by a thin layer of small nerve fibers. A slight delay in the effect of replacing the external solution was recognized in partially cleaned axons; otherwise the results obtained from these axons were similar to those obtained from extensively cleaned axons.

Perfusion chamber

Figure 1 shows the experimental setup which we employed for intracellular perfusion. The nerve chamber was made of Lucite; a pool of external fluid in the chamber was about 3 mm in depth. There were two grooves, one on each side of the chamber. The axon was placed horizontally on the chamber stretching from one groove to the other. Each end of the axon was held by a side piece, which was separated from the main chamber by a distance of several mm. A slight tension was applied to the axon through the two threads attached to the ends of the axon. The other ends of the threads were fixed with wax to the base plate on which both the chamber and the side pieces were mounted. The height of the side pieces was so determined as to maintain a distance of about 0.5 mm between the bottom of the Lucite chamber and the axon. The length of the chamber, namely, the length of the axon immersed in the external solution, was about 35 mm. The portions of the axon outside the chamber were exposed to the air and allowed to dry.

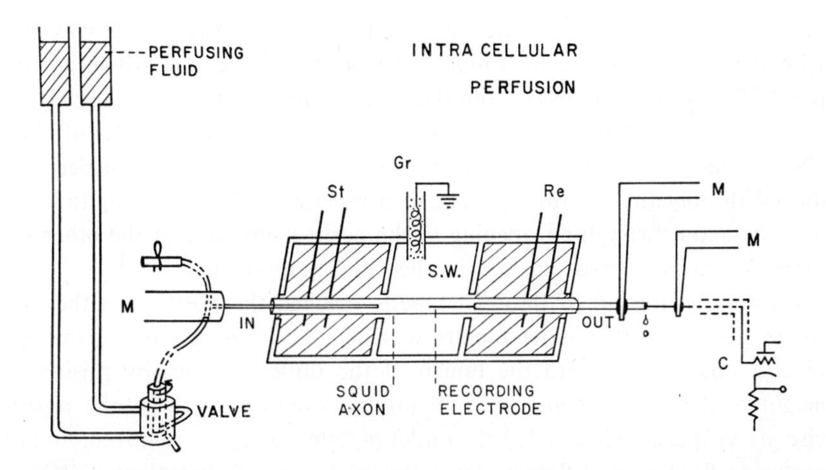

Figure 1 Typical experimental arrangement for intracellular perfusion. By means of micromanipulators (M), glass cannulae (IN and OUT) are inserted into the ends of a squid giant axon within a lucite chamber containing natural sea water (S.W.). The perfusion fluid reservoirs at the left are connected through polystyrene tubing and a glass valve to the inlet cannula (IN); the valve permits rapid alternation between different reservoirs. The intracellular recording electrode is introduced concentrically within the outlet cannula (OUT), by means of a third micromanipulator (M). Extracellular stimulating (St) and recording (Re) electrodes are positioned lateral to the perfusion zone. A large ground (Gr) electrode is placed in the external medium (S.W.). For further details, see text and elsewhere (From Tasaki, *et al. J. Gen. Physiol.* **48**, 1095, (1965)

External electrodes

A pair of metal (silver or platinum) wire electrodes were placed at the bottom of the grooves at each side of the chamber. They were used for stimulation of the axon and for recording action potentials externally. Conducted action potentials were continually monitored with these electrodes before and during insertion of the inlet and outlet pipettes.

Insertion of pipettes

The outlet pipette for intracellular perfusing fluid had a diameter of about 300 μ. With the aid of a Lucite holder and a micromanipulator, the pipette was inserted into the axon along its longitudinal axis: a small opening was made in the axon membrane on one of the side pieces and the pipette was advanced into the axon through the opening. During the process of insertion of the pipette, gentle suction was applied through a polyethylene tubing connected to the outlet pipette. This procedure was performed to remove excess axoplasm thereby reducing the pressure inside the axon during the insertion of the pipette. At the end of this procedure, the pipette was filled with the viscous axoplasm. Finally, the outlet pipette was advanced to a point 25 mm or more away from the point of insertion.

The inlet pipette for intracellular perfusing fluid had a diameter of about 100 μ. This pipette was connected to one of several reservoirs of perfusing fluid with polyethylene tubings. The inlet pipette was inserted longitudinally into the axon through an opening in the nerve membrane at the other side piece. No negative pressure was applied during insertion.

Eventually, the tip of the inlet pipette reached the position in the axon where the tip of the outlet pipette was located. Then the tip of the inlet pipette was inserted into the lumen of the outlet pipette. By raising the height of the reservoir and by applying light suction to the outlet pipette, viscous axoplasm which filled the outlet pipette was easily removed, and the perfusing fluid started flowing from the inlet to the outlet pipette. Because the tips of the two pipettes were overlapping at this moment, the perfusing fluid was not in direct contact with the interior of the axon at this stage. The hydrostatic pressure was maintained at 20 to 30 cm H_2O.

Initiation of intracellular perfusion

Perfusion of the interior of a squid giant axon was initiated when the outlet pipette was retracted and the perfusion fluid was brought in direct contact with the layer of axoplasm remaining in the axon. The length of the zone was usually about 15 mm. In the perfusing zone, the major portion of the protoplasm had been removed by suction through the outlet pipette. The layer of axoplasm on the inner side of the membrane did not as act a barrier for diffusion of ions; this statement is based on the following facts: 1) the effect of changing the internal solution is recognizable within one minute; 2) when the perfusing fluid is stained by adding dye, the color is seen to spread up to the axolemma within 1 min; and 3) when a solution containing a radioisotope is introduced into the chamber the intracellular perfusate (collected at the end of the outlet pipette) becomes readioactive in less than 1 minute.

Removal of axoplasm with protease solution

When intracellular perfusion is made with solution of low ionic strength, the axoplasm swells and it is difficult to maintain flow of the internal perfusion fluid. Apparently, the axoplasm behaves like a typical polyelectrolyte gel (see Michaeli and Katchalsky, 1957). For recommencing the flow, the outlet pipette has to be advanced again to the point were the two tips of the pipettes have been joined previously.

Takenaka and Yamagishi (1966, 1969) found that dilute solutions of certain kinds of proteases, obtained from bacteria, can be used to remove the protoplasm almost completely without altering the properties of the membrane appreciably. For example, when a solution containing 1 mg/ml of "Nagarse" and 400 mM of potassium fluoride is used as a perfusing fluid, most of the axoplasm is removed within five min without bringing about any appreciable change in the shape of the action potential. This finding has a great technical value because continuous flow of perfusing fluid with low ionic strength can be maintained.

In the present studies, pronase, a bacterial protease, was used to remove axoplasm. Concentration of the enzymes was 0.05 mg/ml in most of our recent experiments. Usually, about 0.1 mg/ml of chlorphenol red was added to the perfusing fluid to mark the enzyme solution. Internal perfusion was initiated with the enzyme solution, and 1.0 to 1.5 min later, the perfusion fluid was switched to a test solution which was free of enzyme. To maintain the internal flow, such short exposure to the dilute enzyme solution was found to be sufficient. This procedure brought about no detectable change in the action potential. The survival time under continuous internal perfusion could be more than two hours when suitable external and internal perfusates were employed.

Circulation of external medium

In most experiments described in this article, the external surface of the axon was also perfused continuously with a chemically well-defined salt solution. The extracellular perfusion system was composed of a number of plastic bottles with polyethylene tubings connected to the outlets of the bottles. The bottles were filled with solutions to be used as the external media. The ends of the polyethylene tubings were held above the pool. Flow of the external medium was initiated by raising one of the bottles above the experimental stage. A small plastic receptacle held above the pool and two perforated barriers served to reduce mechanical shocks produced by the rapid flow of the external solution around the axon. The fluid was removed from the other side of the pool by a glass tubing connected to an aspirator through a trap. The trap was used to electrically isolate the nerve chamber from the aspirator.

Exchange of the external solution could be rapidly performed by alternately lowering one bottle and raising another. The fluid in the side grooves was sucked by a pipette connected to another aspirator. During the initial period of about 2 min, the flow rate of the external medium was usually 30 to 50 ml/min. Later, the flow rate of the external fluid was maintained at approximately 20 ml/min.

Intracellular recording and stimulating electrodes

For internal recording of action potentials, a piece of metal wire or a glass capillary filled with electrolyte solution was used. For recording bi-ionic action potentials, a glass pipette of about 100 μ diameter filled with 0.6 M KCl-agar was used. The pipette was connected to the interior of a small plastic container through a hole and was sealed with wax. The container was filled with 0.6 M KCl solution, in which a porous tip of a saturated KCl-calomel reference electrode was inserted. The container was held by a separate micromanipulator. The resistance of the recording electrode was 2 to 10 MΩ.

The internal electrode used for passing stimulating current was made from a piece of 50 μ enameled platinum wire; an approximately 15 mm long portion of the wire was scraped to remove the enamel coating.

Both electrodes were inserted into the axon longitudinally through the outlet pipette. Usually, the stimulating electrode was inserted first by hand and then the recording electrode was inserted by the aid of the micromanipulator. The speed at which these electrodes were inserted had to be sufficiently slow to avoid a rise of pressure inside the axon.

The final position of the tip of the stimulating electrode was near the tip of the inlet pipette. Because the insulating enamel around the wire was removed for the length of separation between the tips of the inlet and outlet pipettes, nearly the entire length of the perfused zone was exposed to the stimulating current more or less uniformly. The tip of the recording pipette electrode was approximately at the center of the perfusion zone.

In the external medium there were two electrodes; one was used as an indifferent electrode for the intracellular recording and the other as the external stimulating electrode. For recording the membrane potential, a calomel half cell was used as a reference electrode. The electrode tip was immersed in the external fluid medium near the glass tubing for draining the fluid. Because the outside medium was continuously flowing at a high rate, contamination of the outside medium by KCl from the salt bridge of the calomel half cell was completely prevented. The recording was made differentially between the two calomel half cells. The input stage of the recording system was a Bak preamplifier which had a high input impedance and a grid current of less than 10^{-9} A.

The external salt solution was grounded with a coiled silver (or platinum) wire electrode. Stimulating current pulses were delivered to the axon between the intracellular wire electrode and the ground. A square pulse generator was used as the source of stimulating currents. The output terminal of the pulse generator was connected to the intracellular stimulating electrode through a resistor of 1 to 10 MΩ. A conventional double-beam oscilloscope was used to record the membrane potential and to monitor the stimulating pulse.

Composition of solutions

External and internal perfusing fluids were mixtures of several "isotonic" stock solutions. For uni-univalent electrolytes (e.g., NaCl), stock solutions were 0.6 M. For di-univalent electrolyte (e.g., $CaCl_2$), 0.4 M was the stock solution concentration. For "isotonic" non-electrolyte solutions, 12 volume percent glycerol solution was usually used. From the standpoint of balancing the osmotic pressure, the above procedure is known to be inaccurate because the osmotic coefficient and the "reflection coefficient" deviate from unity and change with the concentration. However, we did not encounter an irreversible suppression of excitability by the osmotic unbalance with solutions prepared with the above procedure. It seems that excitability of the axon is not sensitive to a slight variation in the osmotic pressure. A small amount of buffer solution was added to the external solution. The buffer was prepared by adding concentrated HCl to an 0.8 M solution of tris-(hydroxymethyl)aminomethane; the pH of the buffer was 8.0 ± 0.1. The final level of Tris concentration was usually 1 mM.

Solutions of fluoride or phosphate saltes were used internally in most of our experiments. These salts are known to be most favorable for maintaining excitability (Tasaki, Singer and Takenaka, 1965). The pH of the internal fluid was adjusted to 7.3 ± 0.1. For phosphate solutions no extra buffer was needed. Fluoride solutions were buffered with a phosphate solution of the same cation species: the ratio of fluoride to phosphate was usually 9 to 1.

Temperature

The experiment was usually performed at room temperature of 19 to 20°C. In one series of experiments, the temperature of the external perfusing fluid was lowered by keeping the container of the external solution in ice water.

III. RESULTS

A. Cesium as internal cation

Demonstration of excitability

When squid giant axons were intracellularly perfused with a dilute cesium fluoride or cesium phosphate solution, they remained excitable in an external medium which contained a salt of calcium, strontium or barium, but no univalent cations. An example of the records obtained in these experiments is shown in Figure 2. The procedure of this experiment was as follows.

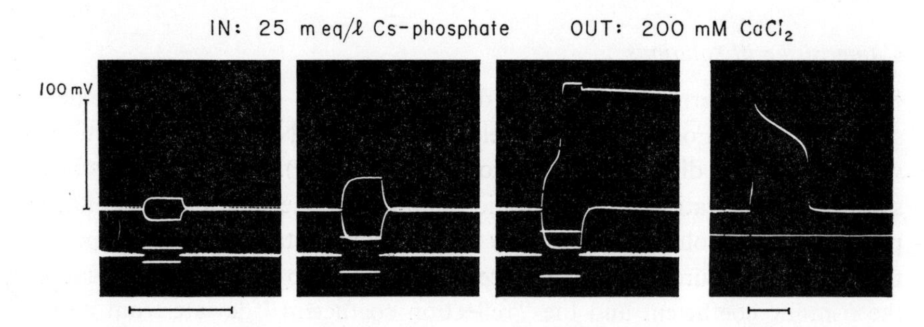

Figure 2 Oscillograph records demonstrating changes in membrane potential (upper traces) in response to square current pulses (lower traces). The axon is perfused internally with 25 mM Cs-phosphate and immersed in 200 mM $CaCl_2$ solution. In the first three records, two sweeps with outward and inward currend are superposed. With increasing stimulus intensity an all-or-none response is observed. The record on the right was obtained after a period of continuous internal and external perfusion. In this record the inward current pulse is applied prior to stimulation with a depolarizing current pulse. The time marker on the left represent 250 msec and that on the right 2 sec. Temperature 21°C

An axon was dissected out and was mounted in the experimental chamber which was filled with natural sea water. Two perfusion pipettes were inserted into the axon and the tip of the small (inlet) pipette was inserted into the lumen of the large (outlet) pipette inside the axon. The action potential observed at this stage was large (about 100 mV) in amplitude and short (about 1 msec) in duration. Then, extracellular perfusion of the axon was started with 200 mM $CaCl_2$ solution. The conducted action potential disappeared within 2 min. Next, one of the reservoirs for the intracellular perfusing fluid was raised to a level of about 40 cm H_2O above the nerve chamber. This procedure immediately initiated flow of the internal perfusing fluid, which contained the enzyme colored with chlorphenol red. The two pipettes were then separated; the final distance between the tips of the pipettes was about 15 mm. Approximately 1.5 min after the onset of perfusion with the enzyme solution, the internal fluid was switched to an enzyme-free perfusing fluid—a 25 mM cesium fluoride solution in this case (with 10% phosphate, see Methods). The red color inside the axon faded away in a few minutes. Finally, both the internal stimulating and the recording electrodes were brought to the center of the perfusion zone.

The oscillograph records in Figure 2 were taken about 11 min after separation of the pipettes. In response to a square current pulse, the membrane potential changed roughly exponentially. The final potential level obtained with an inward current was approximately the same as that obtained with outward current. "Delayed rectification" was not observed at this stage. With a strong outward current a large action potential was observed. In most axons examined, it took approximately 5 to 10 min to develop all-or-none action potentials.

Shape and magnitude of the action potential

The amplitude of the fully developed action potential observed under these conditions was usually between 80 and 100 mV; in a few cases the amplitude was 120 mV. The amplitude changed slightly with the stimulus intensity; with stronger current pulses the peak potential increased by several millivolts.

The duration of the action potential varied in a wide range from axon to axon. Usually when the duration was shorter than 50 msec, the action potential amplitude was graded. The action potential usually becomes longer and larger during continuous internal and external perfusion of the axon. In the final, more-or-less steady stage, the duration was usually 0.5–10 sec. Sometimes the membrane potential was arrested at the plateau level of the action potential, in which case the axon remained inexcitable until the membrane potential was lowered to the resting level either by applying hyperpolarizing current pulse or by increasing the external calcium concentration.

The duration of the action potential was very sensitive to the state of refractoriness. When a stimulaing current pulse was applied a short time after a preceding action potential, the duration of the second action potential was much shorter than that of the first.

The rate of potential change at the end of the action potential is smaller than that in the rising phase. Sometimes the falling phase showed an extra step, which probably indicates a spatial non-uniformity of the axon membrane. Such action potentials were seen more frequently when the perfusing fluid flowed outside of the outlet pipette. Under this circumstance, the axon membrane surrounding the outlet pipette was imperfectly perfused.

Our interpretation of the extra step is as follows: the imperfectly perfused end zone is excited by electric current (local current) arising from the central zone. The action potential of the end zone terminates earlier than that of the central zone. Termination of the action potential in the central zone is brought about by the local current which is inwardly directed in the central zone. This successive termination of the action potentials in two different zones creates and extra step in the falling phase.

Membrane resistance

The membrane resistance of unstimulated axons under these experimental conditions is of the order of 5 to 10 $K\Omega \cdot cm^2$. During excitation, the membrane resistance falls below the resting level. The resistance gradually recovers toward the end of the plateau. At the initial phase of action potential, the membrane resistance is between one-tenth and one-fifth of the resting value, as indicated by the decrease of the amplitude of electronic potential (see Figure 3).

Resting potential

In this article, we define the resting potential operationally as the difference between the potential levels which are observed when the same glass capillary electrode (filled with 0.6 M KCl-agar) is moved back and forth between the interior and the exterior of an axon. The resting potential defined in this manner may be taken as a useful measure of the physiological state of an axon.

Figure 3 illustrates an example of the records of bi-ionic action potentials upon which the potential level of the external medium is superposed. The resting potential is very small and usually slightly negative inside. Therefore, during almost the entire period of action potential, the potential inside the axon is higher than the potential level recorded with the same capillary electrode immersed in the external medium.

Effects of external calcium ion concentration

The most conspicuous effect of changing the calcium concentration in the external medium is on the duration of the action potential. To cite an

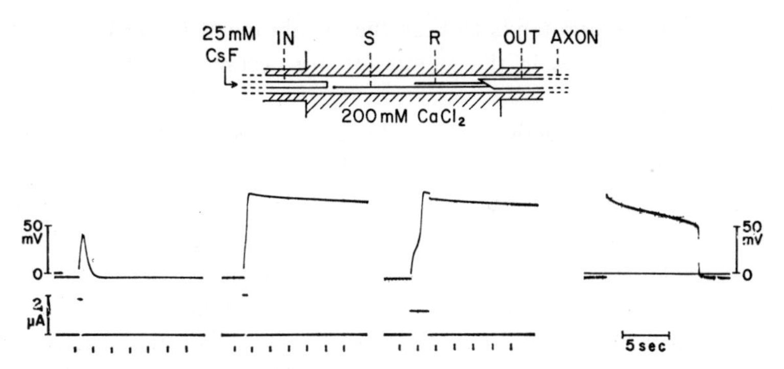

Figure 3 *Top*: Schematic diagram showing the experimental arrangement used to obtain the oscillograph records shown below. The inlet cannula (IN), outlet cannula (OUT), stimulating wire electrode (S), and recording glass capillary electrode (R) are shown. The distance between the tips of the two cannulae was approximately 20 mm. The diameter of the axon was approximately 500 μ. *Bottom*: Records of action potentials observed in the absence of univalent cation in the external fluid medium. In the first record (extreme left), the stimulating current pulse (indicated by the lower oscillograph trace) was subthreshold. The time markers are 50 msec apart. The last record was taken at a slower sweep speed. The zero level for the potential trace indicates the potential observed when the recording electrode (filled with 0.6 M KCl) was placed in the outside fluid medium. 21°C

example, the duration of the action potential was between 1 and 4 sec in 100 mM $CaCl_2$ and it was reduced to 0.1 sec in 200 mM $CaCl_2$ solution. In all the cases examined, an increase in the external Ca-ion concentration resulted in a drastic shortening of the action potential duration.

[The excited state is considered to be a state in which the membrane is occupied predominantly by univalent cations. An increase in the divalent cation concentration in the outside medium increases the probability of replacing the univalent cations in the membrane with divalent ions in the medium, thus facilitating termination of the excited state.]

The membrane resistance did not change appreciably with changes in the external calcium concentration. The threshold for the action potential increased slightly at a high calcium concentration. In the particular example cited above, the rise in threshold resulting from replacement of 100 mM Ca with 200 mM Ca was roughly 4 mV. This finding is consistent with previous findings on squid giant axons (Frankenhaeuser and Hodgkin, 1957) as well as on other excitable tissues (e.g., Weidmann, 1955; Hagiwara and Naka, 1946).

[The so-called "stabilizing action" of the calcium ions can be explained in terms of the two-stable-states theory: an increased calcium concentration in the external medium tends to keep the membrane at the resting state, in which the charged sites in the membrane are occupied predominantly by divalent ions.]

The amplitude of the action potential was increased slightly by an increase in the external Ca-ion concentration in the external medium. Quantitative analysis of this increase is somewhat difficult because the action potential amplitude varies to some extent with the intensity of stimulating current. In the example cited above, the increase was roughly 10 mV for doubling the external Ca-ion concentration: comparison was made with the same current intensity. It should be noted, however, that such an increase is not necessarily an indication of a "specific permeability increase" of the membrane to the Ca-ion. When Ca is the only cation species in the medium, a non-specific cation exchange membrane is expected to behave in this manner (Helfferich and Ocker, 1957).

When an axon is exposed to a concentrated calcium solution for a long period of time, the action potential tends to deteriorate irreversibly. This loss of excitability is possibly due to an influx of $CaCl_2$ into the axon. [At a high salt concentration, co-ion exclusion is not complete.] On the other hand, when the axon is exposed to a dilute calcium solution, the membrane potential often fluctuates, leading to gradual depolarization of the membrane and eventually to an irreversible loss of excitability. The loss of excitability in this case is probably due to a decrease in the divalent cation in the membrane. For these reasons, the range of favorable external Ca-ion concentrations is limited.

Effect of external anions

$CaBr_2$ instead of $CaCl_2$ could be used without affecting the shape and magnitude of the action potential. The resting potential was not affected by this replacement of Cl with Br.

Calcium ethylsulfate, $Ca(C_2H_5SO_4)_2$, could also be used as the sole external electrolyte. The difference between chloride and ethylsulfate externally was small, especially when the action potential in $CaCl_2$ solution was large. Sometimes, ethylsulfate ion produces better action potentials than chloride. In one axon which developed relatively small action potentials, the action potential amplitude increased by about 7 mV when Cl in the external medium was replaced with ethylsulfate; the effect was reversible. The action potential was prolonged in ethylsulfate solution in this axon. Such effects can be explained by assuming that the exclusion of anions be the membrane is more complete with ethylsulfate than with chloride because of the steric effect of the large anions. The anion effect is expected to be greater when the axon is poorer, because we have evidence that under poor conditions the membrane fixed charge density is apparently decreased (Tasaki, Watanabe and Lerman, 1967).

Removal of Tris from the external medium

As described under Methods, the external media contain a small amount of Tris-HCl buffer. Several experiments were conducted to show that the existence of Tris ion was not essential for the production of the bi-ionic action potential. In one of these experiments the buffer was completely removed; the pH was adjusted to about 8 by adding a small amount of $Ca(OH)_2$ solution to the external medium. Since the latter does not posses any buffer action, the pH of the external medium was not stable. In spite of this lability of the pH, typical bi-ionic action potentials could be recorded under these conditions. The amplitude of the action potential of these axons was up to 110 mV. Based on these observations, it was concluded that the existence of Tris ion externally is not essential for production of the so-called bi-ionic action potential.

An alternative method of eliminating the Tris buffer was to mix the external solution with $CaCO_3$ powder. When $CaCO_3$ dissolves, the pH of the solution shifts toward the alkaline side. This method of immersing an axon in a suspension of calcium carbonate in a $CaCl_2$ solution was used on one occasion, yielding somewhat poor, but all-or-none action potentials. The experiment agin shows that Tris is not essential for production of the action potential under the present conditions.

Effect of internal cesium concentration

Bi-ionic action potentials could be evoked with the cesium ion concentration in the internal perfusion fluid at any level between 3 and 200 mM. The most favorable concentration range was between 10–50 mM. When the concentration was higher than 50 mM, the probability of obtaining large action potentials was reduced. When the internal Cs concentration was higher than 100 mM less than 50% of the axons examined produced all-or-none action potentials. With 200 mM of Cs internally, only one axon produced small, but clearly all-or-none bi-ionic action potentials.

Effect of internal anions

It is known that survival time of intracellularly perfused axons is strongly influenced by the difference in chemical species of the anions in the intracellular perfusing fluid (Tasaki, Singer and Takenaka, 1965). Fluoride (with 10% phosphate) and phosphate are the most favorable anions for production of bi-ionic action potentials, and they are almost equally favorable. For other anions our knowledge is still limited.

Role of enzymes

Bi-ionic action potentials are usually recorded after extensive removal of the axoplasm. To maintain a continuous flow of perfusing fluid with low ionic strength, it is almost imperative to initially remove the axoplasm by use of a proper proteolytic enzyme.

After establishing the existence of the bi-ionic action potential, attempts were made to eliminate the use of enzymes. Twenty axons were internally and externally perfused without enzyme and excitability was tested under bi-ionic conditions. Out of 20, 3 axons produced all-or-none responses, 10 axons gave graded responses, and 7 axons did not give any clear regenerative response. The results are certainly unsatisfactory as compared with those obtained with the standard procedure, in which the axoplasm is initially removed by proteases. In all of these cases, there was serious difficulty in maintaining the flow of the internal perfusion fluid.

We believe that all the difficulties in these experiments derive essentially from cessation of the flow of the internal fluid medium under these conditions. Judging from the flow rate of the internal fluid (10–30 μl/min) and the volume of the axoplasm remaining in the perfusion zone (1–3 μl), the flow has to be maintained for more than 5–10 min before the electrolyte concentration in the axoplasm is lowered below the salt concentration in the perfusion fluid. A high potassium concentration in the axon interior is detrimental to production of bi-ionic action potentials. Furthermore, whenever the outlet pipettes are introduced into the axon (to recommence flow), there is a high probability of injuring the axon membrane.

"Nagarse" and "prozyme", which were originally used by Takenaka and Yamagishi (1966, 1969), can also be employed to produce good bi-ionic action potentials. Trypsin was not suitable for the present purpose. Cystein has been described as being usable for removing protoplasm (Huneeus-Cox *et al.*, 1966), but it was not very efficient as compared to the action of proteases on axons of *Loligo pealii.* Rojas and Atwater (1967) recently examined effects of prolonged pronase perfusion on clamping current.

B. Amines as the internal cations

Internal cations other than cesium

Until the summer of 1967, cesium was the only internal ionic species with which bi-ionic action potentials were demonstrated in the squid giant axon. Based on our previous experience, the fluoride or phosphate salt of cesium was considered to be the most favorable internal electrolyte to demonstrate excitability under these experimental conditions. Later, we found that a variety of organic and inorganic cations can serve as the internal cations in a way similar to cesium ion. The ability to produce bi-ionic potentials is now considered a rather common property of many univalent cations.

Table 1 summarizes the present status of our knowledge. It should be emphasized here that negative results are of very little significance. A slight modification of the experimental conditions may convert an inexcitable axon into an excitable one. For example, the chemicals in the second column may produce the all-or-none response when the experimental conditions are modified.

Table 1 List of internal univalent cations which gave rise to all-or-none (left), graded (middle) and no action potentials under bi-ionic conditions. Rb and K-ion gave responses only under anodal polarization.

All-or-none response	Graded response	No response
Cesium	Dimethylamine	Ammonium
Choline	Monomethylamine	Hydrazinium
Ethanolamine	Propylamine	
Guanidium	Benzylamine	TBA
Lithium	(Potassium)	Tris
Methylguanidium		
Monomethylamine		
(Rubidium)		
Sodium		
TEA		
TMA		
TPA		
Trimethylamine		

In this section, experiments are described in which amines were used as the internal cation. The behavior of the following three ion species was examined most extensively: choline, tetramethylammonium (TMA) and tetraethylammonium (TEA). The characteristics of the axons internally perfused with other amines are described briefly in a separate subsection.

Preparation of chemicals. The amines were used as internal cations in the form of either phosphate or fluoride salts. Because neutralization of alkali with hydrofluoric acid was more tedious than with phosphoric acid, phosphate salts were used more frequently than fluoride salts. The procedure of neutralizing amines was a follows. To 100 ml of 0.6 M free amine solution, about 2.8 ml of concentrated phosphoric acid was added. This lowered the pH to about 7.3. The solution was then diluted with 12 volume percent glycerol solution to make a solution of the desired ionic strength. When amines were available only as chloride and bromide salts, the halide salts were first changed into the hydroxyde form by utilizing an ion exchange column. An anion exchange resin, Rexyn 201 (OH) (Reagent Grade, Mesh, 16–50, Fisher Scientific Co.) was washed with distilled water at least five times. The resin was then mounted in a 100 ml buret with a diameter of about 1.7 cm, and again washed for at least ten times, each time with 10 ml of distilled water. The amine salt was dissolved in distilled water to make a 1 M solution and the solution was passed through the column. The outflow from the column was collected into a number of beakers. In each beaker about 10 ml of the fluid was collected. The fractionated solutions were then checked for contamination with chloride ions; this was done by neutralizing the sample with nitric acid and by adding a drop of $AgNO_3$ after dilution. When a trace of turbidity was recognized the collection was discontinued. Usually about 80 ml of about an 0.7 M solution was obtained by this procedure. The solution was then diluted to 0.6 M and was neutralized with concentrated phosphoric acid as described above.

Maintenance of excitability with TMA, TEA or choline internally

With various amine phosphate solutions, we found that most of the amines examined are capable of producing large action potentials under the bi-ionic conditions. With a 200 mM $CaCl_2$ solution externally, the amplitude of the action potential was sometimes larger than those obtained with cesium salts. The action potential appeared usually about 5 min after the onset of intracellular perfusion and then gradually developed to its full size. When the external $CaCl_2$ concentration was 200 mM, however, the action potential started to deteriorate following a short steady period: first, the action potential duration became shorter, then the threshold higher and the amplitude smaller. Eventually, the excitability was permanently lost. The process was always irreversible; switching the internal solution to 25 mM CsF solution did not restore the ability of the axon to develop large action potentials.

A characteristic change in the resting level was also noticed. After initiation of intracellular perfusion, the resting level of the intracellular potential was found to shift in the positive direction. This shift is common to all the observations made with dilute salt solution internally (Tasaki and Shimamura, 1962; Narahashi, 1963). When the action potential started to deteriorate, however, it was found that the resting level of the membrane potential moves in the negative direction. [This negative shift of the resting potential appears to be associated with a loss of the negative fixed charge in the axon membrane; the membrane potential is then determined essentially be the diffusion potential of $CaCl_2$ across the membrane.]

A significant prolongation of the period of maintenance of excitability was achieved by diluting the external $CaCl_2$ solution with isotonic glycerol (or sucrose) solution. In the previous experiments, 200 mM $CaCl_2$ solution was the standard external medium. When 100 mM $CaCl_2$ solution was used externally, the excitability of the axons internally perfused with the phosphate salts of TMA, TEA or choline lasted more than 1 hr without any sign of deterioration.

The reason for this effect of dilution is a matter of conjecture. It is possible that when the external calcium concentration is high, calcium ions invade the axon interior together with chloride ions, because of imperfect co-ion exclusion at a high salt concentration. Such divalent ions inside the cell could disrupt the integrity of the membrane, causing an irreversible loss of excitability.

Shape of action potential

The general appearance of the bi-ionic action potential is very similar to that obtained with cesium as the internal cation. Figure 4 shows several examples. Although some characteristic differences among action potentials do exist with different internal cations, the similarities among these records seem striking.

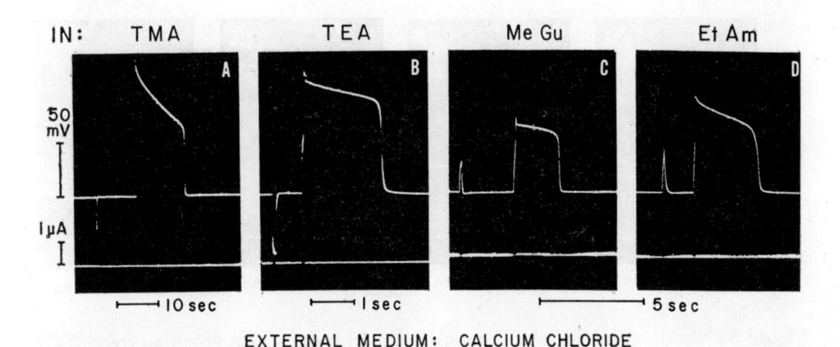

Figure 4 "Bi-ionic" action potentials of squid axons internally perfused with phosphate salt solutions containing 25 mM tetramethylammonium (TMA), tetraethylammonium (TEA), methyl guanidinium (MeGu), and ethylamine (EtAm). The external medium contained 100 mM $CaCl_2$ for the experiments cited in records A, C, and D and 200 mM $CaCl_2$ for B. Upper oscillograph trace represents membrane potential, lower trace represents stimulating currents. In records A and B, a hyperpolarizing current pulse was delivered prior to stimulation with a depolarizing current pulse. In records C and D, subthreshold stimuli were delivered shortly before action potentials were evoked. The length of the internal perfusion zone was between 12 and 15 mm. Axon diameters ranged between 350 and 450 μ. Temp: 18–21°C. (From Tasaki, I., *et al. Am. J. of Physiol.* **216**, No. 1, 130 1969)

The amplitude of the action potentials range between 80 and 100 mV. With TMA or choline, action potentials of 120 mV in amplitude were often encountered. The duration of the action potential varied in a wide range; it was usually between 0.1 and 20 sec. On occasion, it was seen that the membrane was abruptly depolarized spontaneously; in such instances the external $CaCl_2$ concentration had to be raised slightly to restore excitability. The action potential duration was very sensitive to the interval between successive stimuli. When the interval was short, the duration of the following action potential was reduced.

As described in the preceding section, the resting membrane potential was very small under these experimental conditions. The magnitude of the resting potential was usually less than 10 mV and the sign can be positive or negative.

Reversibility of changes brought about by internal perfusion

After the demonstration of prolonged action potentials with choline internally and Ca externally, it was possible to obtain "normal" action potentials of about 1 msec in duration by introducing a K phosphate solution intern-

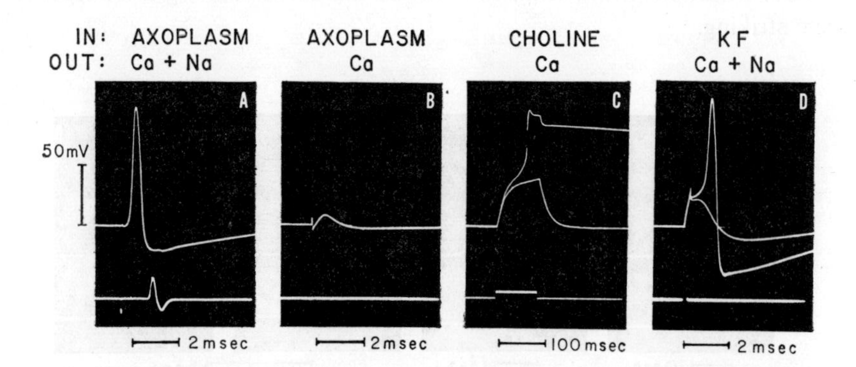

Figure 5 Oscillograph records demonstrating bi-ionic action potential (record C) of an axon immersed in 100 mM $CaCl_2$ solution and internally perfused with 25 mM choline (phosphate) solution. Records A and B were obtained before internal perfusion; the external medium contained *300 mM NaCl* and *100 mM $CaCl_2$* in A and 100 mM $CaCl_2$ in B. The lower trace in these records indicates the presence (A) and absence (B) of impulse propagation along the axon. Record C was obtained 7 min after the onset of internal perfusion with choline phosphate. Record D was obtained from the same axon after switching the internal perfusion fluid to 400 mM KF and the external medium to the solution used in A. In records C and D, responses to supre- and subthreshold stimulating pulses are superposed. The lower trace in C and D represents the intensity of electric current applied to the 15-mm-long perfusion zone. Axon diameter: 450 μ. Temp: 19°C. (From Tasaki, I. *et al. Am. J. of Physiol.* **216,** No. 1, 130 1969)

ally and a solution containing $CaCl_2$ and NaCl externally. An example of such observation is illustrated in Figure 5. This fact indicates that internal perfusion with choline or TEA does not bring about any detectable irreversibile changes in the properties of the membrane.

Effects of Mg, Sr, and Ba in external medium

With the salt of either choline or TMA used internally, a reversible loss of excitability was observed when the external Ca-ion was completely replaced with Mg-ion. Under internal perfusion with the salt of choline or TEA, either Sr or Ba-ions could substitude for the Ca-ion without loss of excitability. Substitution of Ba-ion for Ca-ion was found to reduce the amplitude of the action potential; however, no improvement of action potential amplitude was observed when the external medium was switched back to $CaCl_2$ solution. In one axon, replacement of an external 100 mM $SrBr_2$ solution with a 100 mM $CaCl_2$ solution was found to improve the response; but on switching back to 100 mM $SrBr_2$, no clear reduction in the response amplitude was observed. In conclusion, the difference in favorability among Sr, Ba and Ca was not very clear, although Ca seemed to be the best.

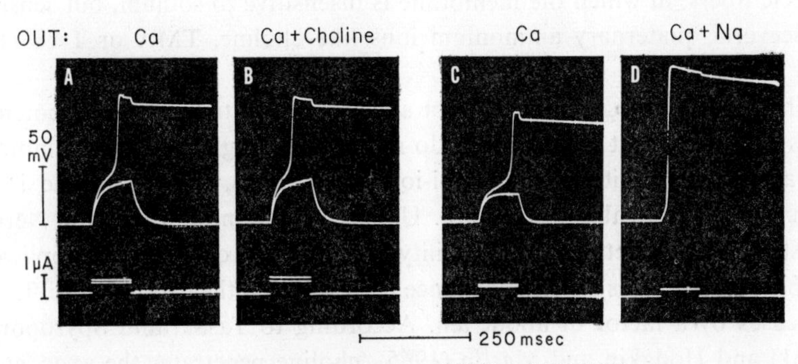

Figure 6 Records A and B: effect of addition of choline chloride to external medium of axon producing action potentials under bi-ionic conditions. Internal perfusate was 25 mM choline (phosphate). External media in A was 100 mM $CaCl_2$. Record B was obtained 3 min after switching the external medium to a solution containing 100 mM $CaCl_2$ and 300 mM choline chloride. Axon diameter: 450 μ. Temp. 19°C. Records C and D: modification of bi-ionic action potential by addition of NaCl to the external medium. Internal perfusate was 50 mM choline (phosphate). External media contained 100 mM $CaCl_2$ in record C. Record D was obtained 4 min after switching the external medium to a solution containing 100 mM $CaCl_2$ and 150 mM NaCl. Axon diameter: 450 μ. Temp. 20°C. Note that, external application of choline produced little, if any, change in the action potential while addition of NaCl externally significantly increased the action potential amplitude and the membrane conductance during excitation. (From Tasaki, I. *et al., Am. J. Physiol.* **216**, No. 1, 130 1969)

Addition of Na to the external medium

An enormous increase in action potential amplitude was observed when sodium ion was added to the external medium (Figure 6). The rate of potential rise was also enhanced, and the impedance loss at the peak of the action potential was increased. The duration of the action potential was also increased by this procedure. After addition of sodium salt to the external medium, it was often necessary to abolish the action potential by application of an inward current or by reduction of the external Na-ion concentration.

Addition of amines to the external medium

With choline inside the axon, the bi-ionic action potential was little affected by addition of the same univalent ion to the external medium. Figure 6 presents one example of such an observation. As can be seen, no appreciable change was produced in the resting potential, membrane resistance, threshold, size and shape of the action potential by addition of 100 mM choline to the external $CaCl_2$ solution. Similar observations were made with TMA. In this respect perfused squid giant are somewhat different from crustacean muscle fibers, in which the membrane is insensitive to sodium, but sensitive to seeveral quaternary ammonium ions like choline, TMA or TAE (Fatt and Katz, 1953).

The fact that these amines do not affect the amplitude of action potential does not imply that these cations do not pass through the axon membrane. In fact, during excitation under bi-ionic conditions, the membrane impedance decreases to about one-tenth. Under the assumption that the Nernst-Einstein relation between ion mobility and diffusion coefficient is applicable to these cations, this fall in impedance indicates that the cation interdiffusion increases by a factor of about ten. According to Tasaki and Spyropoulos (1961) and Hodgkin and Martin (1965), choline penetrates the axon membrane at a considerable rate, indicating that there is no serious steric hindrance of choline molecules penetrating the axon membrane.

Abrupt depolarization produced by depletion of the external Ca-ion

Addition of the salt of various univalent cations to the external medium (containing Ca-ion) produces abrupt depolarization (Tasaki, Takenaka and Yamagishi, 1968). The ability of a univalent cation to depolarize the membrane varies to a great extent with the chemical species of the ion involved. In axons internally perfused it was found that addition of choline to the external medium does not bring about abrupt depolarization as long as the external calcium concentration was kept at 100 mM. Abrupt depolarization could be produced by reducing the external calcium concentration.

An example of an experiment in which abrupt depolarization is produced by variation in the external divalent calcium concentration is as follows: Addition of 300 mM choline to the external 100 mM $CaCl_2$ solution produced no significant change in the membrane potential. On reducing the calcium concentration to 10 mM, the membrane potential started shifting in the positive direction; and after a latency of about 10 sec., an abrupt depolarization took place. The potential jump associated with this depolarization, measured from the original baseline before depletion of Ca, was about 106 mV, which was comparable with the amplitude of the action potential elicited by electrical stimulation (110 mV). This fact suggests that the abrupt depolarization is a phenomenon similar to excitation produced by electrical stimulation. [The experiment is, again, consistent with the basic postulate that an electrical stimulus triggers the process of excitation by removing the Ca-ions in the membrane (Tasaki, 1968).]

Effect of tetrodotoxin

Four axons perfused with choline phosphate solution were used to study the effect of TTX. In two axons perfused internally with 50 mM choline and externally with 100 mM $CaCl_2$, adding 100 µl of 10^{-5} g/ml TTX solution to the external medium blocked the bi-ionic action potential in 4–10 min. The final concentration of TTX was about 2×10^{-7} g/ml in these cases.

In a different series of experiments two axons were perfused internally with 25 mM choline phosphate and immersed in 100 mM $CaCl_2$ solution. TTX was added to the external medium step by step until a final concentration of about 5×10^{-7} g/ml was reached. By this procedure, the action potential amplitude was reduced, the rate of potential rise was decreased, and the impedance loss during action potential was decreased; but the response was still all-or-none. The batch of the poison used in the above series of experiments blocked the conducted spike of unperfused axon immersed in sea water within 3 min at a concentration of 3×10^{-8} g/ml. These experiments suggest that axons perfused with choline and immersed in 100 mM $CaCl_2$ are less susceptible to TTX than the axons in natural ionic environments, or axons perfused externally with 200 mM $CaCl_2$ and internally with 25 mM CsF.

The effect of TTX on the perfused squid axon has been discussed in several recent papers (see Tasaki, Singer and Watanabe, 1966; Kao, 1966; Moore, Anderson and Narahashi, 1966; Watanabe, Tasaki, Singer and Lerman, 1967; Moore, Narahashi and Anderson, 1967; Moore Blaustein, Anderson and Narahashi, 1967; Singer, 1967).

Effect of anions in external medium

In axons internally perfused with dilute choline-phosphate solution and immersed in 100 mM $CaCl_2$ solution, Ca-ethylsulfate could be used in the external medium without suppressing bi-ionic action potentials. Substitution of ethylsulfate for Cl often increased the duration of the action potential. Sometimes the membrane potential stayed at the level of the plateau of the action potential permanently after such substitution. The resting potential shifted in the positive direction by about 5 mV, and the amplitude of the action potential was sometimes increased slightly by this substitution.

As described before, $SrBr_2$ could be used in place of $CaCl_2$ or $SrCl_2$ in the external medium.

Effect of internal cation concentration

In most of the experiments described above, 25 mM solutions of amines were used as the internal perfusate. A 50 mM solution of choline could be used without noticeable difference in excitability. With a 100 mM solution internally, the amplitude of the action potential was reduced.

C. Other amines as internal cations

Tetrapropylammonium (TPA)

Five axons were perfused internally with 25 mM TPA phosphate and immersed in a 200 mM $CaCl_2$ solution, and the ability of the axons to develop action potential was examined. Except in one axon, all-or-none action potentials were obtained under these conditions. The main feature of these action potentials was that the loss of the membrane impedance at the peak of excitation was very small.

Tetrabutylammonium (TBA)

Ten axons were perfused internally with 25 mM TBA phosphate solution under a variety of conditions. As the external solution, a 200 mM $CaCl_2$, 100 mM $CaCl_2$, 100 mM $BaCl_2$, or a solution containing 300 mM NaCl and 100 mM $CaCl_2$ were employed. No action potential was observed under any of these conditions. The effect of internal perfusion with TBA appeared to be irreversible, in the sense that switching to 400 mM KF internally did not restore excitability in an axon immersed in medium containing 300 mM NaCl and 100 mM $CaCl_2$.

Mono-, Di- and Trimethylamine

Among these methyl amines, trimethylamine was most favorable as the internal cation, giving rise to bi-ionic action potentials in axons immersed in 200 mM $CaCl_2$ solution. Out of three axons perfused with 25 mM trimethylamine phosphate, one axon gave an action potential of more than 100 mV; in the other two axons, the action potential amplitude was slightly smaller.

Dimethylamine was examined in two axons and gave rise to poor graded responses. Monomethylamine, examined in three axons, also gave rise to clear, but graded responses. All of these experiments were made with axons immersed in 200 mM $CaCl_2$ solution.

Ethylamine, Propylamine and Benzylamine

In two axons internally perfused with ethylamine, all-or-none bi-ionic action potentials with an amplitude of 70–100 mV were observed in an external medium containing 100 mM $CaCl_2$. Propylamine was examined in one axon immersed in 200 mM $CaCl_2$ solution; the responses observed in this axon were small and graded. Benzylamine was examined in one axon immersed in 100 mM $CaCl_2$ solution; only poor, graded responses were observed in this axon.

Ethanolamine

Two axons immersed in 200 mM $CaCl_2$ solution were perfused internally with 25 mM ethanolamine (in the acetate and phosphate form). In both axons, bi-ionic action potentials with amplitudes of 90–110 mV were observed. The resting potential of these axons was about −10 mV.

Guanidium

Axons internally perfused with 25 mM guanidium phosphate solution developed action potentials of 40 to 60 mV in amplitude in a medium containing 100 mM $CaCl_2$. In all six axons examined, the regenerative power of these responses were found to be limited; strong stimulating currents were needed to evoke these action potentials.

Methylguanidium

Two axons immersed in 100 mM $CaCl_2$ solution were internally perfused with 25 mM methylguanidium phosphate solution and were found to be able to produce action potentials of about 70 mV in amplitude and about 4 sec in duration. The characteristic feature of the responses was that the membrane potential did not fall appreciably in the period between the peak and the abrupt end of the action potential; the time course of the response was nearly rectangular.

Tris(hydroxymethyl)aminomethane (Tris)

Four axons were used to examine excitability under internal perfusion with either 40 or 120 mM Tris phosphate solution. [Note that only about 17%

of the Tris molecules are dissociated at the standard pH of the intracellular fluid, 7.3.] The external medium used was 200 mM $CaCl_2$ solution. No axons retained excitability under these bi-ionic conditions. Axons internally perfused with Tris appear to be in a depolarized state.

D. Sodium and lithium as internal cations

During the course of experiments on bi-ionic action potentials using various amines as the internal cation, we noticed that the external calcium concentration was one of the most critical factors determining the survival time of the axon. In a medium containing 100 mM $CaCl_2$, such cations as guanidium could be used internally to produce bi-ionic action potentials. After recognizing these facts, attempts were made to demonstrate bi-ionic potentials with Na-phosphate internally and $CaCl_2$ externally.

Amplitude and duration of the action potential

With 100 mM $CaCl_2$ in the external medium, bi-ionic action potentials were observed about 5–15 min after initiation of intracellular perfusion with dilute (10–30 mM) sodium phosphate solution. These responses were usually all-or-none and the excitability was maintained at least one hour with little trace of deterioration.

The amplitude of the action potential was in most cases between 50 and 60 mV. The largest action potential observed had an amplitude of 80 mV. In 30 axons perfused internally with 10–100 mM phosphate, and externally with 100 mM calcium chloride, two axons gave rise to graded responses; all other axons had all-or-none action potentials.

The duration of these all-or-none action potentials was in most cases longer than a few seconds. As in the experiments described earlier, the duration of these action potentials was found to vary during the course of internal perfusion. In some axons, it was seen that the membrane potential was arrested at the plateau level for more than 20 sec.

The resting potential of the axon varied between +7 and −51 mV depending on the internal and external ion concentration. There was a reversal of the membrane potential at the peak of these action potentials. In most cases examined, the membrane potential was positive inside throughout the entire plateau period.

The membrane resistance was found to fall during these action potentials. At the onset of these action potentials, the resistance was between about one-tenth to one-twentieth of the resting value. During the later part of the plateau phase the resistance approached a value of about one-third of the resting level. The fall in membrane resistance observed during the action potential in these axons was larger than that seen in axons with cesium, TMA, or TEA internally.

The rate of potential rise following suprathreshold stimulation was found to vary with the intensity of the stimulating current. When the stimulus intensity was barely at the threshold level, the maximum rate of potential rise was often about 0.2 V/sec. The rate increased with the stimulus intensity. The rate observed under these conditions was smaller than that seen in unperfused axons under natural conditions but larger than that observed in axons internally perfused with guanidium.

Reversibility of the state of the membrane

Following the demonstation of all-or-none action potentials with Na internally and Ca externally, the internal solution was switched to 400 mM KF and the external medium to a mixture of 100 mM $CaCl_2$ and 300 mM NaCl. The action potentials observed after this procedure were very similar to those of unperfused axons immersed in normal sea water (Figure 7). This observation indicates that the conformation of the membrane macromolecules have not undergone any detectable irreversible alteration at the time when bi-ionic action potentials are demonstrated.

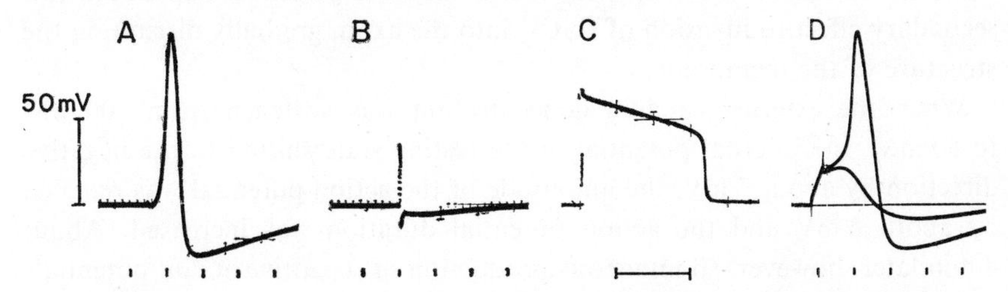

Figure 7 Oscillograph records demonstrating development of an all or none action potential in an axon perfused internally with sodium phosphate and externally with $CaCl_2$ solution. Records A and B were taken before initiation of internal perfussion; the external medium contained 300 mM NaCl and 100 $CaCl_2$ in A and only 100 mM $CaCl_2$ in B. The time markers are 1 msec apart. Record C was taken approximately 12 minutes after the onset of internal perfusion with 10 mM sodium phosphate; the external medium was 100 mM $CaCl_2$. The time marker represents 10 seconds. Record D was obtained from the same axon after switching the internal perfusion fluid to 400 mM KF and the external medium to the solution used in A; sub and suprathreshold responses are superposed. Time markers are 1 msec apart. The stimulus duration was 0.01 msec for A and B, 100 msec for C, and about 0.3 msec for D. Temperature, 21°C. (From Watanabe *et al.* Proc. Natl. Acad. Sci. **58**, 2246 1967)

Effect of external calcium concentration

When concentration of $CaCl_2$ in the external medium was increased, the action potential became larger in amplitude. In most axons, however, excitability was impaired irreversibly on prolonged exposure to media with a high $CaCl_2$ concentration.

The reversibility of the effect of an increase in the external Ca ion concentration is illustrated by the following example of experiment. In an axon perfused wiht 10 mM Na-phosphate and immersed in 100 mM $CaCl_2$ action potentials of about 50 mV in amplitude (10 sec in duration) were elicited upon stimulation. When the external solution was switched to 300 mM $CaCl_2$, the action potentials amplitude increades by about 12 mV; simultneously, the duration shortened to about 2 sec. About one min after this procedure, however, the response became smaller and graded. When the external medium was switched back to 100 mM $CaCl_2$, there was a gradual recovery of excitability, and in about 4 min all-or-none responses with almost the initial amplitude were observed.

This experiment shows that the effect of an increase in the external $CaCl_2$ concentration in the outside medium is essentially an increase in amplitude and a decrease in duration of the action potential. This finding is in agreement with the results of similar experiments with cesium as the internal cation. A complication derives from the secondary effect of a high $CaCl_2$ solution which tends to suppress excitability. At present we attribute this secondary effect to invasion of $CaCl_2$ into the axon, gradially disrupting the structure of the membrane.

When the external calcium concentration was switched from 100 mM to 50 mM, the internal potential in the resting state shifted to the negative direction by about 7 mV, the amplitude of the action potential was reduced by about 6 mV and the action potential duration was increased. About 4 min later, however, spontaneous production of repetitive action potentials was observed.

Effect of other divalent cations

The effect of replacing the external 100 mM $CaCl_2$ solution with 100 mM $MgCl_2$ solution was examined in axons perfused internally with 30 mM Na-phosphate. The result obtained was similar to that of similar experiments with Cs and other univalent cations internally. Replacement of Mg for Ca was found to suppress the bi-ionic action potential produced by reversibly.

Addition of Na to the external medium

Action potentials produced by axons with sodium as the internal cation could be modified by addition of sodium to the external medium. The types of modification brought about by the external Na-ions are: (1) a shift of the resting membrane potential, mostly in the positive direction (i.e., a decrease in negativity), (2) an increase in amplitude of the action potential,

(3) an increase in the rate of potential rise, (4) an increase in impedance loss of the axon membrane during action potential, (5) an increase in tendency toward spontaneous firing, and (6) an increase in action potential duration.

The effects on the resting and action potentials are summarized in Table 2. The "overshoot" of the action potential invariably increased on adding sodium to the external medium. The dependence of the overshoot on the logarithm of the external sodium concentration was, however, smaller than the level expected from the Nernst slope. In the range of concentration above 20 mM, the slope was 20–30 mV per ten-fold change in the external Na concentration.

Table 2 The effect of addition of NaCl to the external medium (containing 100 mM $CaCl_2$) on the resting and action potential

Axon	[Na] in	[Na] out	Resting potential	Action potential	Overshoot
1	10	0	+7	+52	+59
	10	150	+16	+70	+86
2	10	0	−2	+51	+49
	10	150	+17	+64	+81
3	25	0	−25	+45	+20
	25	25	−23	+43	+20
	25	200	−30	+71	+41
4	25	0	−25	+51	+17
	25	25	−28	+55	+27
	25	200	−23	+82	+59
5	25	0	−36	+55	+19
	25	25	−25	+52	+27
	25	50	−21	+65	+44
	25	100	−13	+65	+52
6	25	0	−50	+63	+13
	25	25	−43	+64	+21
	25	50	−29	+78	+49
	25	100	−22	+76	+54
	25	200	−14	+74	+60
	25	0	−30	+58	+28

The action potential duration was markedly increased by addition of sodium to the external medium. In one axon the duration in sodium-free 100 mM $CaCl_2$ solution was about 5 sec. On addition of 50 mM sodium, the duration increased to about 15 sec. In many axons we found that the action potential became infinitely long: the membrane remained depolarized until the external Na salt (mixed with 100 mM $CaCl_2$) was completely removed.

In axons with 25 mM Na-phosphate internally, switching the standard external medium containing 100 mM $CaCl_2$ to a solution containing 400 mM NaCl and 100 mM $CaCl_2$ was found to lead immediately to rapid repetitive firing of action potentials. Eventually the membrane potential was arrested indefinitely at the plateau level of the action potential.

Axons with identical Na-ion concentration in- and outside the membrane

The behavior of squid giant axons with the identical Na-ion concentration in- and outside the axon membrane was a matter of controversy in recent years (Tasaki and Takenaka, 1963; Chandler and Hodgkin, 1965; Tasaki, Luxoro and Ruarte, 1965). For this reason, a series of experiments was conducted to examine the resting and action potential under these conditions.

The results obtained are summarized in Table 3, and an example of the experiments in presented in Figure 8. It is seen that the intracellular potential at the peak of the action potential is always higher than the potential level

Table 3 The resting and action potentials observed in axons with equal Na-ion concentration in- and outside the axon. The overshoot indicates the action potential amplitude less the resting membrane potential. The last column indicates the Na-ion concentration examined. The external medium contained $CaCl_2$ 100 m M in addition to NaCl

Axon No.	Resting potential	Action potential	Overshoot	[Na]
1	31 mV	53 mV	22 mV	30 mM
2	30	58	28	30
3	30	42	12	50
4	38	61	23	50
5	30	50	20	30
6	33	53	20	30
7	23	48	25	30
8	23	48	25	30

in the external fluid medium. Out of 8 axons, one axon gave a 12 mV overshoot; all other axons gave overshoot 20 mV or more. At least under these experimental conditions, there is no doubt that the intracellular potential is well above the "Na-equilibrium" potential: which is zero when the internal Na-ion concentraion is equal to that in the external medium.

[As described in Methods, the recording electrode is a capillary glass tubing of about 100 μ outside diameter, filled with 0.6 M KCl-agar. When freshly made, the resistance is about 2 MΩ. During experiments, the resistance tends to increase. We found a value of 2–5 MΩ in most cases. However, even when the electrode resistance rises to 20 MΩ, the cannula artefact

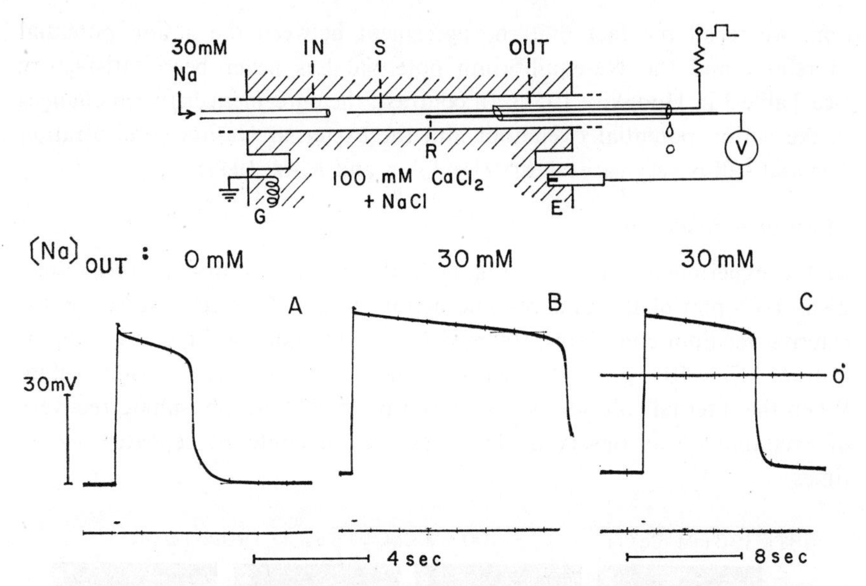

Figure 8 *Top*: Schematic diagram showing the experimental arrangement used to determine the resting and action potentials of axons internally perfused with sodium salt solutions. The inlet cannula (IN), outlet cannula (OUT), stimulating wire electrode (S), recording glass-pipette electrode (R), external calomel electrode (E), and Ag—AgCl ground electrode (G) are indicated. The recording electrodes, (R) and (E), were connected to differential inputs of an oscilloscope via a high input-impedance differential amplifier (designed by A. Bak). The axon is represented by two horizontal lines enclosing the cannulae.
Bottom: Resting and action potentials recorded from an axon internally perfused with a 30 mM sodium salt solution. The external media contained 100 mM $CaCl_2$ for all the records. The external sodium concentrations were varied as indicated. The internal anion was a 1 : 1 mixture of fluoride and phosphate in this case. Stimuli used were 100 msec in duration and approximately 1.5 $\mu A/cm^2$ in intensity (indicated by the lower trace). Note that Record C was taken at a slower sweep sweed. In Record C, the potential recorded when the internal recording electrode was withdrawn and placed in the external medium was superposed on the action potential trace. Axon diameter: approximately 450 μ. 20°C

cannot explain the overshoot. In the bi-ionic situation, the duration of the action potential becomes so long that the "capacitative artefact" can be completely ignored.]

The experimental results just mentioned are not surprising in view of the fact that axons can produce action potentials in the absence of any univalent cation in the external medium. In axons internally perfused with Na-phosphate solution and immersed in a medium containing $CaCl_2$ as the sole electrolyte, the Na-equilibrium potential is negative infinity. Action potentials observed under these conditions gave a finite overshoot. In this connection, we recall the fact that the agreement between the action potential overshoot and the Na-equilibrium potential has never been satisfactory (see Table 3 in Hodgkin, 1951). In contrast, the agreement between changes in the action potential overshoot and the excternal Na-ion concentration was and still is very satisfactory (Hodgkin and Katz, 1949).

Effect of tetrodotoxin

In the experiment shown in Figure 9, the axon was first perfused with 25 mM Cs-phosphate, and bi-ionic action potentials were observed in an external medium containing 100 mM $CaCl_2$. On adding TTX at a concentration 2.7×10^{-7} g/ml the action potential disappeared within 2 min. When the internal solution was switched to 25 mM Na-phosphate recovery of excitability was observed. This observation could be repeated several times.

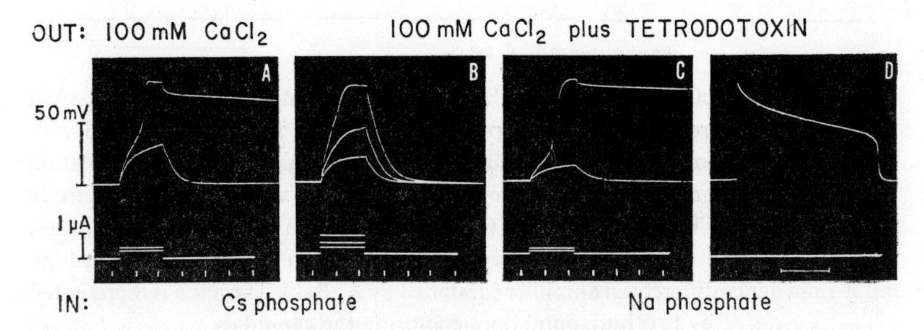

Figure 9 Oscillograph records demonstrating the effect to TTX on bi-ionic action potentials. Record A: Subthreshold response and action potential obtained from axon perfused internally with 25 mM Cs-phosphate and immersed in 100 mM $CaCl_2$ solution. The upper beam represents membrane potential; the lower beam membrane current. Record B: Loss of excitability upon addition of 2.7×10^{-7} gm/ml TTX to external medium. Three sweeps at different stimulus strengths are superposed. Record C: Recovery of all-or-none response after switching internal solution to 25 mM Na-phosphate. Two sweeps with different stimulus strengths are superposed. Dots in Records A, B, and C are 50 msec apart. Record D: Same as in C, but slower time base. Time marker, 10 seconds

The experiment just mentioned shows that susceptibility of the membrane to TTX is dependent on the chemical species of ions in the internal medium. The susceptibility appears to depend also on the external Ca-ion concentration. Several axons which were internally perfused with 25 mM Cs-phosphate and immersed in 100 mM (dilute) $CaCl_2$ solution showed all-or none responses with TTX of 10^{-6}–10^{-7} g/ml externally. The failure of TTX to block action potentials has never been encountered in axons immersed in 200 mM (concentrated) $CaCl_2$ solution.

Voltage clamp experiments

In a series of experiments, the voltage clamp technique was used to study the behavior of axons under bi-ionic conditions with 100 mM $CaCl_2$ externally and 25 mM Na-phosphate internally. The usual inwardly directed membrane currents were observed when the membrane potential was clamped at appropriate depolarizing levels. In the example shown in Figure 10, a maximum inward current of 17 $\mu A/cm^2$ was observed when a clamping pulse of 39 mV (above the resting potential) was applied. The $I - V$ curve showed a typical N-shaped configuration; the membrane resistance at the peak of excitation was about one-tenth of that in the resting state.

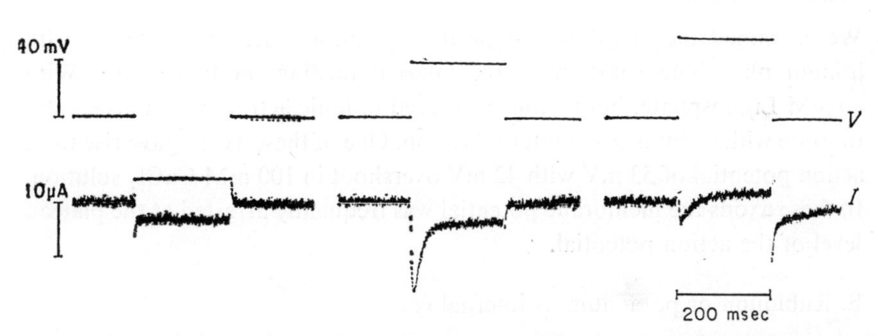

Figure 10 Membrane potentials and currents recorded under voltage-clamp of an axon internally perfused with Na phosphate and immersed in $CaCl_2$ solution. Record 1 is the current-time trace obtained when the voltage pulse used to clamp the axon membrane was hyperpolarizing. Record 2 shows the current-time trace of the membrane current when the clamping pulse was depolarizing and 39 mV in amplitude. In record 3, the curren-time trace observed when the clamping voltage was 56 mV in amplitude and depolarizing in direction is presented. Axon diameter: approximately 450 μ. 19°C

[We believe it is erroneous to regard the inward current as the clacium current. When the membrane potential is clamped at the peak of the action potential, the peak inward current is close to zero; under these conditions the equivalent Ca-influx is equal in magnitude (and opposite in sign) to the Na efflux. The observed inward current is then due to a decrease in the outwardly directed sodium flux and an increase in the inwardly directed calcium flux.]

Effect of changing internal soidum concentration

In the present studies the standard internal Na-ion concentration was 20 to 30 mM. One axon with 5 mM sodium solution internally gave an all-or-none action potential with no sign of rapid deterioration. Between 10 and 50 mM excitability could be maintained. We examined three axons with 100 mM $CaCl_2$ externally and 100 mM Na-phosphate solution internally. One axon gave all-or-none action potentials with an amplitude of about 20 mV. The second axon gave graded response, and the third showed no trace of responses.

The resting membrane potential increased with the internal sodium concentration. In an axon with 200 mM $CaCl_2$ solution externally, alteration of the internal Na-concentration from 10 mM Na to 30 mM changed the resting potential from -21 mV to -29 mV; this efiect was reversible. The collected data from different axons also indicate that the dependence of the resting potentials on the internal sodium concentration is real.

Lithium as the inside cation

We examined the possibility of producing bi-ionic action potentials with lithium phosphate solution as the internal medium in two axons. With 10 mM Liphosphate, both axons produced bi-ionic action potentials similar to those with sodium as the internal cation. One of these axons gave rise to an action potential of 53 mV with 42 mV overshoot in 100 mM $CaCl_2$ solution. In these axons the membrane potential was frequently arrested at the plateau level of the action potential.

E. Rubidium or potassium as internal cation

In a series of experiments, Rb- and K-phosphate were used as the sole internal electrolyte species. When immersed in an aqueous medium containing only $CaCl_2$ and glycerol, axons internally perfused with the salt of Rb- or K-ion never gave rise to claer action potentials (in the absence of polarizing current). Axons in this simple ionic environment showed delayed rectification. A long pulse of outwardly directed membrane current produced a rise in the intracellular potential followed by a rapid fall; but there was no clear sign of "regenerativeness" in this transient positive deflection of the internal potential.

In some axons, particularly when the external $CaCl_2$ concentration was low, clear signs of the presence of "hyperpolarizing responses" are observed. (The properties of hyperpolarizing responses are described by Segal, 1958; Tasaki, 1959; Bennet and Grundfest, 1966.)

It is well known that axonal membranes capable of developing hyperpolarizing responses produce action potentials of the normal sign if excitability is tested in the presence of a steady, hyperpolarizing membrane current (Müller, 1958; Tasaki, 1959). For this reason, attempts were made to demonstrate all-or-none action potentials in axons under these conditions by superposing pulses of outward (stimulating) current on a steady inward (polarizing) current.

In all nine axons internally perfused with Rb-phosphate solution, it was possible to demonstrate clear signs of excitability under these conditions. In two of these axons we were able to demonstrate all-or-none action potentials under steady anodal polarization. In one out of the three axons internally perfused with K-phosphate solution, we could demonstrate the presence of regenerative excitatory processes.

In the experiments described in this article our objective was to demonstrate action potentials in media free of univalent cations. If both KCl and $CaCl_2$ are present in the external medium there is no difficulty in demonstrating hyperpolarizing responses. (An explanation of these facts is presented in a recent paper by Tasaki, Lerman and Watanabe, 1969.)

IV. DISCUSSION

Under Results we have described electrophysiological properties of squid giant axons immersed in a medium containing the salt of a divalent cation and internally perfused with the salt of a univalent cation. We have seen that the ability of the axon membrane to develop all-or-none action potentials is maintained under these "bi-ionic conditions". Although we did not go into detailed explanations of the experimental findings, some of the implications of these findings in relation to the theory of nerve excitation is almost self-evident. As in many other excitable tissues, there is in squid giant axons no absolute requirement for the presence of univalent cations in the external medium to maintain excitability. On the other hand, the requirement for divalent cation in the external medium appears to be absolute. At present, no tissue is known to develop all-or-none action potentials in the presence of Ca-chelating agents (e.g., Na-salt of EDTA).

Although univalent cations are dispensable in the external medium, addition of certain univalent cations to the medium does bring about a drastic increase in the amplitude of the action potential. This amplitude-augmenting effect is strong with such cations of the hydrophilic nature as Na, Li, hydrazinium, guanidinium, ammonium, etc. The effect is absent or weak in polyatomic cations with hydrophobic side-chains. Such behavior of external univalent cations is observed in all axons (irrespective of the chemical species of the internal univalent cations), indicating that this behavior derives from the property of the outer layer of the axon membrane.

The experimental findings described in this article indicate that the peak level of the action potential does *not* in general coincide with the so-called Na-equilibrium potential (see Figure 8). The voltage clamp studies under these conditions contradict the notion that a depolarizing (clamping) voltage pulse applied across the axon membrane opens up the "sodium channels" in the membrane (see Figure 10). On the contrary, these findings provide strong support for the theory in which cooperative ion-exchange processes involving di- and univalent cations are assumed to be responsible for production of action potentials. Since, however, the present article is designed to present "raw" experimental data, and not to discuss theories of nerve excitation, the readers who are interested in theoretical implications of these data are referred to our original papers cited in the Reference.

V. SUMMARY

Squid giant axons under intracellular perfusion with the salt of various univalent cations are shown to maintain their ability to develop action potentials in a medium containing solely the salt of a divalent cation (Ca, Sr, Ba). The conformational state of the axon membrane under these "bi-ionic conditions" are within the normal, reversible range in the sense that normal action potentials can be obtained when K-phosphate solution is introduced internally and a mixture of NaCl and $CaCl_2$ externally. Bi-ionic action potentials were demonstrated with Na-phosphate as the sole internal electrolyte and $CaCl_2$ as the sole external electrolyte.

REFERENCES

Baker, P. F., A. L. Hodgkin, and T. I. Shaw (1961). *Nature*, **190**, 885.
Bennett, M. V. L., and H. Grundfest (1966). *J. Gen. Physiol.* **50**, 141.
Chandler, W. K., and A. L. Hodgkin (1965). *J. Physiol.* **181**, 594.
Fatt, P., and B. Katz (1953). *J. Physiol.* **120**, 171.
Frankenhaeuser, B., and A. L. Hodgkin (1957). *J. Physiol.* **137**, 218.
Hagiwara, S., and K. Naka (1964). *J. Gen. Physiol.* **48**, 141.
Helfferich, F., and H. D. Ocker (1957). *Z. Phys. Chem. N. F.* **10**, 213.
Hodgkin, A. L. (1951). *Biol. Rev.* **26**, 339.
Hodgkin, A. L., and B. Katz (1949). *J. Physiol.* **108**, 37.
Hodgkin, A. L., and K. Martin (1965). *J. Physiol.* **179**, 26P.
Huneeus-Cox, F., H. L. Fernandez, and B. H. Smith (1966). *Biophys. J.* **6**, 675.
Kao, C. Y. (1966). *Pharm. Rev.* **18**, 997.
Michaeli, I., and A. Katchalsky (1957). *J. Polymer Sci.* **23**, 683.
Moore, J. W., N. Anderson, and T. Narahashi (1966). *Federation Proc.* **25**, 569.
Moore, H. W., M. Blaustein, N. Anderson, and T. Narahashi (1967), *J. Gen. Physiol.* **50**, 1401.
Moore, J. W., T. Narahashi, and N. Anderson (1967). *Science* **157**, 220.
Muller, P. (1958). *J. Gen. Physiol.* **42**, 137.
Narahashi, T. (1963). *J. Physiol.* **169**, 91.
Oikawa, T., C. S. Spyropoulos, I. Tasaki, and T. Teorell (1961). *Acta Physiol. Scand.* **52**, 195.

Rojas, E., and I. Atwater (1967). *Nature*, **215**, 850.
Segal, J. R. (1958). *Nature*, **182**, 1370.
Singer, I. (1967). *Nature*, **215**, 852.
Takenaka, T., and S. Yamagishi (1966). *Proc. Jap. Acad.* **42**, 521.
Takenaka, T., and S. Yamagishi (1969). *J. Gen. Physiol.* **53**, 81.
Tasaki, I. (1959). *J. Physiol.* **148**, 306.
Tasaki, I. (1968). Nerve excitation, a macromolecular approach, Thomas, Springfield.
Tasaki, I., L. Lerman, and A. Watanabe (1969). *Am. J. Physiol.* **216**, 130.
Tasaki, I., M. Luxoro, and A. Ruarte (1965). *Science*, **150**, 899.
Tasaki, I., and M. Shimamura (1962). *Proc. Nat. Acad. Sci.* **48**, 1571.
Tasaki, I., I. Singer, and Takenaka (1965). *J. Gen. Physiol.* **48**, 1095.
Tasaki, I., I. Singer, and A. Watanabe (1966). *Am. J. Physiol.* **211**, 746.
Tasaki, I., and C. S. Spyropoulos (1961). *Am. J. Physiol.* **201**, 413.
Tasaki, I., and T. Takenaka (1963). *Proc. Nat. Acad. Sci.* **50**, 619.
Tasaki, I., T. Takenaka, and S. Yamagishi (1968). *Am. J. Physiol.* **215**. 152.
Tasaki, I., A. Watanabe, and I. Singer (1966). *Proc. Nat. Acad. Sci.* **56**, 1116.
Watanabe, A., I. Tasaki, and L. Lerman (1967), *Proc. Nat. Acad. Sci.* **58**, 2246.
Watanabe, A., I. Tasaki, I. Singer, and L. Lerman (1966). *Science*, **155**, 95.
Weidmann, S. (1955). *J. Physiol.* **129**, 568.

*Some Relations between External Cations and the Inactivation of the Initial Transient Conductance in the Squid Axon**

WILLIAM J. ADELMAN, JR. and YORAM PALTI

The Marine Biological Laboratory
Woods Hole, Massachusetts, 02543,
The Department of Physiology,
University of Maryland,
School of Medicine,
Baltimore, Maryland, 21201,
and
The Department of Physiology,
Hebrew University School of Medicine,
Jerusalem, Israel

INTRODUCTION

The initial transient conductance in the giant axon of the squid, which normally enables sodium ions to rapidly permeate the membrane, has been described in quantitative detail by Hodgkin and Huxley (1952c). This description was based on the analysis of the membrane potential dependency of membrane currents obtained with the voltage clamp method introduced by Cole (1949, see also Cole, 1968). According to the Hodgkin and Huxley formalism, inactivation of the initial transient conductance is determined by $1 - h$, where h is a dimensionless parameter which is time and voltage dependent. The time dependency of h is given by: $dh/dt = \alpha_h(1 - h) - \beta_h h$, where α_h and β_h are voltage dependent rate constants. Thus, the h process

* This work was supported by USPHS research grant No. NS-04601, by USPHS Special Research Fellowship No. 1F 10 NS-02204, and by USPHS International Research Fellowship No. 1 F 05 TW 01220.

varies with a voltage dependent time constant, τ_h, and reaches a steady-state value, h_∞, within msecs. When first formulated, α_h and β_h were assumed to be only voltage dependent. However, Frankenhaeuser and Hodgkin (1956) showed that both the absolute magnitude of h_∞ and the position of the h_∞ vs membrane potential (E_M) curve on the voltage axis were functions of the external [Ca]. In addition, perfusion experiments have shown that the inactivation mechanism is dependent both on internal [K] (Narahashi, 1963; Adelman, Dyro and Senft, 1965a; Adelman, 1971) and on external [K] (Adelman, Dyro and Senft, 1965b). Recently, the amplitude of the initial transient current has been related to the reciprocal of the external [K], $1/[K_0]$, in intact axons (Adelman and Senft, 1968); Adelman and Palti, 1969a and b). The purpose of this paper is to consider possible mechanisms for producing changes in the concentrations of potassium and calcium ions close to the outside of the excitable membrane and to correlate these changes with ionic inactivation of the initial transient conductance.

A model system is proposed to describe a mechanism for producing alterations in [K] in close proximity to the external surface of the axon membrane, as a function of membrane currents. The model is derived from anatomical and physiological characteristics of the axon and axon sheath (Frankenhaeuser and Hodgkin, 1956; Adelman and Palti, 1969b). On the basis of this model, we will attempt to correlate the kinetics of inactivation of the initial transient conductance as a function of $[K_0]$ (Adelman and Palti, 1969a and b) with the kinetics of K ion changes in the space just outside the axolemma.

A somewhat similar model is proposed to describe a mechanism for producing changes in [Ca] in the space just external to the membrane, as a function of membrane currents. These [Ca] changes will also be correlated with inactivation of the initial transient conductance (Frankenhaeuser and Hodgkin, 1957).

RELATION BETWEEN SODIUM INACTIVATION AND EXTERNAL POTASSIUM ION CONCENTRATION

In order to correlate such a model system with inactivation of the sodium conductance, we determined experimentally the functional relation between external [K] and the inactivation parameters. Values of h_∞, α_h and β_h were measured at different values of $[K_0]$ (Adelman and Palti, 1969a).

Single giant axons obtained from *Loligo pealei* were prepared and voltage clamped as described in Adelman and Palti (1969a). Membrane currents in the voltage clamp were measured as a function of external potassium ion concentration. External sodium concentration was kept constant at 230 mM in

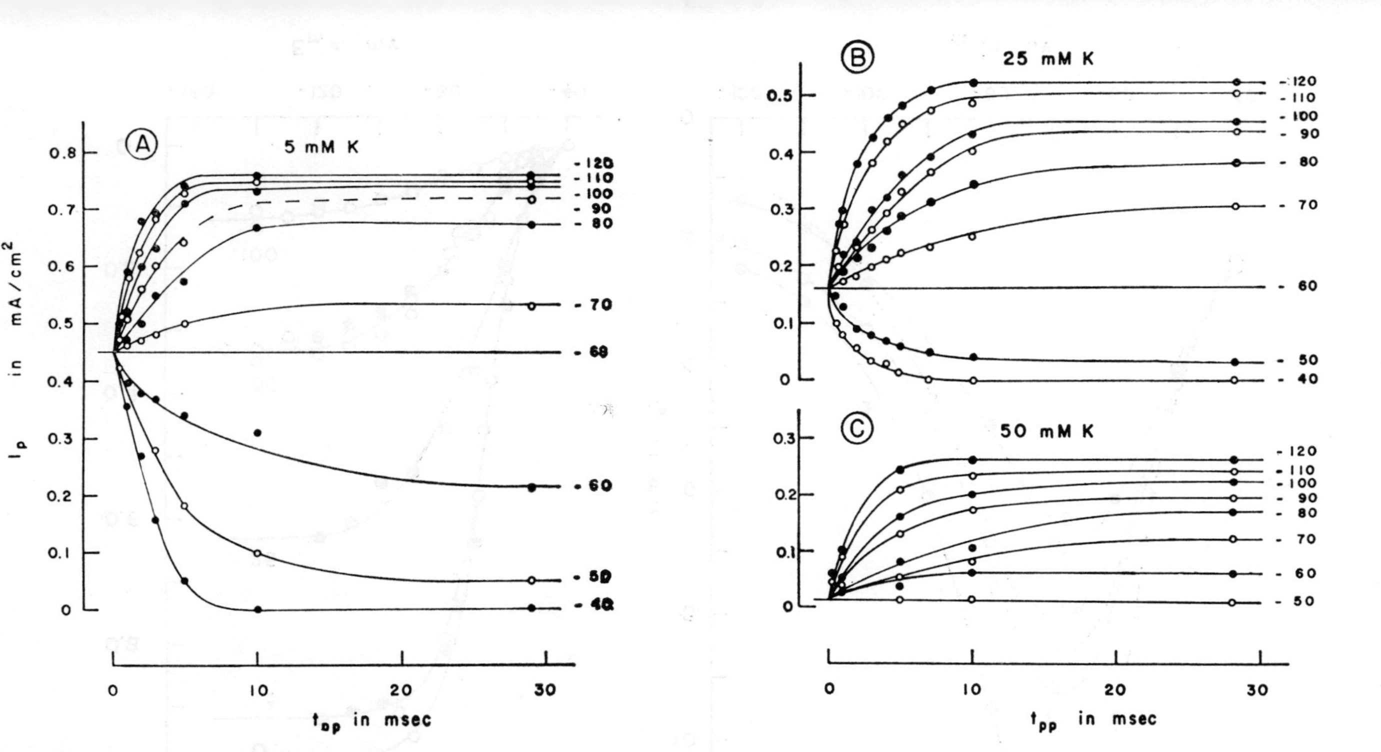

Figure 1 Time course of the variation in the peak value of the initial transient membrane current, I_p, as a function of the duration, t_{pp}, of a conditioning prepulse potential step. The values of the prepulse step potential are given to the right of each curve. I_p was always measured upon stepping from the prepulse potential to the test potential of zero millivolts. Axon 68–50. A, axon in 5 mM K, 230 mM K, 230 mM Na ASW, $E_{HP} = -68$ mv. B, axon in 25 mM K, 230 mM Na ASW, $E_{HP} = -60$m v. C, axon in 50 mM K, 230 mM Na ASW, $E_{HP} = -44$ mv. E_{HP} is the value of the holding potential. (Reprinted by permission of the Rockefeller University Press from *The Journal of General Physiology*, 1969, 53, 689)

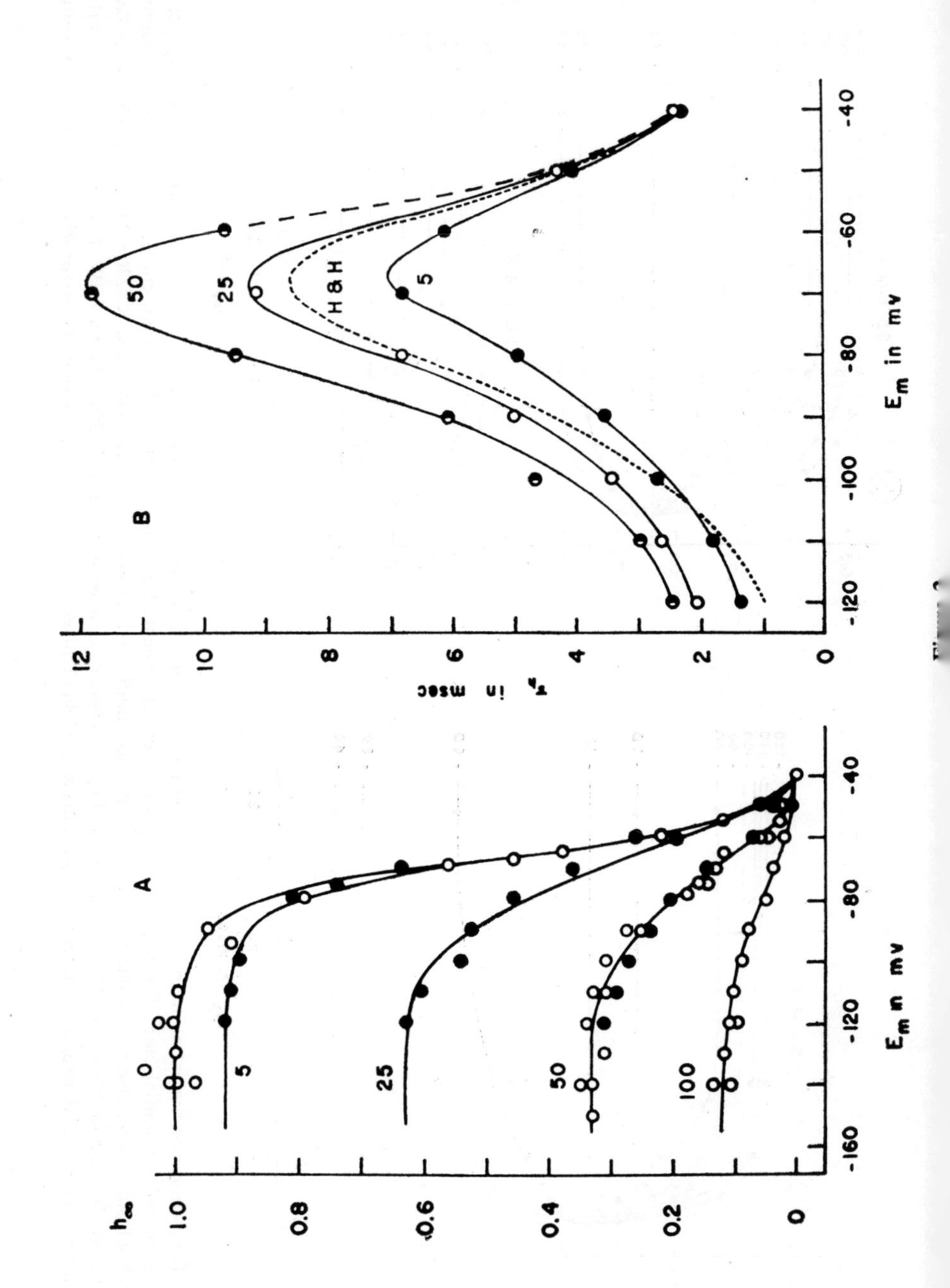
A
h∞
1.0
0.8
0.6
0.4
0.2
0
0
5
25
50
100
-160
-120
-80
-40
E_m in mv
B
τ_h in msec
12
10
8
6
4
2
0
50
25
H & H
5
-120
-100
-80
-60
-40
E_m in mv

Figure 2 (A) Steady-state relations between h and E_m as a function of $[K_O]$. Numbers on each curve refer to the $[K_O]$ in 230 mM Na ASW. Zero mM [K]: axons 68–17 and 68–33, 5 mM K: axon 68–50, 25 mM K: axon 68–50, 50 mM K: axons 69–50, 68–17, 68–34, 100 mM K: axon 68–34. h_∞ = steady-state values of h normalized with respect to the maximum values of h obtained in K-free, 230 mM Na ASW. (B) Values of the time constant of the h process, τ_h, plotted against E_m as a function of $[K_O]$. Experimental ponts obtained from the exponential portions of the curves shown in Figure 2. Open circles, experimental points in 50 mM K, 230 mM Na ASW. Solid circles, experimental points in 25 mM K, 230 mM Na ASW. Half-filled circles experimental points in 5mM K, 230mM Na ASW. Axon 68–50. Dashed line, normalized digital computer solutions of the Hodgkin-Huxley equations (1952c) for ASW containing 10 mM K plotted to indicate only the shape of the H and $H\tau_h$ vs. E_m relation. The amplitude of the H and H curve for τ_h was arbitrarily scaled so as to fall between our τ_h curves for 25 and 5 mM $[K_O]$. (Reprinted by permission of the Rockefeller University Press from *The Journal of General Physiology*, 1969, **53**, 690 and 691)

---

the artificial sea water (ASW) solutions. Trishydroxymethylaminomethane chloride (Tris Cl) was added to the solutions keeping osmolarity approximately equal to that of 430 mM sodium sea water as indicated by freezing point depression measurements. External potassium ion substitutions were made for given fractions of the Tris concentration. Tris Cl was prepared by titrating Tris base with HCl to a pH of 7.4 at 3.5 ± 0.5°C, the temperature at which the experiments were done. At this pH and temperature, the conductivity of a 200 mM Tris, 230 mM Na ASW solution was measured to be within 10% that of 430 mM Na ASW. Experiments using all Tris ASW solutions showed no inward initial transient currents in the voltage clamp.

Figure 1 plots the initial peak current, I_p, against the duration, t_{pp}, of a conditioning pre-pulse for various amplitudes, E_{pp}, of the pre-pulse at three different concentrations of K_0. Test pulses were to zero potential and potentials were measured internally with respect to earth. Analysis of these results was similar to that of Hodgkin and Huxley (1952b). As shown in Figure 1, inactivation is increased with depolarization and decreased with hyperpolarization. At pre-pulse durations greater than 20 msec., inactivation, and consequently I_p, tends toward a steady state value. In experiments not illustrated in this figure, in three axons exposed to different concentrations of K_0, the steady state level of inactivation was maintained over the range of t_{pp} from 20 to 100 msecs.

We will define h_∞ as the ratio of the I_p value in the steady-state to the maximum I_p value, $I_{p_{max}}$. $I_{p_{max}}$ was obtained at zero $[K_O]$ following hyperpolarizing pre-pulse potentials. Typical values of h_∞ vs E_M for various $[K_O]$ are illustrated in Figure 2A. Figure 2A indicates the relation between

h_∞ and $[K_O]$, showing a general decrease in the values of h_∞ with increasing $[K_O]$. For each value of $[K_O]$, a plateau value, $h_{\infty_{plat}}$, is obtained with hyperpolarizing pre-pulses. Each curve has a shape similar to the K-free curve. $h_{\infty_{plat}}$ is inversely related to the log of $[K_O]$. Unlike the curves determined by Frankenhaeuser and Hodgkin (1957) for Ca_O, the curves of h_∞ vs E_M for different $[K_O]$ cannot be made to superpose, as they have different slopes even when their $h_{\infty_{plat}}$ values are normalized. For $[K_O] \geqq 10$ mM, even 30 msec. -40 mV pre-pulses could not bring $h_{\infty_{plat}}$ to the maximum value of $h_{\infty_{plat}}$ found in K-free, 230 mM Na ASW. This effect and the change in slope of h_∞ vs E_M with $[K_O]$ imply that the K^+ effect on h is voltage dependent.

Values of τ_h were obtained from the exponential portion of curves such as those shown in Figure 1. Figure 2B gives values of τ_h vs E_M for various $[K_O]$. Each of the curves has the same general shape. However, increasing external potassium generally increases the values of τ_h. It is possible to obtain relations between τ_h and the log $[K_O]$ for various values of the pre-pulse potential. These relations are approximately linear. Deviations from linearity increase slightly in the depolarizing direction. The rate constants, α_h and β_h, were determined in the E_M range from -120 mV to -50 mV as follows (Hodgkin and Huxley, 1952c):

$$\alpha_h = \frac{h_\infty}{\tau_h}, \tag{1}$$

and

$$\beta_h = \frac{1}{\tau_h} - \alpha_h = \frac{1 - h_\infty}{\tau_h}. \tag{2}$$

Figure 3A gives values of α_h vs E_M for artificial sea water solutions containing 5, 25 and 50 mM K and 230 mM Na. α_h is inversely related to $[K_O]$. The points for 5 mM K_O follow a locus somewhat similar to that published by Hodgkin and Huxley (1952c) for normal sea water (when a temperature correction is made with $Q_{10} = 3$).

The experimental points plotted in Figure 3A were fit by curves obtained from

$$\alpha_h = (0.126 - 0.028 \ln [K_O]) \exp [-(E_M + 60)/27.4]. \tag{3}$$

When $[K_O] = 10$ mM, Equation (3) reduces to

$$\alpha_h = 0.06 \exp [-(E_M + 60)/27.4]. \tag{4}$$

Equation (4) is very similar to Hodgkin and Huxley's relation for α_h as a function of E_M. The experimental α_h vs E_M relationship can also be expressed in a purely exponential form [cf. Adelman and Palti, 1969a, Equations (5) and (6)].

Figure 3B is a plot of β_h vs E_M for 5, 25, and 50 mM K_0, 230 mM Na ASW solutions. Unlike the α_h curves, the β_h curves have minima between $E_M = -70$ and $E_M = -90$ mv. For depolarizing potentials, the values of β_h are similar to those reported by Hodgkin and Huxley (1952c). The increasing values of β_h observed for the hyperpolarized membrane were in a voltage range not considered by Hodkin and Huxley (1952b). The curves drawn to fit the points in Figure 3B were computed from:

$$\beta_h = 1/[\exp[-(E_M + 36)/9] + 1] + 0.01 \exp[-E_M B_{(K)}], \qquad (5)$$

where

$$B_{(K)} = [K_0]/(32.5\,[K_0] + 185). \qquad (6)$$

The first term in equation (5) is independent of $[K_0]$ and is almost identical to the equation for β_h given by Hodgkin and Huxley (1952c). Thus, while α_h is inversely related to $[K_0]$, β_h is directly related to $[K_0]$ for hyperpolarizations, and is virtually independent of $[K_0]$ for depolarizations. The β_h vs E_M relations imply that the potassium effect on β_h is a function of membrane potential. Both equations 3 and 5 contain more information than the original Hodgkin and Huxley equations as they incorporate, in a general way, the $[K_0]$ effect on the *h* process.

It is apparent that both h_∞ and τ_h are proportional to log $[K_0]$ in the concentration range dealt with here. This fact may imply that the K^+ effect on the *h* process is mediated through a corresponding change in resting potential. However, the change in slope of h_∞ vs E_M curves with $[K_0]$ necessitates that external potassium ions affect the inactivation process in other ways than those brought about by simple shifts of the *h* parameter on the voltage axis.

As the values of I_p in Figure 1 reach a plateau within 28 msec., changes in h_∞ produced by conditioning the membrane with long hyperpolarizing pulses (cf. Narahashi, 1964; Adelman and Palti, 1969b) in axons bathed in high $[K_0]$ should be regarded as removals of inactivation. In the above experiments, greater amplitude sodium currents were obtained with depolarizing pulses if the axon membrane was previously conditioned with hyperpolarizing potentials of a few seconds than if conditioned with pre-pulses of approximately 50 msec. Narahashi (1964) called the effect of these long duration hyperpolarizing conditioning potentials, "removal of inactivation". Figure 4 shows an example of the kinetics of this process (Adelman and Palti, 1969b). An axon is bathed in 50 mM K, 230 mM Na for about 15 min and then is voltage clamped to its resting potential (−40 mV). A hyperpolarizing pre-pulse to −120 mV is applied which is immediately followed by a depolarizing pulse to a potential of zero mV.

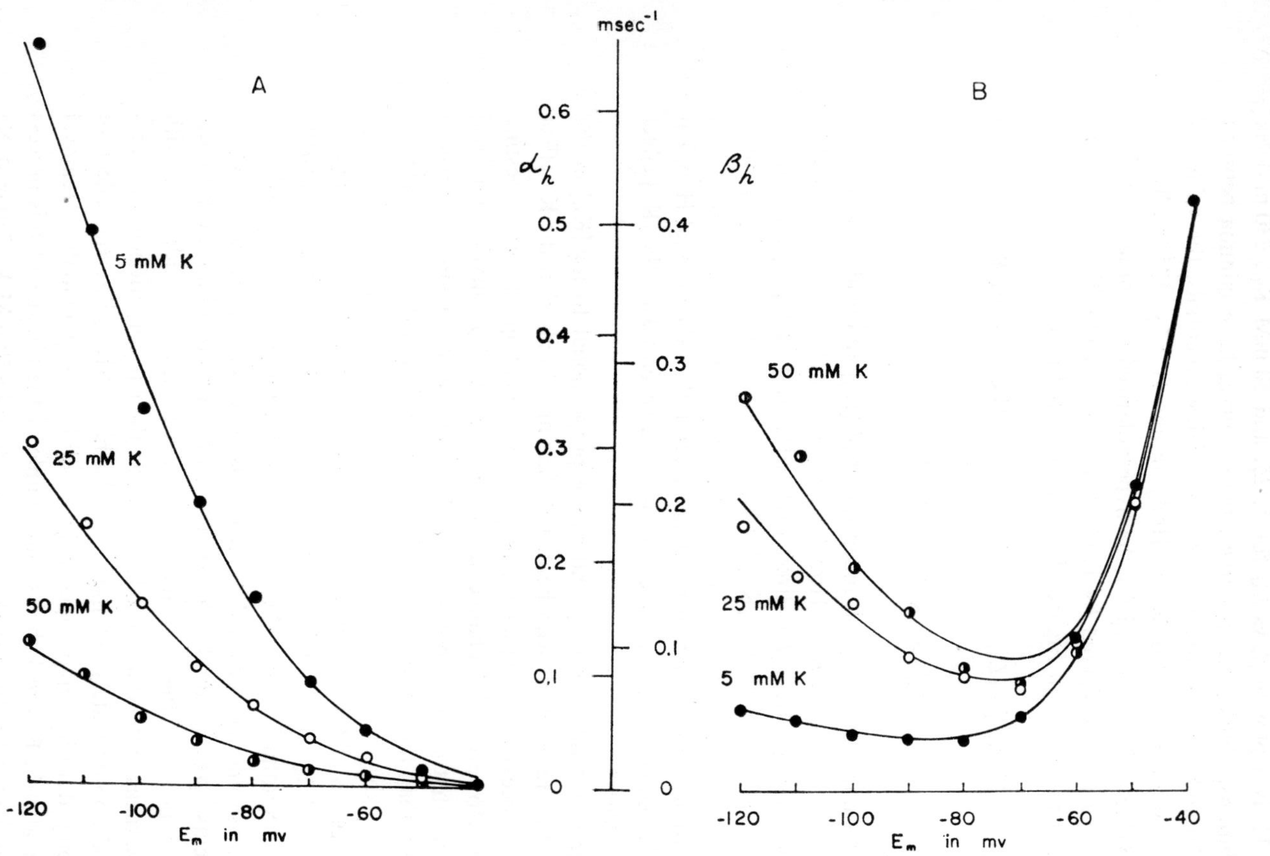

Figure 3 (A) Values of the rate constant, α_h, of the *h* process plotted against E_m as a function of $[K_0]$. Values of $[K_0]$ in 230 mM Na ASW are given above each curve. Points obtained from experimental values of h_∞ and τ_h. Curves plotted from equation 3. Axon 68–50. (B) Values of the rate constant, β_h, of the *h* process plotted against E_m as a function of $[K_0]$. Values of $[K_0]$ in 230 mM Na ASW are given above each curve. Points obtained from experimental values of h_∞ and τ_h. Curves plotted from equation 5. Axon 68–50. (Reprinted by permission of The Rockefeller University Press from *The Journal of General Physiology*, 1969, **53**, 695 and 698)

The duration of the pre-pulse was varied between 0 and 16 msec (A), between 30 msec and 2 sec (B) and between 1 sec and 3 min (C). It can be seen that the values of I_p for a standard depolarizing pulse increase to three plateaus as a function of the duration of the conditioning hyperpolarizing pre-pulse. Each of the subsequent plateaus is reached with prepulse durations, t_{pp}, which are roughly 100-fold longer than those required to achieve the previous plateau. The first process seen in A has a time constant equivalent to the classical Hodgkin and Huxley τ_h. The time constant of the process in B is about 200 msec and the time constant of the process in C is almost 1 min. Adelman and Palti (1969b) demonstrated that in the absence of external K^+, application of pre-pulses achieved plateau level 2 with the msec time constant. In other words, hyperpolarizing pre-pulse durations from 10 msec to 2 sec did not further increase the values of I_p beyond the I_p value obtained following a 10 msec pre-pulse. However, when $[K_0] = 0$ mM, I_p values obtained following a 10 msec pre-pulse were greater than those obtained following a 1 sec pre-pulse when $[K_0] = 50$ mM.

On the basis of these observations, Adelman and Palti (1969b) concluded that the effects of hyperpolarizing pre-pulses of seconds duration were somehow related to removing the effects of potassium ion on the sodium inactivation mechanism. The third plateau achieved with time constants approaching one minute was independent of $[K_0]$ for hyperpolarizing pre-pulse potentials equal to or more negative than -120 mV. The external potassium ion and membrane potential dependency of these three plateaus are shown in Figure 5. The figure plots $I_p/I_{p_{max}}$ against the pre-pulse potential for the three pulse durations corresponding to the plateau levels seen in Figure 4. Notice that $I_{p_{max}}$ in A, B and C is not the same, but is the value corresponding to each of the three plateau maxima achieved in K-free 230 mM Na ASW with increasing hyperpolarizing pre-pulses. The curves in Figure 5A correspond to those shown in Figure 2A.

Adelman and Palti (1969b) concluded that the two slow restoration processes only indirectly affected the inactivation mechnism and proposed that the intermediate time constant (highly K_0^+ dependent) process involved removing potassium ion from physical proximity to the external membrane surfase. They also suggested that the even slower time constant restoration might be related to a membrane current dependent process involving a redistribution of Ca ion just external to the membrane surface.

In the following section, models are developed for providing means by which 1) K ion could accumulate, 2) K ions could be removed and 3) Ca ions could be varied in concentration in a space or unstirred lay just external to the axon membrane.

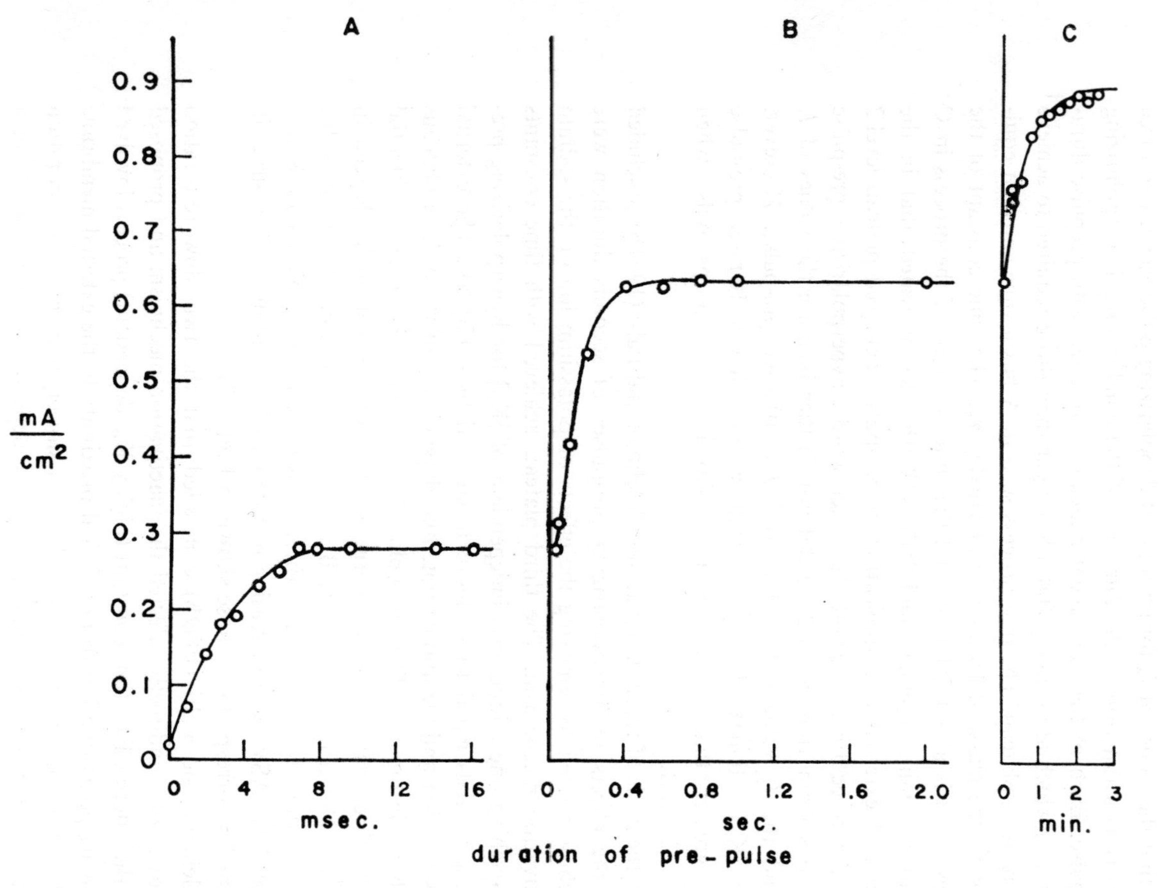

Figure 4 Typical I_p values in ma/cm² plotted as a function of the duration of a prepulse having a constant E_{pp} of −120 mV. Axon 68–17 exposed to 50 mM K, 230 mM Na ASW. A, prepulse durations varied between 0 and 16 msec. B, prepulse duration varied between 30 msec and 2 sec. C, prepulse durations varied between 1 sec and 3 min. (Reprinted by permission of the Rockefeller University Press from *The Journal of General Physiology*, 1969, **54**, 595

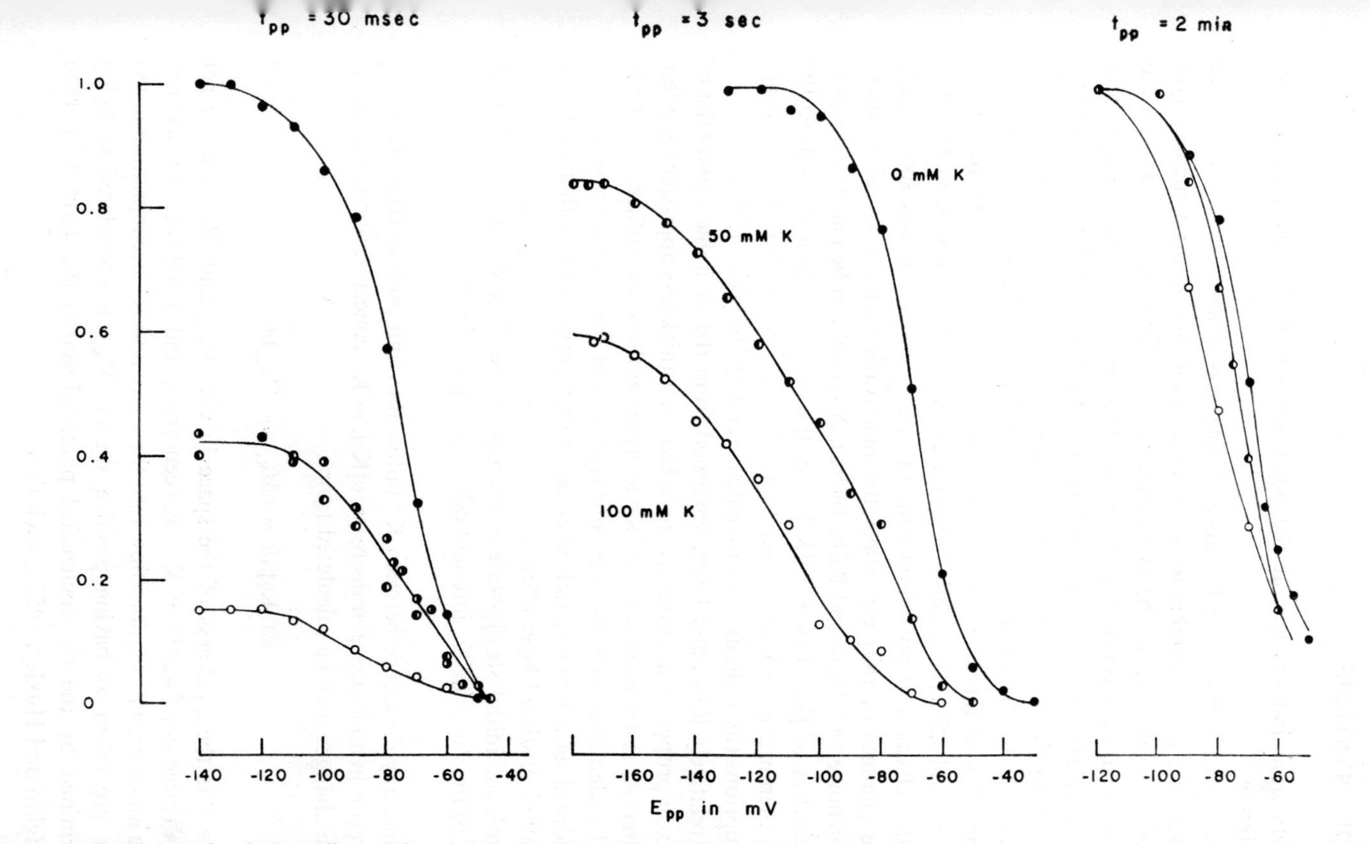

Figure 5 Comparison of relative effects of conditioning prepulses of three different durations, t_{pp}, on $I_p/I_{p_{max}}$ ratios plotted as a function of prepulse potential and $[K_0]$. $I_{p_{max}}$ was taken as unity for $[K_0] = 0$ and E_{pp} -120 mV for each prepulse duration. Filled circles, K-free, 230 mM Na ASW, half-filled circles, 50 mM K, 230 mM Na ASW, open circles, 100 mM K, 230 mM Na ASW. A, $t_{pp} = 30$ msec, axons 68–17 and 68–34. B, $t_{pp} = 3$ msec, axon 68–11. C, $t_{pp} = 2$ min, axon 68–38. (Reprinted by permission of the Rockefeller University Press from *The Journal of General Physiology*, 1969, **54**, 596)

MODEL SYSTEMS

a. Potassium ion changes at the outer surface of the squid giant axon membrane.

The model is based on the general idea that when currents are passed through the axon membrane, ionic concentrations in the aqueous layer immediately external to the membrane may be different than those in the bulk external solution (Frankenhaeuser and Hodgkin, 1956). This difference may be attributed to insufficient mixing between the ions immediately adjacent to the membrane (usually termed unstirred layer) and the bulk solution, or to the different ionic permeability of the Schwann layer as compared with the axolemma. In this last case, potassium (or other ion) influxes and effluxes resulting from membrane potassium currents must produce changes in potassium concentration in the Schwann space, $[K_s]$. These changes in $[K_s]$ can affect the inactivation of the initial transient conductance (Adelman and Palti, 1969a). A model can be made to describe the kinetics of [K] changes, $\delta[K_s]_{(t)}$, in the aqueous space external to the axon membrane (Adelman and Palti, 1969b) by making the following assumptions (hypothesis 1, Frankenhaeuser and Hodgkin, 1956):

1) A relatively thin outer layer, separated from the excitable membrane of the axon proper by an aqueous space having a thickness approaching a few hundred Å, restricts diffusion of K^+ away from the axon surface. This layer may be identified with the Schwann layer but in the general sense, may be considered as a hypothetical layer of proper permeabilities that gives the measured unstirred layer effects.

2) Both the initial steady-state concentration $[K_{s_0}]$, and the integral, over time, of the fluxes of K^+ into and out of the space determine $[K_s]_{(t)}$ in mole/cm^3.

Thus, any difference between K^+ inflow into and outflow from the space will result in an increase or decrease, $\delta[K_s]$, of K^+ concentration in the space. These changes may be calculated from:

$$d\delta\,[K_s]/dt = (M_{K_{sa}} + M_{K_{so}})/\theta \qquad (7)$$

where θ is the thickness of the space in cm, $M_{K_{sa}}$ and $M_{K_{so}}$ are the net fluxes (mole cm^{-2}sec^{-1}) of K^+ between space and axoplasm and between space and external solution (flow into the space and K^+ accumulation in the space are taken as having positive signs.). $M_{K_{sa}}$ is considered as being determined by the electrochemical potential across the axon membrane (Hodgkin and Huxley, 1952c) such that

$$M_{K_{sa}} = G_K(E_M - E_K)/F \qquad (8)$$

where G_K is the potassium ion conductivity in ohm^{-1} of the axon membrane, E_M is the potential difference across the axon membrane, E_K is the potassium current reversal potential and $F = 96{,}500$ coul/mole. It is apparent that for depolarization, G_K is a function of membrane potential and has the form $\bar{g}_K n^4$ (Hodgkin and Huxley, 1952c). Using appropriate values of the rate constants α_n and β_n, it is possible to solve for n as a function of time and voltage. For hyperpolarizations, values of n tend toward zero and the potassium current is carried mainly through the leakage conductance. Under these conditions, G_K is virtually independent of time, but the apparent G_K is a function of $[K_0]$ [see equation (13)].

Frankenhaeuser and Hodgkin (1956) give the following relation for the diffusion of K ions, M_{K_d}, from the space to the outer layer:

$$M_{K_d} = -[K]_s P_{K_s} \tag{9}$$

where P_{K_s} is the K^+ permeability of the outer layer in cm sec^{-1}.

The K^+ flux carried between the space and the external solution by an electric current flowing to the external electrodes is given by:

$$M_{K_e} = -I_{total}\, t_K/F \tag{10}$$

where I_{total} is the sum of all ionic currents through the axon membrane, and t_K is the transport number of K^+ in the solution used.

Upon assuming that the values of the resistances in series with the axon membrane are small compared with the membrane resistance, the potential difference between the internal and external current electrode approaches E_M.

Note that M_{K_e} has a positive sign for an influx of K ions into the space. As the flux through the outer layer is given by the following relation:

$$M_{K_{so}} = M_{K_d} + M_{K_e} \tag{11}$$

from equations (7) through (10) we get:

$$\delta[K_s] = \int_0^t [(G_K(E_M - E_K) - G_t E_M t_K - \delta[K_s]\, P_K F)/F\theta]\, dt \tag{12}$$

As $G_t = \bar{g}_K n^4 + g_L + \bar{g}_{Na} m^3 h$, and $G_K = \bar{g}_K n^4$, these may be readily solved for by means of the Hodgkin and Huxley (1952c) equations (cf. Palti, 1971).

The integral of equation (12) was solved numerically by the finite increment method using either an IBM 360/44 or an SDS Sigma 7 digital computer. As $[K_s] = [K_{s_0}] + \delta[K_s]$, the values of $[K_s]_{(t)}$ can be readily obtained from iterative solutions of equation (12). If $\delta[K_s]$ is positive, then $[K_s]$ increases with time; if $\delta[K_s]$ is negative, $[K_s]$ decreases with time.

Figure 6 plots calculated values of $[K_s]$ as a function of time at a given depolarization for two values of θ, the space thickness, and three values of P_{K_s}, the Schwann layer permeability. The initial conditions were $[K_0] = 10$ mM, $[K_i] = 350$ mM, $E_H = -60$ mV and $E_p = 0$ mV. It can be seen that $[K_s]$ doubles within the first few msec and tends toward values approaching or even exceeding 100 mM. If θ is taken as 100 Å then $[K_s]$ is five times the initial value within 5 msec for any of the three values of P_{K_s} chosen.

Now let us calculate the effects of axon membrane hyperpolarization on $[K_s]$ values. In this case, the values of parameter n tend to zero so that the the current carried by potassium ion is mainly through the leakage conductance. Therefore, G_K is practically independent of time. Under these conditions, the values of G_K and G_t can be determined under voltage clamp in real axons (cf. Adelman and Palti, 1969b) and the apparent G_K was found

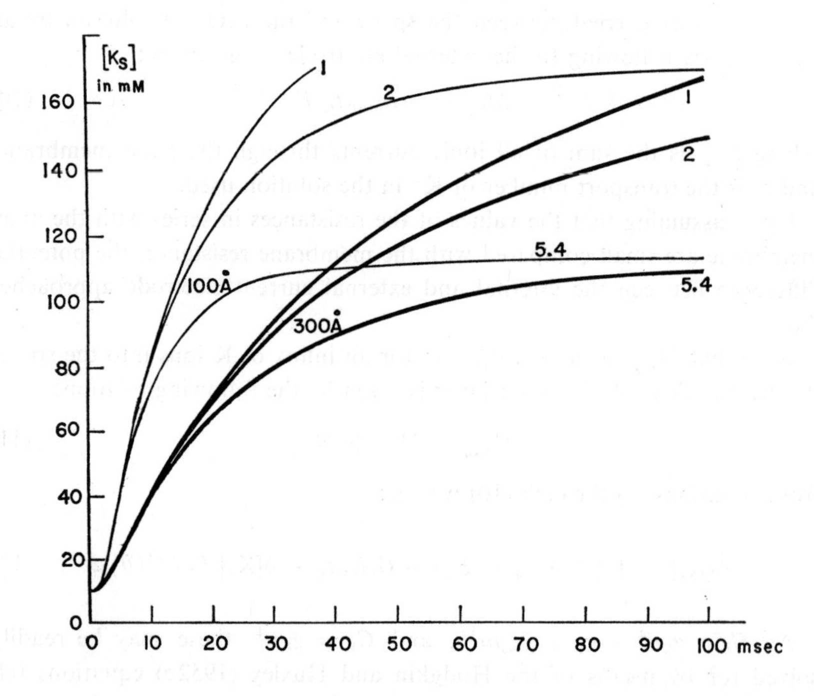

Figure 6 Potassium ion concentration, mM, in the space between the axolemma and Schwann cell layer, $[K_s]$, calculated by means of a digital computer from equation 12 as a function of time after onset of a depolarizing pulse to a membrane potential of zero mv. $[K_s]$ values plotted as a function of space thickness, θ, and the outer layer pemeability, P_{K_s}. Thin lines: $\theta = 100$ Å; thick lines: $\theta = 300$ Å. Numbers adjacent to each curve represent values of P_{K_s} multiplied by 10^{-5} cm/sec. Initial values of $[K_s]$ were set at 10 mM. See text.

to be a function of $[K_0]$. The following relation was found between G_t and $[K_0]$:

$$G_t = 10^{-3} (0.094 + 0.336 \log [K_0]) f \quad (13)$$

where $200.0 \geqq [K_0] \geqq 5.0$ and f is a conductance factor (ohm^{-1}) specific for each axon (values of f were found experimentally to very between 0.5 and 2).

Equation (12) was solved by varying the values of G_t in accord with equation (13) as $[K_s]$ varied with time. Figure 7 illustrates solutions of equation (12) obtained for a given set of P_{K_s} values in the range from $1 \cdot 10^{-5}$ to $5.4 \cdot 10^{-5}$ cm sec^{-1} (see Frankenhaeuser and Hodgkin, 1956) and for a set of space thicknesses from 100 to 2000 Å. In all cases, $[K_s]$ changes were computed as a function of time after onset of a -120 mv hyperpolarizing potential.

Figure 7A plots the changes in $[K_s]$ as a function of time for different apparent space thicknesses (θ) when $P_{K_s} = 1$—$2 \cdot 10^{-5}$ cm sec^{-1}. It is seen that $[K_s]$ values decrease exponentially to about half the original values (100 mM)*, with time constants of the order of 100 msec. The larger θ is, the slower are the changes in $[K_s]$. For any given P_{K_s} the steady-state values of $[K_s]$ are very little affected by θ. The rate of decline and steady-state values of $[K_s]$ calculated for a 300 Å thick space and $P_{K_s} = 2 \cdot 10^{-5}$ cm sec^{-1} are in good agreement with the experimentally observed sodium conductance restorations (see Figure 4B and Adelman and Palti, 1969b). Figure 7 (B, C and D) illustrates the changes of $[K_s]$ as a function of time for different P_{K_s} and f values as well as different initial concentrations of $[K_s]$.

b. Calcium ion changes at the outer surface of the squid axon membrane.

The changes in calcium concentration ($\delta[Ca_s]$) in the so-called Schwann space resulting from axon membrane hyperpolarization can be calculated using the same model, assumptions and methods used for calculation of $\delta[K_s]$. To a good approximation, the excess or deficieny in calcium ions, $\delta[Ca_s]$, above the steady-state level $[Ca_{s_0}]$ is given by:

$$\delta[Ca_s] = \int_0^t [(G_{Ca}(E_M - E_{Ca}) - G_t E_M t_{Ca} - \delta[Ca_s] P_{Ca_s} F)/F\theta \, dt \quad (14)$$

where G_{Ca} is the axon membrane Ca^{++} conductance, E_{Ca} is the calcium current reversal potential (as calculated from the Nernst equation). t_{Ca}

* As the axon was allowed to equilibrate for about 10 minutes in any solution before each pulse, it may be assumed that in practice when $[K_0]$ was 100 mM, the initial values of $[K_s]$ were also 100 mM.

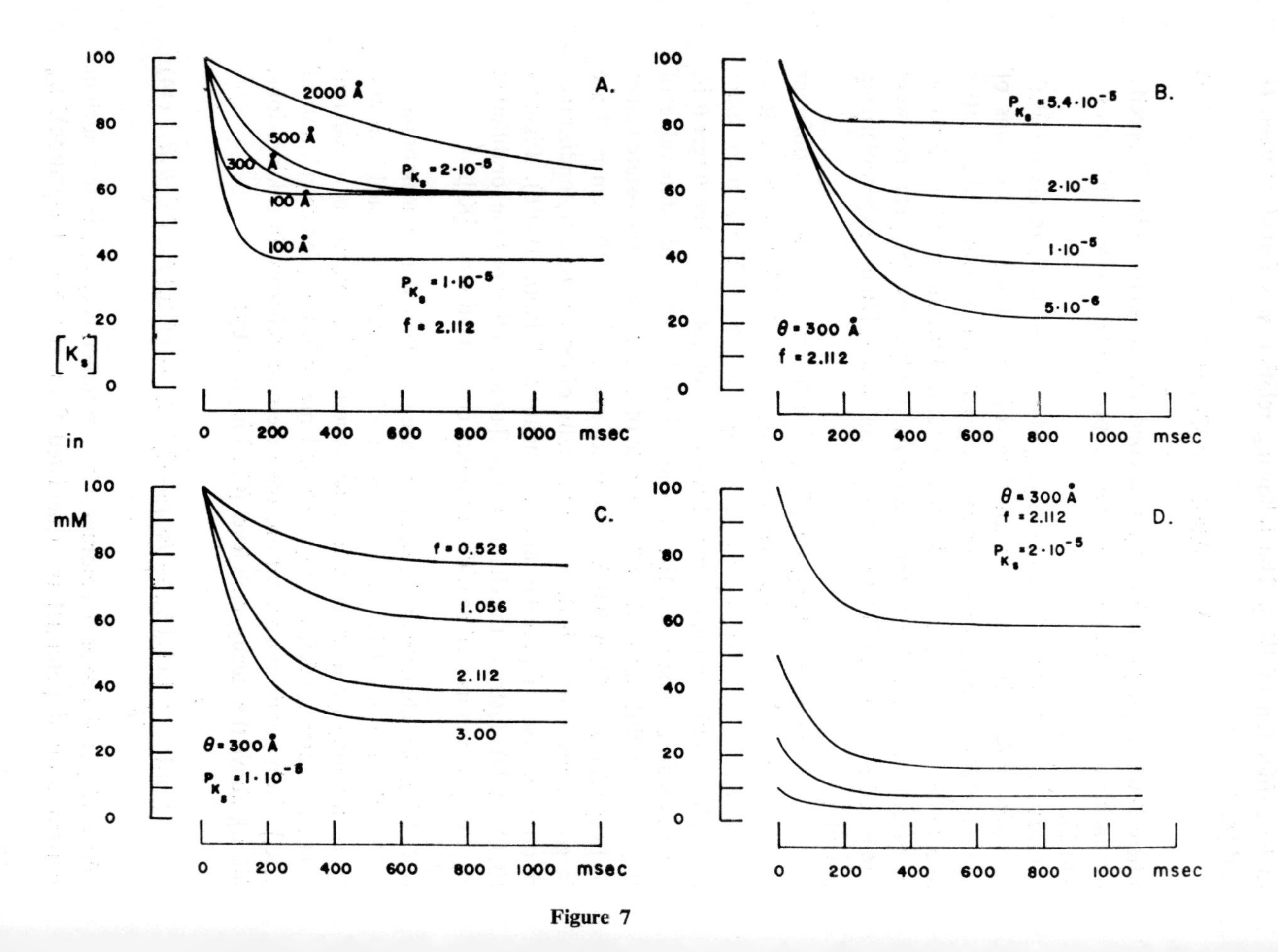
A.
B.
C.
D.
$[K_s]$ in mM
msec
2000 Å
500 Å
300 Å
100 Å
100 Å
$P_{K_s} = 2 \cdot 10^{-5}$
$P_{K_s} = 1 \cdot 10^{-5}$
f = 2.112
$P_{K_s} = 5.4 \cdot 10^{-5}$
$2 \cdot 10^{-5}$
$1 \cdot 10^{-5}$
$5 \cdot 10^{-6}$
θ = 300 Å
f = 2.112
f = 0.528
1.056
2.112
3.00
θ = 300 Å
$P_{K_s} = 1 \cdot 10^{-5}$
θ = 300 Å
f = 2.112
$P_{K_s} = 2 \cdot 10^{-5}$
0
20
40
60
80
100
200
400
600
800
1000

Figure 7

Figure 7 Potassium ion concentration, mM, in the space between the axolemma and Schwann cell layer, $[K_s]$, calculated by means of a digital computer from equation (12) as a function of time after onset of a —120 mV hyperpolarizing pulse. Note $[K_s]$ depletion which reaches a plateau level with time constants in the order or 50–250 msec. A, $[K_s]$ values plotted as function of space thickness, θ. The value of nerve conductance factor, f, used in the computation was 2.112. The outer layer K permeability, P_{K_s}, was 2×10^{-5} cm sec^{-1} for the upper plateau and 1×10^{-5} cm sec^{-1} for the lower plateau. The lower plateau level illustrates the steady-state $[K_s]$ obtained for any θ with a smaller P_{K_s} value. B, $[K_s]$ values as a function of outer layer K permeability, P_{K_s}. $\theta = 300$ Å and $f = 2.112$. C, $[K_s]$ values plotted as a function of nerve conductance factor f [see equation (13)]. $\theta = 300$ Å, $P_{K_s} = 1 \times 10^{-5}$ cm sec^{-1}. D, $[K_s]$ plotted as a function of time for the four different initial $[K_0]$ used experimentally: 100 mM, 50 mM, 25 mM, and 10 mM. (Reprinted by permission of The Rockefeller University Press from *The Journal of General Physiology*, 1969, **54**, 604)

is the transport number of Ca ions in the solutions used and P_{Ca_s} is the outer layer permeability to Ca^{++}. Following Hodgkin and Keynes (1957), we assumed, for a given fiber, that $G_K/G_{Ca} \approx 1000$. Assuming also that roughly the same conductivity ratio holds for the outer layer (no other information regarding this ratio is available), equation (14) was numerically solved for $\delta[Ca_s]$. In the solution, it was assumed that all Ca^{++} influx into the space remains in the free ionic form.

Figure 8 plots the calculated $[Ca_s]$ as a function of time after onset of a hyperpolarizing conditioning potential of -120 mV when the space thickness is 300 Å. $[Ca_s]$ is given for different values of G_{Ca} and P_{Ca_s}. It is seen that when a proper combination of values is taken, the $[Ca_s]$ values increase exponentially to about 3–5 times the initial concentrations with time constants in the order of 50 sec. This time course of the changes in $[Ca_s]$ is similar to that of sodium conductance restoration by long hyperpolarizing conditioning potentials.

Frankenhaeuser and Hodgkin (1957) have shown that by increasing $[Ca_0]$ the relation between sodium inactivation and membrane potential is changed such that at a constant membrane potential the proportion of the sodium-carrying system in the inactive state is decreased (Frankenhaeuser and Hodgkin, 1957, Figure 8). The slow K^+ independent restoration of sodium conductance, described above (Adelman and Palti, 1969b) may thus be related to sodium inactivation curve alterations caused by changes in $[Ca_s]$ brought about by long duration hyperpolarizations.

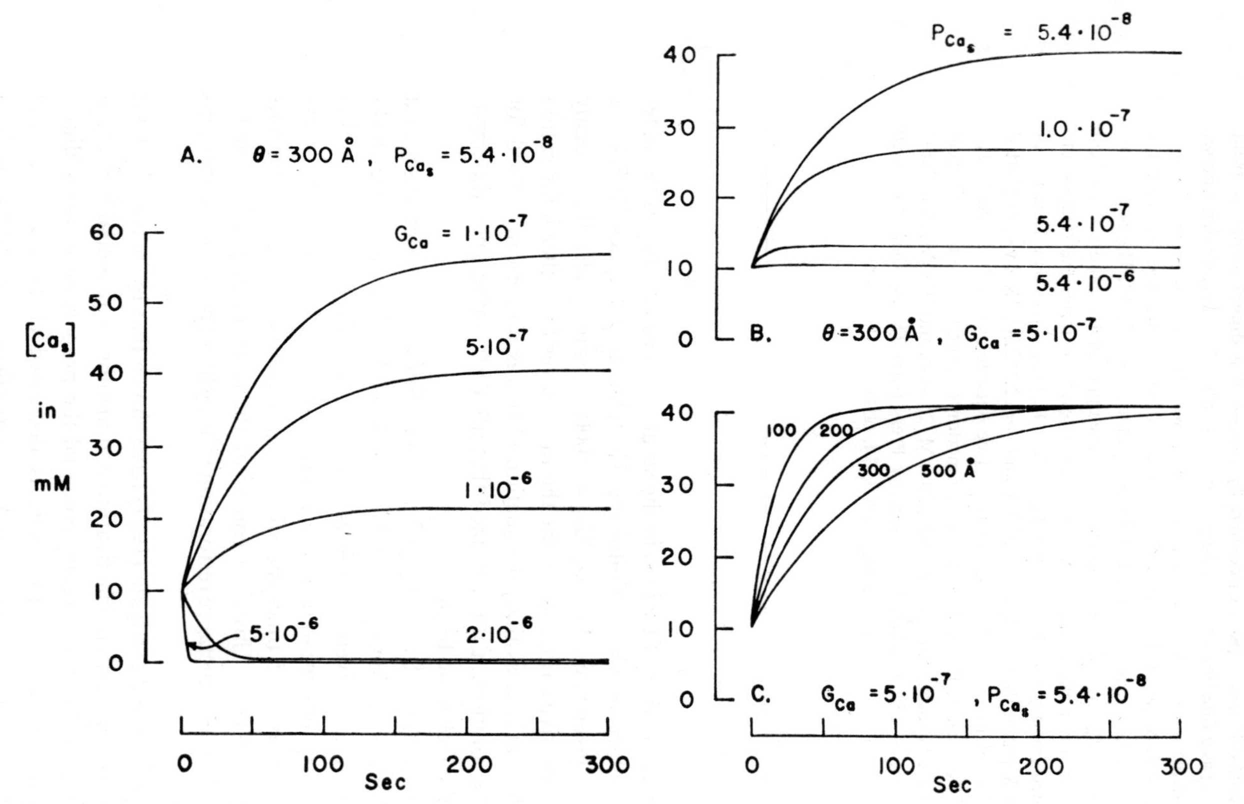

Figure 8 Calcium ion concentration, mM, in the space between axolemma and Schwann cell layer, [Ca_s], calculated by means of a digital computer from equation (14) as a function of time after the onset of a hyperpolarizing pulse to a membrane potential of —120 mv. Note [Ca_s] accumulation in space with time constants in the order of 30 to 100 sec. A, [Ca_s] values plotted as a function of axon membrane permeability to Ca, G_{Ca}. Outer layer Ca permeability used for the computation was 5.4×10^{-8} cm sec^{-1} and space thickness, $\theta = 300$ Å. B, [Ca_s] values plotted as a function of outer layer pemeability to Ca ions. $G_{Ca} = 5 \times 10^{-7}$, $\theta = 300$ Å. C, [Ca_s] values plotted as a function of space thickness, θ. $G_{Ca} = 5 \times 10^{-7}$, $P_{Ca_s} = 5.4 \times 10^{-8}$ cm sec^{-1}. Note that the rate of rise of [Ca_s] decreases with increasing θ values. However, the final [Ca_s]

EXPERIMENTAL TESTS OF THE MODEL SYSTEM FOR POTASSIUM ACCUMULATION EXTERNAL TO THE AXON MEMBRANE

One would expect to find that the reversal potential for the delayed current in response to voltage clamping the axon membrane reflects the activity gradient of potassium ions across the axon membrane (Hodgkin and Huxley, 1952a). The membrane current and voltage records shown in Figure 9 may be used as an example illustrating this point. The upper record in Figure 9 was recorded from a squid giant axon voltage clamped initially to the resting potential, $E_{RP} = -66$ mV. The clamped potential was then stepped to $E_p = +6$ mV and held constant for 44 msec. At the end of this period, the potential was stepped to $E_{\text{post}} = -84$ mV. E_{post} was slightly more positive than the estimated steady-state value of $E_K = -85$ mV [from $E_K = (RT/F) \ln (10/350)$. As $I_K = g_K(E_M - E_K)$], the "tail" current at the end of the 44 msec depolarizing pulse is expected to be transiently outward, and should decline with a characteristic time constant [Hodgkin and Huxley, (1952a) Figure 13] toward an inward small constant value.

Notice in the lower record in Figure 9 that the "tail" current at the end of the 44 msec depolarizing pulse is both large and inward. Also notice that the outward current flow during the depolarizing pulse has a maximum value occurring about 10 msec after the onset of the pulse. Following this maximum, the outward current declines or "droops". Frankenhaeuser and Hodgkin (1956, Figure 14b) analyzed similar records, and interpreted both the reversed "tail" and the outward current "droop" as resulting from an accumulation of K^+ in the unstirred space external to the excitable membrane. Such an accumulation would change the effective E_K as a function of the amplitude and time course of the outward potassium current. Thus, E_K would become less negative and the driving force $(E_M - E_K)$ for the outward current would decrease with time. Upon repolarization, the newly established E_K would be more positive than the initial E_K and the current tail would be inward if E_{post} was more negative than the new E_K.

Figures 10 and 11 illustrate the voltage and time dependency of potassium current "tails" obtained upon voltage clamping squid axons, as well as the method used to estimate the variation of E_K with pulse potential and duration. Values of membrane outward current determined 29 msec after the onset of depolarizing pulses to a variety of membrane potentials, E_p, are plotted in Figure 10 (above the zero current axis). Values of the inward tail current peaks obtained upon stepping the membrane potential from E_p to $E_{\text{post}} = -90$ mV are plotted below the zero current axis at $E_M = -90$ mV. These tail current maxima were obtained by extrapolating the

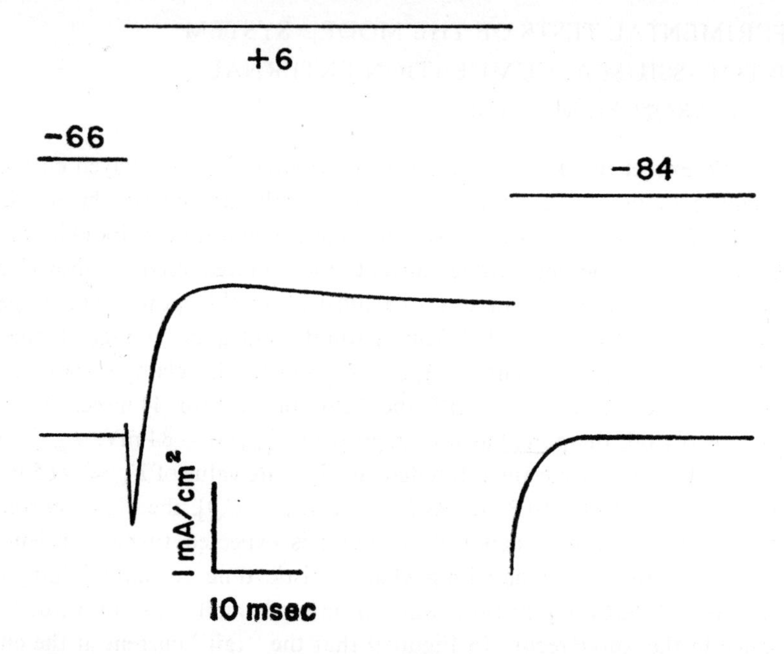

Figure 9 Records of membrane potential (upper trace) and current (lower trace) obtained from a voltage clamped squid giant axon exposed externally to artificial sea ater, ASW. Initial potential was the holding potential, E_H, which was set equal to the resting potential of -66 mV. Membrane was pulsed to $+6$ mV for 44 msec and then stepped to a post-pulse potential of -84 mV. Following the transient inward current, membrane current is outward through out the remainder of the depolarizing pulse. A transient inward tail current is seen upon repolarization. Inward currents are plotted downward, and outward currents are plotted upward. The initial horizontal segment of the current record at the left is the zero current baseline. Axon 69–5; temperature = 5°C

exponential portion of each tail current back to the time of the repolarizing jump. This was done so as to correct for the capacity current surge and to avoid overestimating the instantaneous value of the tail current at the jump in membrane potential. The dashed line in Figure 10 represents the best estimate of the leakage current for each membrane potential. The peak values of sodium current are shown by the continuous line. It is expected that the contribution of the initial transient current and the leakage current to the delayed outward current and the tail current after 29 msec of depolarization should be small. The straight lines connecting the outward and inward current pulses for each potential step should then cross the zero current axis at an E_M value very close to the reversal potential for the potassium current, E_K. Notice in Figure 10 that as E_p becomes more positive, these

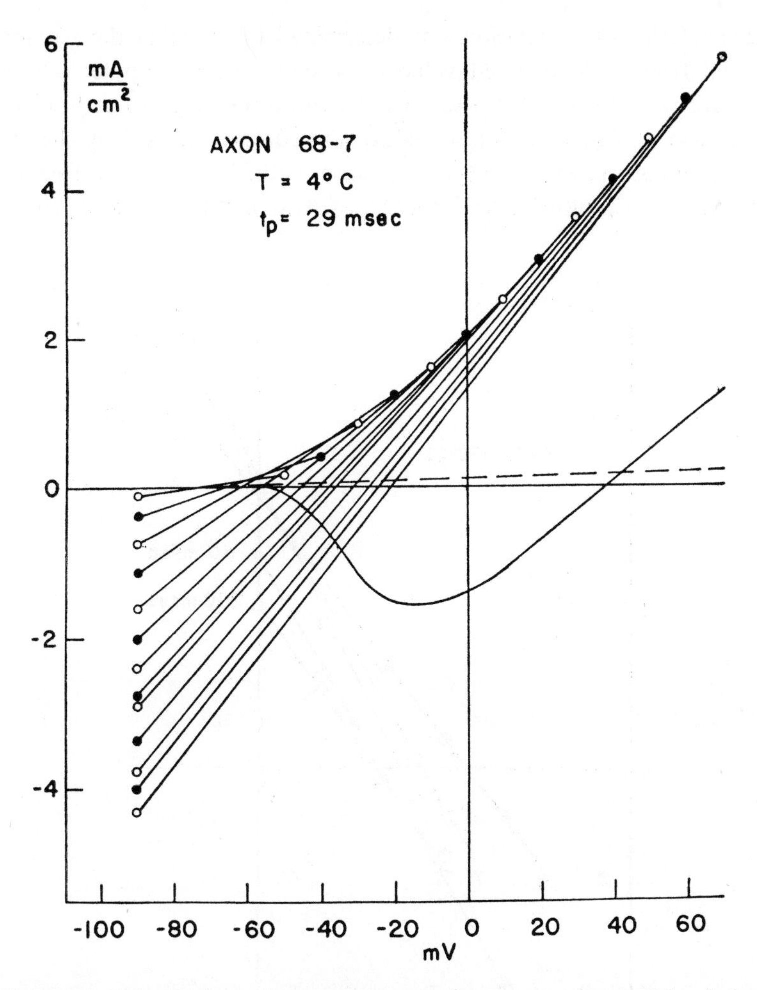

Figure 10 Voltage clamp instantaneous membrane current-potential relations (straight lines between points) recorded at the end of a 29 msec depeolarizing pulse upon stepping the membrane potential from various membrane potentials to a post-pulse potential of —90 mV. Dashed line: leakage current estimate. Continuous curvilinaer line: peak value of the initial transient current recorded during the depolarizing pulse. Outward currents plotted positively; inward currents negatively. Compare with figure 7 and see text

crossover points tend toward more positive values of E_M. It is concluded that the effective E_K is a function of the amplitude of outward potassium current.

Variations in E_K as a function of the duration of I_K were obtained by using the zero current cross-over points of the instantaneous current-voltage relation. This relation was determined by stepping the membrane potential from a depolarized value to several post-pulse potentials. This procedure was repeated for a set of pulse durations so as to determine the time course of E_K. In order to reduce the currents carried by the initial transient conductance, squid axons were bathed in sodium free ASW (430 mM Tris Cl substituted for 430 mM NaCl). While such a procedure

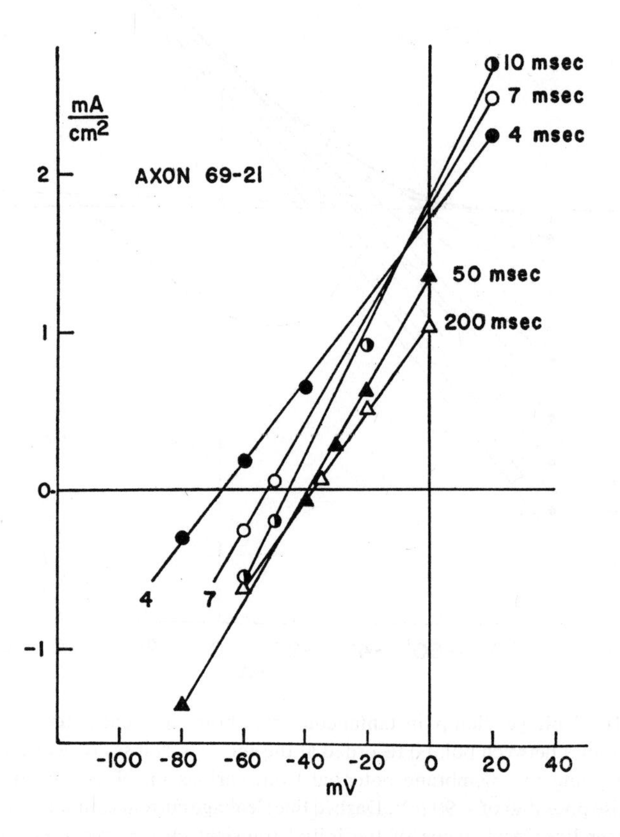

Figure 11 Voltage clamp instantaneous membrane current-potential relations recorded from a squid axon at the end of a depolarizing pulse upon stepping the membrane potential to a variety of post-pulse potentials. Effect of the duration of the depolarizing pulse on the reversal potential for the delayed membrane conductance. Points at the extreme right were recorded at the end of the depolarizing pulse; sets of points to the left were recorded at the beginning of the post-pulse. Filled circles: depolarizing pulse of 4 msec to $E_M = +20$ mV; open circles: depolarizing pulse of 7 msec to $E_M = +20$ mV; half-filled circles: depolarizing pulse of 50 msec to $E_M = 0$ mV, and open triangles: depolarizing pulse of 200 msec to $E_M = 0$ mV. Temperature 7°C. Compare with Figure 10 and see text.

eliminated the inward sodium currents, outward currents through the initial transient conductance were not reduced. While these outward initial transient currents were small, their influence could not be neglected.

Therefore, voltage clamp record of outward membrane currents obtained from axons bathed in Na-free ASW were compared with records obtained with the same voltage pulses upon bathing axons in ASW containing 25 nM TTX (cf. Cuervo and Adelman, 1970). It was found that for depolarizing the membrane from the holding potential, $E_H = E_{RP} = -64$ mV to $E_p = 0$ mV, the fraction (I_K/I_T) of total outward membrane current, I_T, identified as I_K approached 0.95 about 2.5 msec after the onset of the depolarizing pulse. I_K/I_T was 0.60 at 1 msec after the depolarizing step.

Therefore, the shortest pulse duration which was used to obtain an accurate estimate of the effective E_K was 4 msec. In order to increase the accuracy of the estimate and to demonstrate that the current jumps represented an instantaneous conductance, several post-pulse potentials were used and the values of the maximum tail current corresponding to each post-pulse were plotted on a current-voltage diagram. These relations were linear and passed through the current point obtained just before the repolarizing potential step.

Figure 11 is typical of data obtained from 10 squid axons under these conditions. The linearity of the current-voltage relations is apparent and it can be seen that as the depolarizing pulse duration (and I_K) is increased, the zero current crossover points move to the right. Thus, the effective E_K values become more positive as the duration of I_K increases. Notice that the effective E_K values reach a steady-state for I_K durations greater than 50 msec.

Figure 12A plots values of E_K (obtained from I/V plots such as those shown in Figure 11) as a function of the duration of the delayed outward current. The points were obtained from two voltage clamped axons upon pulsing the membranes from the resting potential to zero and then stepping the membrane potential to a variety of post-pulse potentials. The lines are computer read-outs for $[K_s]$ and E_K, as based on solution of equation (12) for $\delta[K_s]$; $[K_s] = [K_{s_0}] + \delta[K_s]$, and $E_K = (RT/F) \ln([K_s]/[K_i])$. The value of P_{K_s}, the outer layer permeability to potassium ion, was taken as 1×10^{-4} cm/sec, in order to obtain a good fit to the experimental data.

From the results shown in Figure 11 and 12, we conclude that the model for K^+ accumulation in the unstirred space external to the axon membrane is a good description of the changes in effective E_K as a function of amplitude and duration of I_K seen under voltage clamp conditions.

Figure 12B shows the time course of parameter h calculated for an initial value, h_0, of 0.600 and for a steady-state value of $h_\infty = 0.0068$. τ_h was

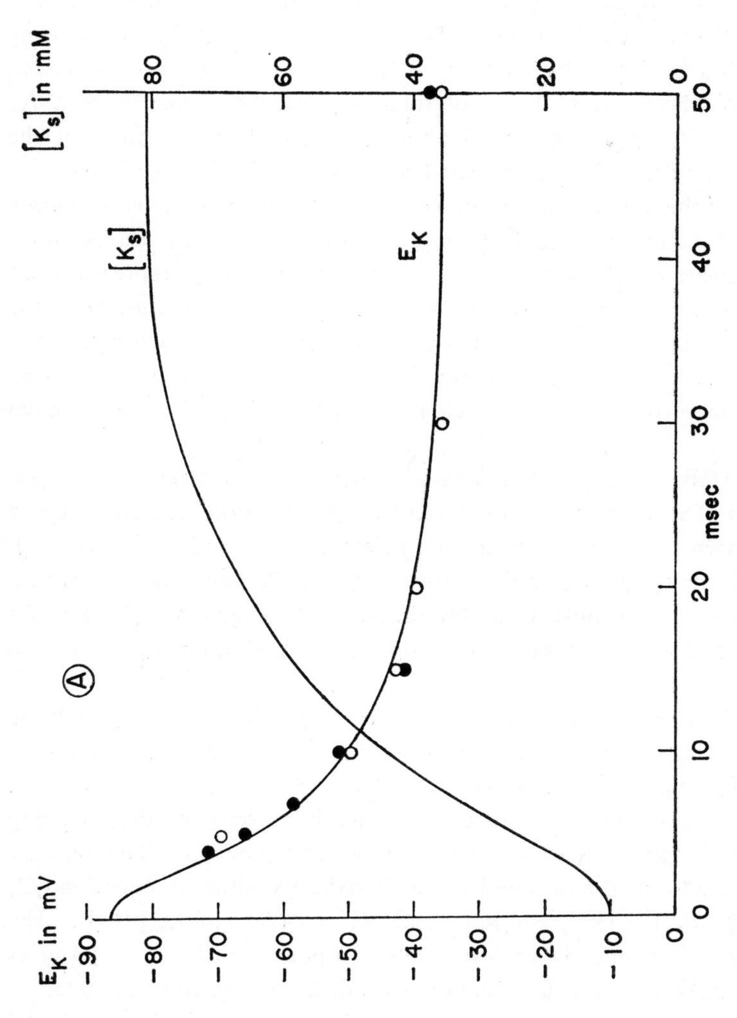

Figure 12 (A) Comparison between recorded values of the reversal potential, E_K, of the delayed membrane conductance, g_K, as a function of depolarizing pulse duration with computed values of $[K_s]$ and E_K obtained by means of digital computer solutions of equation (12). Open circles: values of E_K obtained from axon 69–5, filled circles: from axon 69–21.

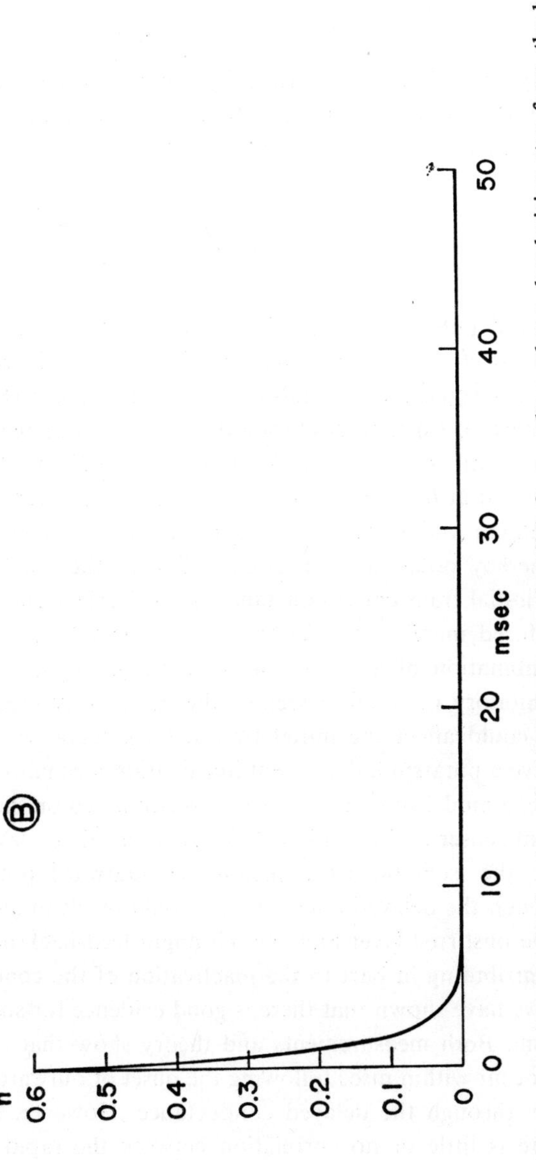

Figure 12 (B) Time course of parameter ***h*** values calculated from equation (15) for a voltage clamp depolarizing step from the holding potential ($E_H = E_{RP}$) to a pulse potential of zero mv. See text.

taken at $E_M = 0$ mV as 1.00 msec. Using the following equation (Hodgkin and Huxley, 1952c):

$$h = h_\infty - (h_\infty - h_0) \exp(-t/\tau_h) \tag{15}$$

the curve in Figure 12B was calculated. By comparing the [K_s] curve in Figure 12A with the h curve in Figure 12B, it can be seen that the decrease in h can only be partially to the locus of [K_s] in time.

DISCUSSION

Adelman and Palti (1969a) concluded that there was ample evidence to demonstrate an effect of external potassium ions on the initial transient conductance in the squid axon membrane. In terms of the Hodgkin and Huxley hypothesis (1952c), this effect took the form of an increase in τ_h and a decrease in h_∞ with increasing [K_0]. However, the K effect on τ_h was primarily manifested at membrane potentials more negative than -40 mV.

Whereas Adelman and Palti (1969a) considered a number of possible mechanisms whereby potassium ion could influence the inactivation mechanism of the initial transient conductance, a mechanism operating solely through the reduced membrane potential brought about by the [K_0] was rejected. A combination of a membrane potential dependent mechanism and a direct inhibitory mechanism seemed the most likely means whereby potassium ions could affect the initial transient conductance.

In order to give a potassium dependent inactivation mechanism physiological significance, a model system has been proposed based on the considerations of Frankenhaeuser and Hodgkin (1956) re an unstirred layer or space just external to the excitable axon membrane. Outward potassium ion movements through the delayed conductance would result in an accumulation of K^+ in the unstirred layer and as such might feed-back on the initial conductance contributing in part to the inactivation of the conductance.

In this work we have shown that there is good evidence for such accumulation mechanism. Both measurements and theory show that a significant rise in [K_s] can occur within msec following the onset of outward potassium ion current flow through the delayed conductance. However, it has been shown that there is little or no correlation between the rapid kinetics of inactivation development with depolarization and these changes in [K_s]. Recovery from inactivation upon repolarizing the membrane following a depolarization of 5 or more msec should depend on the rate of wash-out of K_s as well as on the value of τ_h at the new membrane potential. In this case, the values of α_h and β_h would not be set once repolarization occurs, but would vary somewhat as $[K_s]$ declines toward the value of ambient $[K_0]$.

We have also shown that the inactivating effects of high $[K_0]$ can be overcome by long duration hyperpolarizations (hundreds of msec to secs). A model system for desaturating an external unstirred layer or space has been proposed to account for this phenomenon. This model is similar in kind to that proposed by Frankenhaeuser and Hodgkin (1956) for the accumulation of potassium ions external to the membrane with membrane depolarization. In this respect, the time constant for the removal of inactivation in the presence of high $[K_0]$ is similar to the time constant for removal of K_s by inward current flow during membrane hyperpolarization.

We have presented evidence (Adelman and Palti, 1969 b) that there is an additional restoration of the initial transient conductance which can be brought about by long duration (secs to mins) hyperpolarizations. We have attempted to show that a model system for redistributing Ca^{++} in the space just external to the axolemma can lead to changes in the values of $[Ca_s]$ with time constants in the same order of magnitude of the time constant for the very slow removal of inactivation. These changes in $[Ca_s]$ are sufficient to induce shifts in both the initial transient current—membrane potential and the h_∞ vs membrane potential relations (Frankenhaeuser and Hodgkin, 1957). This model system has not been tested thoroughly and therefore should be considered as only a tentative suggestion.

REFERENCES

Adelman, W. J., Jr. (1971). Electrical studies of internally perfused squid axons. *in* Biophysics and Physiology of Excitable Membranes, W. J. Adelman, Jr., editor. Van Nostrand Reinhold Co. New York.

Adelman, W. J., Jr., F. Dyro, and J. P. Senft (1965 a). Long duration responses obtained from internally perfused axons. *J. Gen. Physiol.* **48,** No. 5, part 2 1.

Adelman, W. J., Jr., F. Dyro, and J. P. Senft (1965 b). Polonged sodium currents from voltage clamped internally perfused squid axons. *J. Cell.* and *Comp. Physiol., supp. 2* **66,** 55.

Adelman, W. J., Jr., and Y. Palti (1969 a). The influence of external potassium on the inactivation of sodium currents in the giant axon of the squid, *Loligo pealei. J. Gen. Physiol.* **53,** 685.

Adelman, W. J., Jr., and Y. Palti (1969 b). The effects of external potassium and long duration voltage conditioning on the amplitude of sodium currents in the giant axon of the squid, *Loligo pealei. J. Gen. Physiol.* **54,** 589.

Adelman, W. J., Jr., and J. P. Senft (1968). Dynamic asymmetries in the squid axon membrane. *J. Gen. Physiol.* **51,** 102.

Cole, K. S. (1949). Dynamic electrical characteristics of the squid axon membrane. *Arch. Sci. physiol.* **3,** 253.

Cole, K. S. (1968) Membranes, Ions and Impulses. Univ. of Calif. Press. Berkeley and Los Angeles, California.

Cuervo, L., and W. J. Adelman, Jr. (1970). Equilibrium and kinetic properties of the interaction between tetrodotoxin and the excitable membrane of the squid giant axon. *J. Gen. Physiol.* **55**, 309.

Frankenhaeuser, B., and A. L. Hodgkin (1956). The after-effects of impulses in the giant nerve fibres of *Loligo*. *J. Physiol.* (London). **131**, 341.

Frankenhaeuser, B., and A. L. Hodgkin (1957). The action of calcium on the electrical properties of squid axons. *J. Physiol.* (London). **137**, 218.

Hodgkin, A. L., and A. F. Huxley (1952a). The components of membrane conductance in the giant axon of *Loligo*. *J. Physiol.* (London). **116**, 473.

Hodgkin, A. L., and A. F. Huxley (1952b). The dual effect of membrane potential on sodium conductance in the giant axon of *Loligo*. *J. Physiol.* (London). **116**, 497.

Hodgkin, A. L., and A. F. Huxley (1952c). A quantitative description of membrane current and its application to conduction and excitation in nerve. *J. Physiol.* (London). **117**, 500.

Hodgkin, A. L., and R. D. Keynes (1957). Movements of labelled calcium in squid giant axons. *J. Physiol.* (London). **138**, 253.

Narahashi, T., (1963). Dependence of resting and action potentials on internal potassium in perfused squid giant axons. *J. Physiol.* (London). **169**, 91.

Narahashi, T. (1964). Restoration of action potential by anodal polarization in lobster giant axons. *J. Cell. Comp. Physiol.* **64**, 73.

Palti, Y. (1971). Digital computer solutions of membrane currents in the voltage clamped giant axon. *in* Biophysics and Physiology of Excitable Membranes, W. J. Adelman, Jr., editor. Van Nostrand Reinhold Co., New York.

PAPER 6

K^+ and Na^+ Transport and Macrocyclic Compounds

D. C. TOSTESON

Department of Physiology and Pharmacology
Duke University Medical Center Durham, North Carolina

INTRODUCTION

Kenneth S. Cole is the father of membrane biophysics in the United States. He promoted the growth of this important twig of science not only through his own incisive investigations, but also through his contributions to the education of his many students. It is a pleasure and an honor for me to join with them in offering this volume to celebrate his 70th birthday.

This paper takes its text from Kenneth Cole's elegant monograph *Membranes, Ions and Impulses* (1). His book records clearly and sequentially the properties of excitable biological membranes that have been revealed by the application of quantitative electrical techniques. It is the story of a triumph built, in large part, on his own pioneer work. He describes the evidence supporting the idea that Na^+ and K^+ are the ions that carry current across most excitable biological membranes and the precise description of the voltage and time dependence of the conductances for these ions. Yet, at the end, he has the courage and candor to conclude: "Even if the structure of a single membrane were well defined, from one aqueous solution to the other, there might still not be an adequate basis to decide where and why it is so difficult for any ion to cross in either direction. Thus any theory of ion permeability must assume and support both a structure and a mechanism. It is most unfortunate, yet most challenging, that a possible and

* This work was supported by grant # 5 P 01-HE 12157 from the National Heart Institute, National Institutes of Health, USPHS.

reasonable explanation of the most powerful facts of ion permeabilities may be completely wrong. And the concepts gained from artificial models may be entirely misleading.

So it is a struggle in the dark except for a few occasional glimmers of light until the approach of one or more attempts is so close that it cannot be denied." This paper seeks to provide another "glimmer of light". I wonder if K. C. was thinking of the sea of Vineyard, agitated by a swimmer or a boat, on a dark night in July.

This paper describes the results of part of a program of research in our laboratories directed toward understanding better the mechanism of sodium and potassium transport across biological membranes. Specifically, this-work deals with the interactions between certain macrocyclic polyesters (e.g. the nonactin series) and depsipeptides (e.g. valinomycin and its analogs) with alkali metal cation in hydrocarbon solvents, thin lipid bilayer membranes, and in red cell membranes.

Since the discovery in 1965 by Pressman and his colleagues[1] that valinomycin produces a potassium-dependent stimulation of respiration in mitochondria, work in many laboratories[2–11] has developed a reasonably coherent but still rather sketchy picture of how this compound and related agents produce a selective increase in the potassium permeability of thin lipid bilayer and biological membranes. One rendition of this picture is shown in Figure 1. The main features of this scheme are well documented and accepted by most workers in the field. A potassium ion from the aqueous solution bathing one side of the membrane interacts with a valinomycin molecule at the interface between membrane and aqueous solution. During this interaction, valinomycin undergoes a conformational change to form a complex with the ion. Since valinomycin itself is uncharged, the complex has a charge of +1. In the process of formation of the complex, potassium is apparently completely dehydrated, the water oxygens in its hydration shell being replaced by carbonyl oxygens from the ester bonds in the macrocyclic compound. The details of the conformation of the complex are different for different macrocyclic compounds[8,10,15] but all possess the property that the exterior surface is composed almost entirely of hydrophobic side chains, for example methyl and isopropyl groups in valinomycin. According to this scheme the potassium complex once formed can diffuse through the hydrocarbon core of the membrane to the opposite surface where it undergoes a reverse series of transformations in order to release the ion.

Despite the progress which has made it possible to form this picture, a number of important questions about the mechanism of selective cation transport in the presence of macrocyclic polyesters and depsipeptides remain unanswered. The work described in this paper attempts to answer

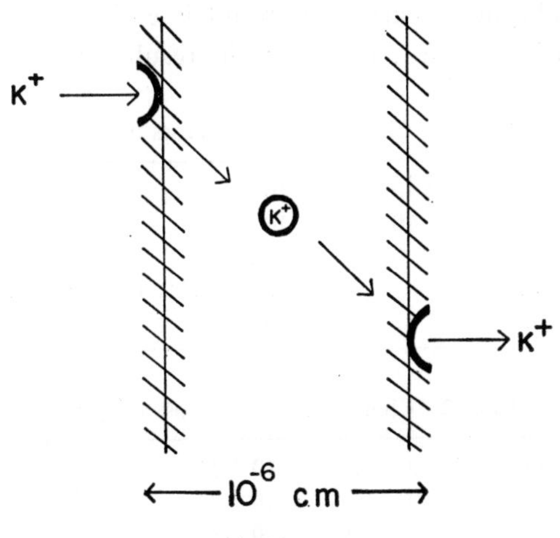

Figure 1

four of these questions. First, is it genrally true, as suggested by this diagram, that complexes between cation and macrocyclic compound are the only charge carriers in these systems? Second, what is the relation between the primary, secondary and tertiary structure of the macrocyclic compounds and their capacity to act as selective cation carriers across bilayer and biological membranes? Third, what kind of interactions occur between positively charged complexes and various anions which might be present in the system? And fourth, what are the similarities between these interactions and the effects of anions on red cell membranes?

CHARGE CARRIERS

We have attempted to identify directly the charge carriers across lipid bilayer membranes containing monactin, dinactin or valinomycin[12]. Our approach has been to measure simulataneously the total electrical current and the net flux of different ions. These and all other bilayer experiments described in this paper were performed on membranes prepared from lipids extracted from sheep red cells. Part of these results are summarized in Table 1. In these experiments, the potassium current was estimated from the difference between the two unindirectional fluxes of potassium each measured separately with tracer ^{42}K. The potassium chloride concentration was 0.1 M on the left and 0.01 M on the right. The electrical potential difference across the membrane was maintained zero by an appropriate external circuit. Note that under these conditions the current carried by potassium ions was not significantly different from the total current passing through the membrane. Clearly, this result is entirely consistent with the picture presented in Figure 1. However, the results shown in Table 2 show

Table 1 Charge transport across bilayers

Monactin-dinaction	10^{-6} M		10^{-6} M
KCl	10^{-1} M		10^{-2} M
Voltage	0 mV		0 mV
Total current		→	$1.3 \pm 0.5 \times 10^{-5}$ amps/cm^2
K^+ current		→	$1.6 \pm 0.3 \times 10^{-5}$ amps/cm^2

Table 2 Charge transport across bilayers

Monactin-dinactin	10^{-6} M		10^{-6} M
KCl	10^{-1} M		10^{-1} M
Voltage	+60 mV		0 mV
Total current		→	$6.1 \pm 0.2 \times 10^{-5}$ amps/cm^2
K^+ current		→	$1.8 \pm 0.03 \times 10^{-5}$ amps/cm^2

that this is not always the case. In this experiment, the concentration of KCl was 0.1 M on both sides and current was driven by an external battery which produced a voltage difference of 60 mV across the membrane. Under these conditions, less than one-half of the total current was carried by potassium ions. In general, whenever current is passed from one phase across the membrane into another phase with an equal or higher potassium concentration, the potassium current is less than the total current. On the other hand, when current is passed from one phase into another phase with a lower potassium concentration, all of the current is carried by potassium ions. Similar results were obtained when valinomycin instead of monactin-dinactin was used as the macrocyclic compound. What is the missing charge carrier? The most obvious possibility is the anion, in this instance chloride. However, direct measurements of anion fluxes in systems which showed the discrepancy between total and potassium current revealed no net current carried by anions. Next, we considered the possibility that hydrogen or hydroxyl ions are charge carriers. Many of the effects of varying the pH of the bathing solutions were consistent with this interpretation. However, we could not be satisfied with this conclusion until we had measured directly hydrogen or hydroxyl ion transport in the system. The result of an experiment designed to make such a measurement is shown in

Figure 2. A micro, pH sensitive glass, electrode was inserted into a closed chamber containing the phase bathing one side of a bilayer membrane containing monactin dinactin. The bathing solutions on both sides of the membrane contained 0.1 M KCl and were free of dissolved CO_2. The application of a 60 mV pulse across the membrane produced no appreciable change in the hydrogen ion content of the chlosed phase. Addition of sufficient HCl to bring the pH of the opposite open phase to 3, also failed to produce significant net hydrogen ion transport. The possibility that the system was insufficiently sensitive to detect a change in hydrogen ion concentration of the magnitude expected was ruled out by adding picric acid to the opposite phase. Under these circumstances, prompt accumulation of hydrogen ions in the closed phase was observed. Parenthetically, these results confirm and extend the observations of Dr. Thompson *et al*[13] that the proton permeability of bilayers is increased by nitrophenols like picric acid. However, the experiment leaves an important problem unsolved.

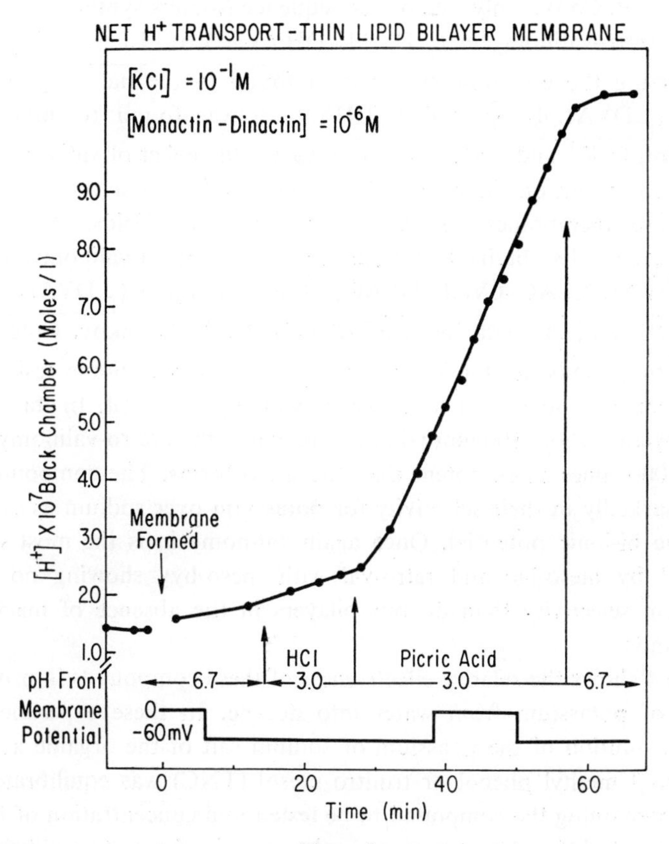

Figure 2

Apparently, under certain conditions, more than half of the current across a membrane containing monactin or valinomycin is carried by some as yet unspecified component. By contrast, measurements of the zero-current potential difference across a valinomycin containing lipid bilayer in the presence of different concentrations of potassium and other ions in the bathing solutions suggest that the transference number for potassium ions is unity.

STRUCTURE-FUNCTION RELATIONS IN VALINOMYCIN ANALOGUES

Shemyakin *et al*[14] have synthesized and characterized many analogues of the naturally occurring depsipeptide. This paper reports the results of synthesis of sequence isomers of valinomycin by the solid method by B. Gisin[15] and the characterization of their interactions with K^+ and Na^+ by M. T. Tosteson and P. Cook. Only one of six sequence isomers synthesized thus far, namely retro-valinomycin ($\underline{\text{DVAL–DHYV–LVAL–LLAC}_3}$) prepared by reversing the direction of acylation found in the naturally occurring isomer ($\underline{\text{DVAL–LLAC–LVAL–DHYV}_3}$) was found to show activity in relation to K^+ and Na^+. Table 3 compares the effect of valinomycin and retro-valinomycin on the electrical resistance and bi-ionic potential of thin lipid bilayer membranes prepared from sheep red cell lipids. Also shown in Table 3 are results obtained with the two meso forms of valinomycin, meso-lac ($\underline{\text{DVAL–LLAC–LVAL–DLAC}_3}$) and meso-hyv ($\underline{\text{DVAL–LHYV–LVAL–DHYV}_3}$) kindly supplied to us by Y. Ovchinnikov. Note that all of the compounds are far less potent at reducing membrane resistance in the presence of potassium ions than is valinomycin itself. In this respect, valinomycin is about 100 more times more potent than retro-valinomycin and about 1000 times more potent than the meso forms. The compounds also differ markedly in their selectivity for potassium over sodium as estimated from the bi-ionic potential. Once again valinomycin is the most selective followed by meso-lac and retro-val with meso-hyv showing no greater potassium selectivity than do our bilayers in the absence of macrocyclic compounds.

Table 4 shows the relative effectiveness of these compounds in promoting uptake of potassium from water into decane. In these experiments, an aqueous solution of the poassium or sodium salt of the organic acid 2, 4, 6 trinitro-3 methyl phenol or trinitro cresol (TNC) was equilibrated with decane containing the compound to be tested in a concentration of 10^{-5} M. The amount of K^+ or Na^+ present in the organic phase at equilibrium was

Table 3 Effects of valinomycin analogs: transport thin lipid bilayer membranes

Compound 10^{-7} M	Resistance KCl (10^{-1} M) ohm cm^2	Bi-ionic potential mV
None	2×10^8	50–80
Valinomycin	3×10^3	180
Meso-LAC	1×10^6	175
Meso-HYV	5×10^6	75
Retro-VAL	3×10^5	120

Table 4 Effects of valinomycin analogs: equilibrium

Water (pH 7.0) Na or KTNC (10^{-2} M) Compound (10^{-5} M)	Decane Mac. Compound (10^{-5} M)		
	$(K)_D$ 10^{-7} M	$(Na)_D$ 10^{-7} M	A'_K/A'_{Na}
None	<0.01	<0.01	–
Valinomycin	56	24	4.1
Meso-LAC	47	19	3.8
Meso-HYV	0.7	45	0.01
Retro-VAL	10	4.3	2.4

estimated with ^{42}K or ^{22}Na. Once again, marked differences between the properties of the several compounds are evident. For example, while approximately one half of the valinomycin or meso-lac present in the decane phase forms a complex with potassium ions under these conditions, only about ten percent of the retro-val and less the 1% of the meso-hyv become associated with potassium ions. On the other hand, meso-hyv is the most potent compound in producing sodium uptake into decane in this system. It should be noted that no uptake of ^{42}K or ^{22}Na was observed in the absence of macrocyclic compounds in this system.

Clearly, there are marked differences between these compounds in their interactions with sodium and potassium ions both under equilibrium conditions as in the phase distribution studies and in transport across bilayers. Apparently, these functions are exquisitely sensitive to the primary structure of the macrocyclic compound. An understanding of this sensitivity depends on greater insight into the relationship between the primary structure of the macrocyclic compounds and their capacity to assume conformations suitable for complexation with ions. For example, the related ineffectiveness of the meso-compounds as ion carriers may be related to the absence of assymmetry in the probable bracelet-like form of the cation-complexes of these substances[16]. Retro-valinomycin may be less active than valinomycin itself because steric interactions between the side chains of hydroxy and amino acids hinder the formation of the intra-molecular hydrogen bonds necessary for stabilization of the bracelet conformation of the K^+ complex.

INTERACTIONS WITH ANIONS

In evaluating the effect of valinomycin and its isomers and analogs, we have studied their relative potency to produce uptake of sodium or potassium ions from aqueous solutions into decane. In order to carry out such extraction experiments, it is obviously essential to use an anion which can enter the organic phase to neutralize the positively charged complex of cation with macrocyclic compound. The experiments reported thus far were carried out with trinitrocresolate. I will now describe the series of investigations which led to the choice of that particular anion for such experiments. In all cases, uptake of cation into the organic phase was measured with an appropriate radio-active tracer.

We began with the hypothesis that lipid solubility would be a major determinant of the effectiveness of the anion in promoting alkali metal cation uptake from water into decane. Table 5 shows the results of experiments with anions of acids with different solubilities in ether. The ether-water partition coefficients are taken from the work of Collander[17] and refer to the undissociated acid form of the organic anion. Note that there is little correlation between relative lipid solubility and the effectiveness of the anion to promote K^+ uptake into decane in the presence of valinomycin. In fact, in this series the most effective anion was picrate which is relatively insoluble in ether. A systematic study of analogues of picrate revealed that potency as a counter ion to K^+ or Na^+ complexes of valinomycin in decane dependend on the nature of the acidic function ($OH > —SO_3H > —COOH$), on the nature of the 2, 4, 6 substituents ($—NO_2 > —CF_3 > —I > —Cl > —C_3$) on the number of $—NO_2$ groups ($3 > 2 < 1$), and on the substituent at the 3 position ($—CH_3 > —H > —OH$) (Table 6).
The most potent anion which we have found to date is TNC^-.

In the course of these experiments, we noted that trinitrocresol appears to have a dramatic effect on the selectivity of valinomycin for potassium over sodium. This point is illustrated by the results shown in Table 4. Note that approximately 56% of the valinomycin became associated with potassium ions and about 24% with sodium ions. The ratio of the apparent association constant for potassium to that for sodium is about 4. This result is in marked contrast to the high degree of selectivity which valinomycin

Table 5 Equilibrium distribution of K^+

Water, pH 7.0 KCl 10^{-2} M KA 10^{-3} M	Decane Valinomycin 10^{-4} M	
Effect of lipid solubility of anion		
Anion (A)	Partition coefficient Ether/Water	$\frac{(\text{K Val}^+)_D}{(\text{Val})_D}$
Pentanoate	23	1.4×10^{-4}
Hexanoate	85	2.6×10^{-4}
Benzoate	74	3.0×10^{-4}
Picrate	3.7	1.5×10^{-1}

Table 6 Equilibrium distribution og K^+

Water, pH 7.0 KCl 10^{-2} M KA 10^{-3}	Decane Valinomycin 10^{-4} M	
Effect of anion: substituted trinitro phenols		
	X	$\frac{(\text{K Val}^+)_D}{(\text{Val})_D}$
O— ; O_2N, NO_3, H, X, NO_2	—OH	4.5×10^{-4}
	—H	1.5×10^{-1}
	$-CH_3$	4.8×10^{-1}

displays in modifying the ionic permeability of bilayers (Table 7). In the presence of 10^{-7} M valinomycin with chloride as the only anion, the ratio of the conductance of a lipid bilayer in potassium to that in sodium is greater than 10^4. Addition of 10^{-3} M trinitrocresol at pH 7 has relatively little effect on membrane resistance in the presence of potassium but dramatically reduces membrane resistance in the presence of sodium. Thus in the presence of 10^{-3} M TNC the ratio of conductance in potassium to that in sodium is less than 10. Addition of TNC in the absence of valinomycin reduces membrane resistance from its control value of 2×10^8 ohm cm^2 to a value of about 10^6 in the presence of either K^+ or Na^+. It is important to note that valinomycin increases markedly the conductance of TNC^+ in bilayers. Table 8 shows the results of experiments in which conductance for Na^+ and TNC^- in bilayers were estimated from measurements of membrane resistance and zero current potential in the presence of concentration differences of each ion. Note that addition of valinomycin systems containing

NaTNC increases both TNC^- and Na^+ conductance. In the absence of TNC^-, valinomycin has no effect on Na^+ conductance.

We have been concerned to try and find an explanation for the remarkable effect of trinitrocresol anions on the selectivity of valinomycin for potassium as compared with sodium ions. One possibility is that the anion forms a complex with valinomycin which alters the distribution of conformational states of the cyclic depsipeptide. Direct evidence in favor of the formation of such a complex is shown in Figure 3 which shows the ultra-violet spectrum

Table 7 K^+—Na^+ selectivity of valinomycin thin lipid bilayer membranes

Bathing solutions			Membrane	G_m^K / G_m^{Na}
10^{-1} M Salt	(Val) M	($TNC^=$) M	Resistance ohm-cm^2	
NaCl	10^{-7}	0	2×10^8	4×10^4
KCl	10^{-7}	0	5×10^3	
NaCl	10^{-7}	10^{-3}	2×10^4	4
KCl	10^{-7}	10^{-3}	5×10^3	
NaCl	0	10^{-3}	1×10^6	1
KCl	0	10^{-3}	1×10^6	

Table 8 Ionic selectivity: Valtinomycin-NaTNC thin lipid bilayer membranes

NaCl (0.1 M)		Conductances	
TNC	VAL M	TNC^- ohm^{-1} cm^{-2}	Na^+
0	0	0	5×10^{-9}
0	10^{-7}	0	5×10^{-9}
10^{-3}	0	5×10^{-7}	2×10^{-7}
10^{-3}	10^{-7}	2×10^{-5}	1×10^{-5}
10^{-2}	0	5×10^{-6}	14×10^{-6}
10^{-2}	10^{-7}	8×10^{-4}	4×10^{-4}

of hexane containing 10^{-4} M valinomycin and either 10^{-4} M KTNC or NaTNC. The trinitrocresol salts were added dry to the hexane-valinomycin. Essentially similar spectra were obtained if hexane containing valinomycin was equilibrated with aqueous solutions of KTNC or NaTNC. Note that there is a significant difference in the spectrum of the two systems. A peak at 260 nm is evident with Na but not with KTNC. The peak at 340 nm observed when the cation is Na^+ moves to 355 nm when the cation is K^+. The characteristics of the shoulder in the region of 400 nm are also altered. In the absence of valinomycin, no TNC^- could be detected in hecane exposed to either KTNC or NaTNC either dry or in aqueous solution. Furthermore, valinomycin alone does not absorb in this region of the spectrum. These strongly suggest the formation of a complex between TNC^- and the cation complexes of valinomycin. The complexes are clearly different if the cation is K^+ or Na^+. Entirely similar results were obtained when decane rather than hexane was used as the solvent. Hexane was chosen for the for the spectroscopic experiments because if the availability of this solvent in spectroscopically pure form. However, the complexes are very sensitive to the solvent. The spectra are different in $CHCl_3$ or CH_3OH as compared with hexane. In CH_3OH there was no difference in the spectra with K^+ as compared with Na^+ (Figure 4). Not only are the complexes different depending upon the cation and the solvent with which valinomycin is combined, they are also different for different macrocyclic compounds. The ultra violet absorption spectrum of hexane solutions of KTNC dissolved in hexane containing meso-lac is distinctly different from that observed in the presence of valinomycin itself. Interactions also occur between TNC^- and its complexes with macrocyclic compounds and lipids. The ultra-violet absorption spectrum of HTNC dissolved in hexane containing valinomycin is different in the presence or in the absence of sheep red cell lipids in a

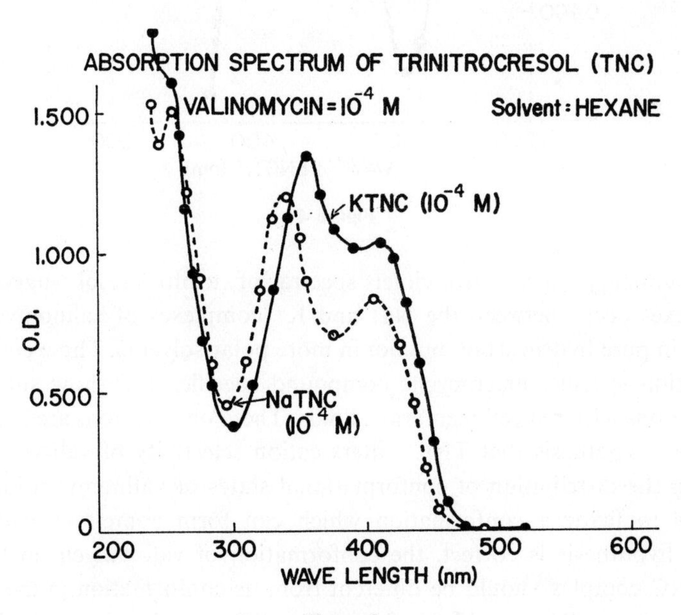

Figure 3

concentration of 1 mgm/ml. Such interactions can also be seen between the K^+-Val and Na^+-Val forms of trinitrocresol and sheep red cell lipids. However, the lipid interactions do not eliminate the spectral difference between the sodium-valinomycin-TNC as compared with the potassium-valinomycin-TNC complexes.

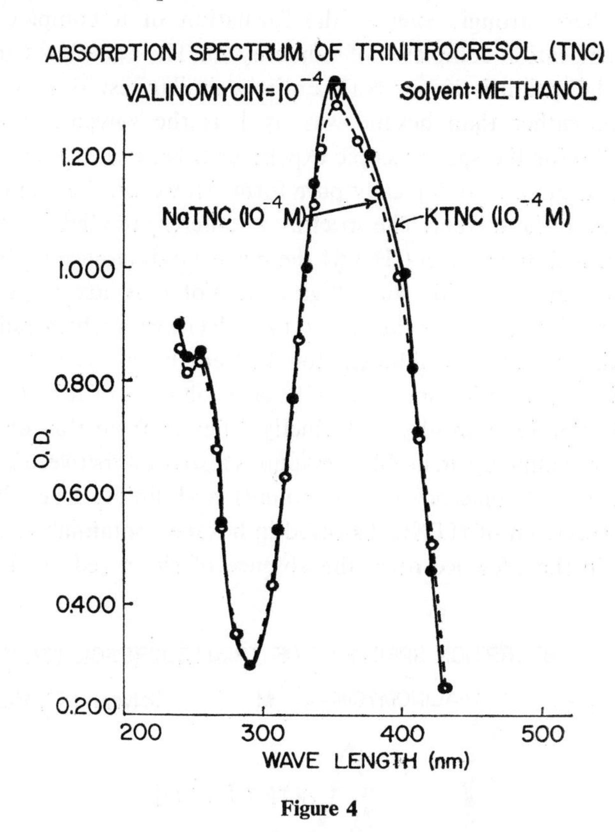

Figure 4

In summary then, ultra-violet spectra of trinitrocresol suggest that complexes occur between the Na^+ and K^+ complexes of valinomycin and TNC^- in pure hydrocarbon but not in more polar solvents. These complexes are cation specific, macrocyclic compound specific, and show substantial interactions with red cell membrane lipids. These observations are consistent with the hypothesis that TNC^- alters cation selectivity of valinomycin by altering the distribution of conformational states of valinomycin in such a way as to favor a conformation which can form complexes with Na^+. If this hypothesis is correct, the conformation of valinomycin in the Na-Val-TNC complex should be different from its conformation in the K-Val-TNC complex. Direct evidence for such a difference has been obtained by

D. Davis in our laboratories through measurements of the proton NMR spectra of free Val, K-Val-TNC and Na-Val-TNC in CD Cl_3 an dhexane. Figure 5 shows that the spectrum of the amide protons of these three forms of valinomycin are all distinctly different.

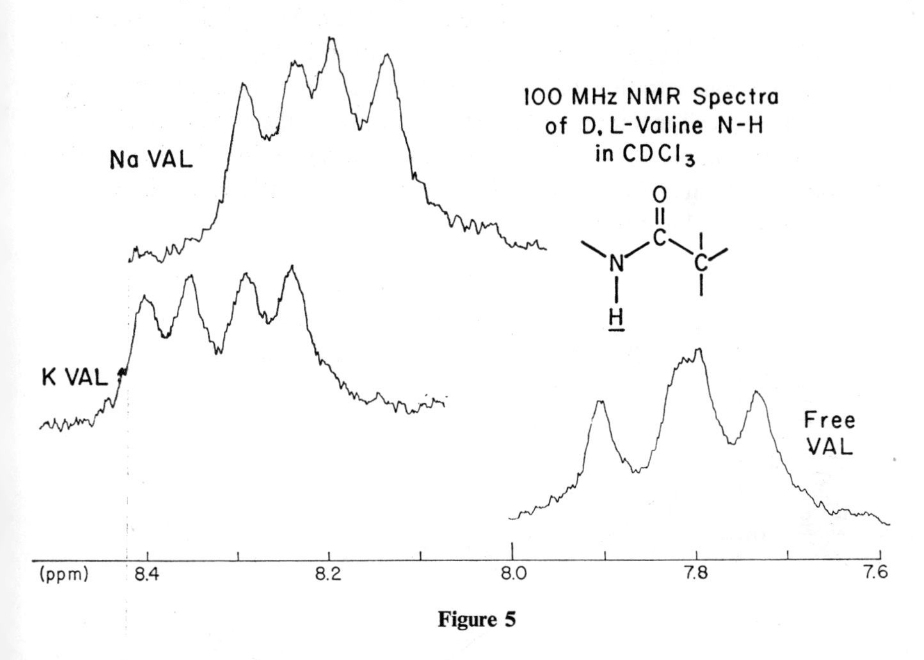

Figure 5

TNC^- AND RED CELL MEMBRANE PERMEABILITY

The membranes of mammalian red cells are known to be 10^6 times more permeable to Cl^- than to small cations like K^+ or Na^+. This permselectivity has often been attributed to the presence in the membrane of fixed positive charges. Because of the discovery that TNC^- forms complexes with K-Val^+ and Na-Val^+, we asked whether this anion would also form complexes with the hypothetical positively charged groups which regulate passive cation transport in sheep red cells. R. Gunn, in our laboratories measured the effect of TNC^- on the permeability of sheep red cell membranes to Cl^-, $SO_4^=$, K^+, and Na^+[19]. We found that TNC^- reduces anion and increases cation permeability. In 10^{-2} M TNC^- at 0°C the half time for exchange of $^{36}Cl^-$ was found to be 1 hour! Under these conditions, the sheep red cell membrane is as permeable to Na^+ as to Cl^-. Na^+ and K^+ transport in the presence of TNC^- shows the same U shaped dependence on temperature as has been reported by Wieth[20] for cation transport in human red cells the presnece of SCN^- or salicylate (Figure 6). This unusual temperature dependence may be related to the effect of temperature on the partition of K-

Val-TNC and K-Val SCN water and organic solvents containing 10^{-4} M valinomycin. The K^+ concentration ratio (organic solvent/water) was 20 times greater at 0°C than at 37°C when TNC was the anion and decane the solvent. This large negative temperature coefficient was not observed with TNC^- and $CHCl_3$ or with SCN^- in either solvent. Thus, it is possible to rationalize the U shaped temperature curve shown in Figure 5 in terms of an increased concentration of TNC^- at the cation permeability regulating sites in the red cell membrane at low temperature. If these sites are similar to the cation complexes of valinomycin, the data shown in Table 9 suggests that they are located in a region of the membrane that is more similar to decane than to chloroform.

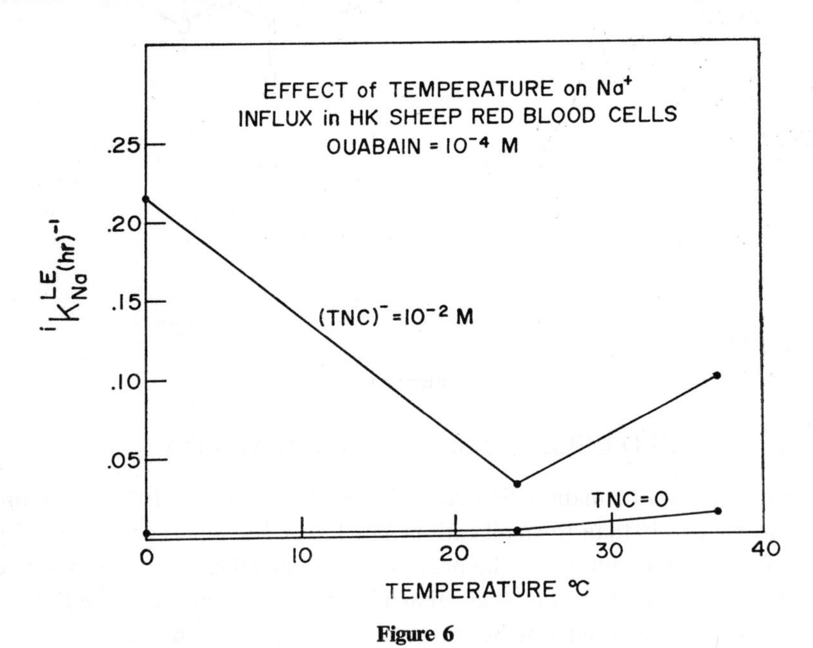

Figure 6

Table 9 Partition of K^+ between water (W) and non-aqueous solvents (S) containing valinomycin at three temperatures

Solvent	Anion	K^+ Concentration Ratio (S/W) 0°C	23°C	37°C
Decane	TNC^-	1.4×10^{-2}	3.5×10^{-3}	8.1×10^{-4}
Decane	SCN^-	1.3×10^{-5}	2.7×10^{-5}	2.9×10^{-5}
$CHCl_3$	TNC^-	3.8×10^{-2}	3.0×10^{-2}	4.1×10^{-2}
$CHCl_3$	SCN^-	4.6×10^{-3}	1.4×10^{-3}	4.1×10^{-3}

RELEVANCE

In conclusion, I would like to review briefly the relevance of these results to the problem of the mechanism of ion transport across biological membranes.

First, our results indicate that it is not, in general, possible to deduce what ions are carrying charge across bilayer membranes containing macrocyclic compounds from measurements of zero current potential or of membrane resistance in the presence of external solutions of different salt composition. When direct comparison is made between total membrane current and ionic current measured by tracer technique, discrepancies appear under circumstances which cannot be predicted from the electrical measurements. Such electrical measurements have provided almost the only estimate of the relative importance of different ions as charge carriers across biological membranes during the unsteady states associated with excitation and propagation in excitable cells. Out results would urge caution in the interpretation of experiments of this kind.

Our studies of the relationship between the structure of macrocyclic compounds and their capacity to function as ion carriers may or may not be relevant to the problem of identifying the ligands which actually interact selectively with potassium or sodium ions in biological membranes. It is evident that the properties of macrocyclic compounds which are necessary for them to act as ion carriers are altered dramatically by small changes in their primary structure. It is also clear that we will soon learn the rules which relate primary structure to membrane activity, presumably through the role of primary structure in determining the characteristics of and the rate of transitions between the preferred conformations of the compounds. Whether these rules are useful in defining more precisely the basis for cation selectivity of biological membranes depends, in large part, on whether the ligands which interact with sodium and potassium in biological membranes are on adjacent or on distant monomers. If they are on adjacent monomers, for example amino acids in a single protein molecule, the rules learned from valinomycin and its analogs may be useful. If on the other hand, the interacting ligands are on distant monomers, as is often the case with the different groups which make up the active site of enzymes, the rules are unlikely to be helpful. Again more work is required. My expectation is that the interacting ligands in the biological membrane will turn out to be on distant monomers so that the rules learned from valinomycin will not be directly applicable to this problem. However, the rules may be of great help in designing compounds which interact selectively with alkali metals and other ions. Such compounds could be useful both analytically and pharmacologically.

Finally, the specific interactions of valinomycin with aromatic anions such as trinitrocresol suggests several points of relevance to the study of biological membranes. First, TNC^- is a member of class of organic anions which are known to uncouple phosphorylation from oxidation in mitochondria. The action of uncouplers is often interpreted in terms of the capacity of these agents to increase the proton permeability of the mitochondrial membrane. The fact that TNC^- can form complexes with the cation complexes of valinomycin suggests that alternative interpretations should be borne in mind. This is particularly true when uncouplers like dinitrophenol are used in combination with cation carriers such as valinomycin. Second, the fact that TNC^- can markedly alter the selectivity of valinomycin for potassium as compared with sodium, apparently without. breaking any covalent bonds, raises interesting possibilities for the mechanism of changes in the sodium-potassium selectivity associated with physiological events in biological membranes. For example, transmitters at synapses are known to modify the relative permeability of the post synaptic membrane to sodium and potassium. A compound with properties like trinitrocresol could function in this manner. Furthermore, the alternating potassium and sodium selectivity of the active transport system on the two sides of the plasma membrane respectively could, in principle, be due to the effect of an anion on a valinomycin-like configuration of a macro molecule without requiring that the necessary changes in this configuration involve the making and breaking of covalent bonds. Finally, the comparison between the effects of anions like TNC^- on biological membrane permeability on the one hand and on the properties of macrocyclic compounds in organic solvents and thin lipid bilayers on the other hand, may help to elucidate the environment through which ions pass during transport. These considerations lead us to hope that investigations of the mechanism of action of macrocyclic compounds will continue to provide "Glimmers of Light" in membrane physiology. But we bear in mind KC's wise admonition "And the concepts gained from artificial models may be entirely misleading."

REFERENCES

1. Pressman, B. C. (1965). Induced active transport of ions into mitochondria. *Proc. Nat Acad. Sci.* (U.S.) **53**, 1076.
2. Mueller, P. and D. O. Rudin (1967). Development of K^+—Na^+ discrimination in experimental bimolecular lipid membranes by macrocyclic antibiotics. *Biochem. Biophys. Res. Comm.* **26**, 398.
3. Lev, A. A., and E. P. Bujinsky (1967) Cation specificity of bimolecular phospho lipid membranes, containing valinomycin. *Tsitologiya* (USSR) **9**, 102.
4. Andreoli T. E., M. Tieffenberg, and D. C. Tosteson (1967). The effect of valinomycin on the ionic permeability of thin lipid membranes. *J. Gen. Physiol.* **50**, 2527.

5. Tosteson, D. C. (1968). Effect of macrocyclic compounds on the ionic permeability of artificial and natural membranes. *Fed. Proc.* **27**, 1269.
6. Shemyakin, M. M., Yu. A. Orchinnikov, V. T. Ivanow, V. K. Antonov, E. I. Vinogradova, A. M. Shkrob, G. G. Malenkov, A. V. Evstratov, I. A. Laine, E. I. Melnik, and I. A. Ryabova (1969). Cyclodepsipeptides as chemical tools for studying ionic transport through membranes. *J. Membrane Biol.* α. 402.
7. Pioda, L. A. R., H. A. Wachtel, R. E. Dohner, and W. Simon (1967). Komplexe von Nonactin und Monactin mit Na^+, K^+ und NH_4^+. *Helv. Chim. Acta* **50**, 1373.
8. Kilbourn, B. T., J. D. Dunitz, L. A. R. Pioda, and W. Simon (1967) Structure of K_4^+ complex with nonactin, a macrotetralide antibiotic possessing specific K^+ transport properties. *J. Mol. Biol.* **30**, 559.
9. Onnishi, M., and D. N. Urry (1970) Solution conformation of Val-K^+ complex. *Science* **168**, 1091.
10. Pinkerton, M., L. K. Steinrauf, and P. Dawkins (1969). The molecular structure and some transport properties of valinomycin. *BBRC* **35**, 512.
11. Szabo, G., G. Eisenman, and S. Ciani (1969). The effects of macrotetrolide antibiotics on the electrical properties of phospholipid bilayer membranes. *J. Memb. Biol.* **1**, 346.
12. Tosteson, D. C., M. Tieffenberg, and P. Cook (1968). The effect of macrocyclic compounds on ionic permeability of HK and LK sheep red cell membranes and on artificial thin lipid membranes. *Metabolism and Membrane Permeability of Erythrocytes and Thrombocytes.* E. Deutsch, E. Gerlach, and K. Moser, Editors. Georg Thieme Verlag Stuttgart. pp. 424.
13. Hopper, Ulrich, Albert L. Lehninger, and Thomas E. Thompson (1968). Protonic conductance across phospholipid bilayer membranes induced by uncoupling agents for oxidative phosphorylation. *Proc. Nat. Acad. Sci.* **59**, 484
14. Shemyakin, M. M., E. I. Vinogradova, M. Yu Feigina, M. A. Aldanova, N. F. Loginova, I. D. Ryabova, and I. A. Pavleno (1965). The structure-activity relation for valinomycin depsipeptides. *Experientia* **21**, 548.
15. Merrifield, R. B., B. F. Gisin, D. C. Tosteson, and M. Tieffenberg. (1969). The synthesis of depsipeptides. *The Molecular Basis of Membrane Function.* D. C. Tosteson, Editor. Prentice Hall, Englewood Cliffs, N. J. U.S.A., pp. 211.
16. Shemyakin, M. M., V. K. Antonov, L. D. Bergelson, V. T. Ivanov, G. G. Malenkov, Yu A. Orchinnikov, and A. M. Shkrob (1969) Chemistry of membrane-affecting peptides, depsipeptides and depsides (structure-function relations). *The Molecular Basis of Membrane Function.* D. C. Tosteson, Editor. Prentice Hall, Englewood Cliffs, N. J. pp. 173.
17. Collander, R. (1949). Die Verteilung organicher Verbindungen zwvischen Äther and Wasser. *Acta Chem. Scand.* **3**, 717.
18. Davis, D. Manuscript in preparation.
19. Gunn, R. B. and D. C. Tosteson (1970). The effect of 2, 4, 6 trinitro-m-cresol on cation and anion transport in sheep red blood cells. *J. Gen. Physiol.* (submitted for publication).
20. Wieth, J. O. (1970). Paradoxical temperature dependence of Na and K fluxes in human red cells. *J. Physiol.* **207**, 563.

PAPER 7

Temperature Dependence of Excitability of Space Clamped Squid Axons

RITA GUTTMAN

Department of Biology, Brooklyn College of the City University of New York and the Marine Biological Laboratory, Woods Hole, Mass.

Kenneth S. Cole has long believed that it will be possible to discover the characteristics of living membranes in general by studying the electrical parameters of excitable membranes. An excitable membrane is, after all, a membrane. It is, however, one for which we have a tremendous literature. No small reason for this was the discovery or rediscovery of the giant axon of the squid by John Z. Young and its utilization by Dr. Cole who was the first to exploit this preparation and to delight in the fact that a whole laboratory could now be put inside an excitable living cell.

It is Dr. Cole also who has long pointed out that it is in the temperature area that the Hodgkin Huxley formulation is probably weakest and that any physical model will have to account for the discrepancies between this formulation and the temperature data. It was his constant encouragement, his insistence that data and theory go hand in hand, his feeling that the temperature area will prove of fundamental importance in elucidating function, that inspired these modest studies and so I am very pleased to have them included in this tribute to him.

The work which I shall report here consists of a series of eight papers published between 1962 and 1970 on the effect of temperature on the excitation process in space clamped squid axons. In this work I have had the loyal, valued technical assistance of Mr. Robert Barnhill over the years and more recently of Mr. Lon Hachmeister.

The earliest paper in the temperature series (Guttman, 1962) seems rather crude now. Many refinements of technique have been developed in the interim. However, though the instrumentation was primitive, certain facts were nevertheless established. Utilizing the space clamped condition and the double sucrose gap technique, which had only recently been developed by Julian, Moore and Goldman (1962) from an earlier version of Stämpfli (1954), it was possible to establish a Q_{10} of 2.3 for threshold (rheobasic) current and 2.0 for threshold voltage, Table I. The former agreed well with the computations of FitzHugh (cf. Guttman 1962) based on the HH equations, which also yielded a figure of approximately 2. There was a discrepancy between my finding for the Q_{10} of the threshold voltage and that of Sjodin and Mullins (1958) who reported an increase in membrane potential threshold when the temperature was lowered while we found a *decrease*. This was troubling until FitzHugh called attention to the possibility that the difference in stimulus duration (they used 1 msec pulses while I used 100 msec pulses) might make the results not comparable.

This indeed proved to be the case when I subsequently set myself the task with the help of Robert Barnhill (who was to assist me loyally for many years to come and to contribute immensely to the refinement and accuracy of the technique) to investigate the specific effect of stimulus duration upon the temperature dependent strength duration curve. Dr. Cole pointed out

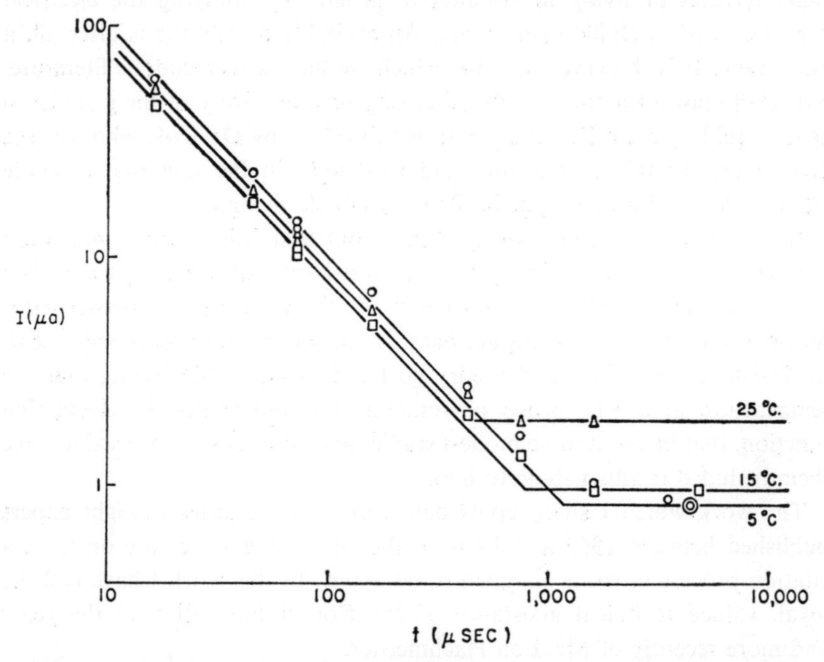

Figure 1 Temperature dependence of threshold membrane current when stimulating pulses of various durations are utilized. Intensity of the current in microamperes is plotted against duration in microseconds both on log scales. Runs were carried out at 5°C, 15°C, and 25°C on the same xaon

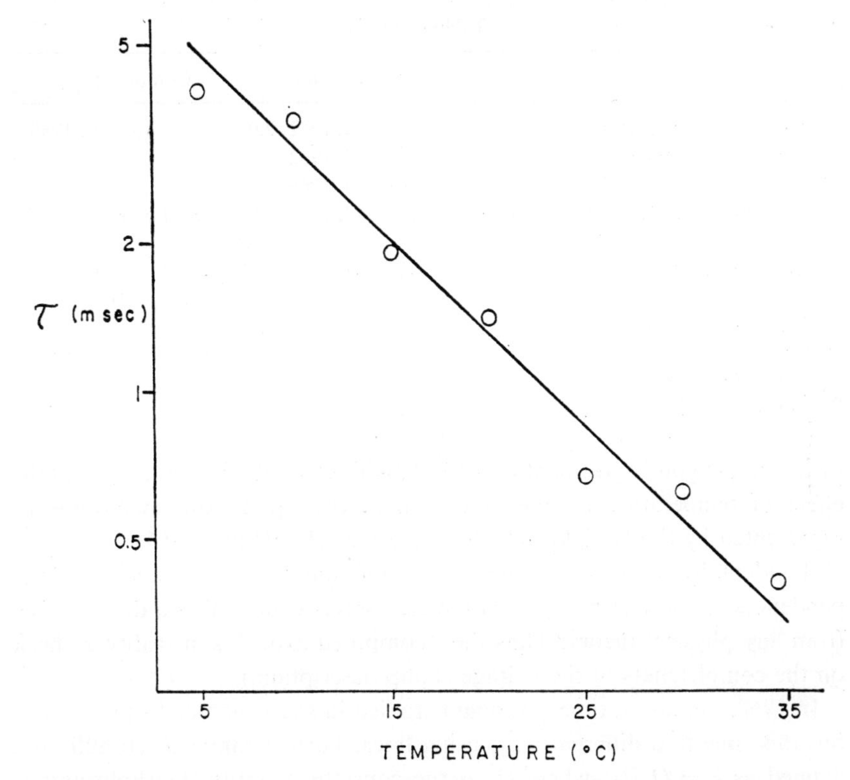

Figure 2 Temperature dependence of time constants of excitation in four axons. Average τ in milliseconds plotted logarithmically vs. temperature in degrees Centigrade

the utility of expressing the results on a log log scale (Figure 1), for when this is done it is apparent that at short durations there is a constant quantity threshold and at long durations, a constant current (rheobasic) threshold. The constant quantity threshold is relatively independent of temperature but the constant current (rheobasic) threshold is strongly temperature dependent. The threshold currents of intermediate duration and at all temperatures are less than both the single and two factor exponential expressions for excitation of Blair and Hill. These characteristics are in good agreement with the 1947 space clamped excitation data of Cole and Marmont (Bonhoeffer, 1953; Cole, 1955) and with the first calculations of the Hodgkin-Huxley equations (Hodgkin and Huxley, 1952). Where the constant quantity threshold line intersects the constant current line, Figure 1, we have τ, the time constant of excitation. The average τ at 20°C was found to be 1.2 msec. It was then possible to establish the effect of temperature upon τ (Figure 2) and a Q_{10} of 0.44 was demonstrated (Table I). These results are,

Table I Q_{10}'s

	Experimental		Computed	
Threshold current (I_0)	2.3	G, 1962	2.0	F, 1966
Threshold voltage (V_t)	2.0	G, 1962		
	1.15	G, 1966		
Time constant of excitation (τ or k)	1.58	G, 1968	1.89 when $A = 1$)	C & D*
			1.92 when $A = 4$)	
Accommodation time (λ)	2.27	G, 1968	1.92 when $A = 1$)	C & D*
			2.07 when $A = 4$)	
Subthreshold oscillation	2.25	G, 1969	2.25 F**	

* In Guttman, 1968 a
** In Guttman, 1969

in general, in good agreement with FitzHugh's computer investigations of the effect of temperature on the excitation of the squid axon membrane as represented by the Hodgkin-Huxley equations (FitzHugh, 1966).

(Incidentally, it might be useful to point out that the temperature dependences of the HH parameters are themselves empirical and do not come from any physical theory. Thus the "computed axon" is in reality a check on the completeness of the voltage-clamp description.)

In 1968, excitation time was again studied in space clamped squid axons, but this time in a different way, using linear current ramps as stimuli, and defined as $k = Q_0/I_0$, where Q_0 is the constant quantity threshold and I_0 is rheobase. The value obtained, 0.62 msec at 15°C, is probably more reliable than that obtained on 1966 using a cruder method, Table II. This value is considerably below the computed value obtained by Cooley and Dodge (cf. Guttman, 1968 a), Table II. On the other hand, accommodation time, also determined by using linear current ramps as stimuli and defined as the ratio of rheobase to the minimum gradient of a linearly rising current that will just excite: $\lambda = I_0/(dI/dt)$, was found to average 3.2 msec at 15°C (Table II), not too different from the computed values of Cooley and Dodge.

The values of Q_{10} for accommodation and excitation are practically identical in the computed axon (about 2). In the experimental axon, however, the Q_{10} of accommodation is 44% higher than the Q_{10} of the excitation process, Figure 3. Thus the HH equations probably need some modification in this area.

Next Fred Dodge suggested to me that he would be interested in learning the effect upon excitation of lowering the concentration of the sodium ion bathing the axon. This was investigated and in a paper published in 1968 (Guttman, 1968 b) it was shown that if the concentration of external sodium

Table II Threshold electrical characteristics of space clamped squid axons

	Experimental		Computed
Threshold current (I_0)	3.2 μA at 26°C to 1 μA at 7.5°C	(G, 1962)	
Threshold potential (V_t)	50 mv at 26°C to 6 mv at 7.5°C	(G, 1962)	
Threshold current density at rheobase	12 μA/cm² at 20°C	(G, 1966)	
Threshold charge	1.5×10^{-8} coul/cm²	(G, 1966)	
Excitation time (τ or k)	0.62 msec at 15°C	(G, 1968a)	1.7 msec when $A = 1$) C & D*
	1.2 msec at 20°C	(G, 1966)	.0 msec when $A = 4$) both at 15°C
Accommodation time (λ)	3.2 msec at 15°C	(G, 1968a)	3.4 msec when $A = 1$) C & D*
			2.2 msec when $A = 4$) both at 15°C

* In Guttman, 1968a

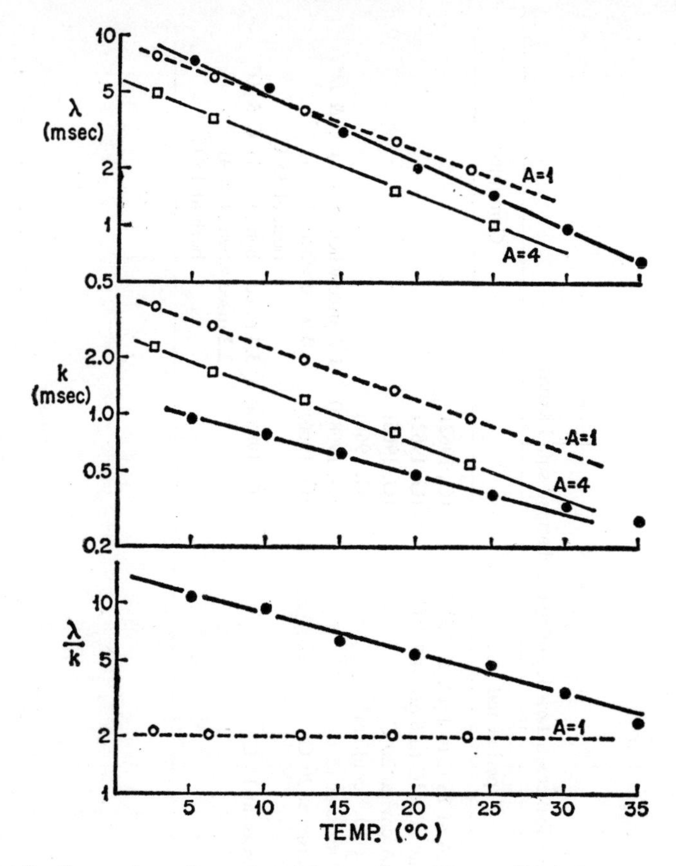

Figure 3 Comparison of experimental axon (heavy lines, solid circles, representing average values) with calculated axon (open circles for $A = 1$ condition; open squares for $A = 4$ condition) for λ, accommodation time; k, excitation time and λ/k, all on logarithmic scale vs. temperature in °C on a linear scale. (For discussion of conditions $A = 1$, $A = 4$, see text.)

was lowered, the shape of the strength-duration curve remained unaltered but that both the constant quantity threshold, Q_0, and the rheobase, I_0, were raised (Figure 4). Cooley and Dodge were able to show that the same held true for the computed axon but that the theoretical axon is about twice as sensitive to sodium reduction as the average experimental axon, (cf. Appendix, Guttman 1968 b). Low sodium did not affect the temperature dependence of the strength-duration relation ship.

Next the effect of applying TTX (tetrodotoxin) externally was investigated since it had been demonstrated that TTX specifically affects the inward sodium current in voltage clamped axons. (However it should be mentioned that lowered sodium and TTX have different effects theoretically because the

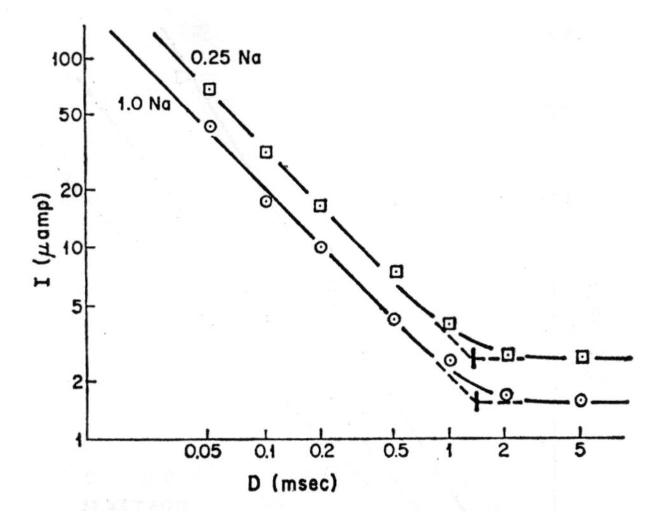

Figure 4 Strength-duration curves for axon in seawater (⊙), and in seawater in which 75% of the sodium has been replaced by choline (⊡). Duration of square wave pulses in milliseconds, D, is plotted against the current in microamperes, I, both on log scales. All points taken on the same axon at 5°C. Vertical bars represent τ, as defined in the text for the two different concentrations of sodium

former produces changes primarily in V_{Na} while the latter changes only g_{Na}.)

The excitability of TTX treated real axons was found to be more temperature dependent than that of normal real axons. Also the blocking effect of TTX was found to have a positive temperature coefficient, Figure 5. Narahashi and Anderson (1967), on the other hand found a negative temperature coefficient for blocking by allethrin, a potent blocking agent which affects both sodium and potassium conductance. Yamasaki and Narahashi (1953) showed that the effectiveness of DDT in developing poisoning symptoms also has a negative temperature coeficient.

Analysis of the data on dosage-response to TTX of real axons reveals that they fit the dose-response reltionship of a hypothetical system in which one TTX ion binds reversibly to its receptor to produce a fraction of the inhibitory effect, the curve being identical to a simple adsorption isotherm, Figure 6.

Although both of us had only the few precious summer months at Woods Hole for these projects, by now Robert Barnhill had been able to snatch sufficient time from the experimental work to develop high precision methods which much increased our accuracy and enabled us to observe and measure phenomena until then impossible to study. These instrumentation improvements were not completed in time for the first study of the temperature

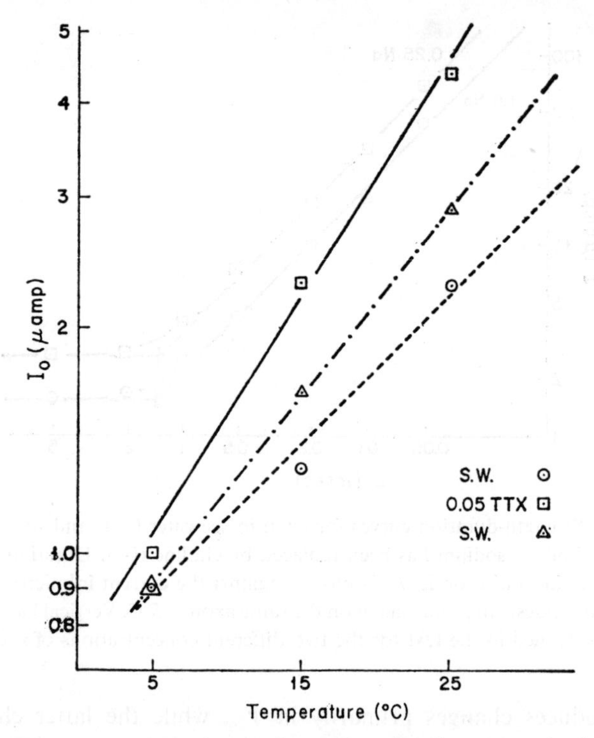

Figure 5 Temperature dependence of tetrodotoxin effect upon rheobase. Rheobasic current, I_0, in microamperes, is plotted against temperature in °C. First data were taken with the axon in seawater (⊙). Then the axon was treated on 0.05 μg/ml solution of tetrodotoxin (⊡). Then partial recovery in seawater (△) is shown. All data taken on the same axon. One of two silmiar experiments.

dependence of oscillation (Guttman, 1969) but they proved invaluable in the subsequent investigations of oscillation and repetitive firing (Guttman and Barnhill, 1970; Guttman and Hachmeister, in Ms.).

One refinement consisted of a base line stabilization circuit, incorporating synchronization and track-and-hold memory circuits and an adder. The track-and-hold memory circuit was synchronized with the stimulus in such a way that the base line was sampled for 133 msec, ending about 16 msec prior to each stimulation. The negative of this potential was held for the remainder of the 1 second repetition period and was added to the unaltered response signal. The sum produced the trace on the oscilloscope. The effect of this circuit was to set the oscilloscope base line to zero immediately before each stimulation, thus practically eliminating the effect of changes in resting potential while introducing no distortion to the active response. In this way, large amplifications (up to 1 mV/cm) of interesting portions of the response became practical.

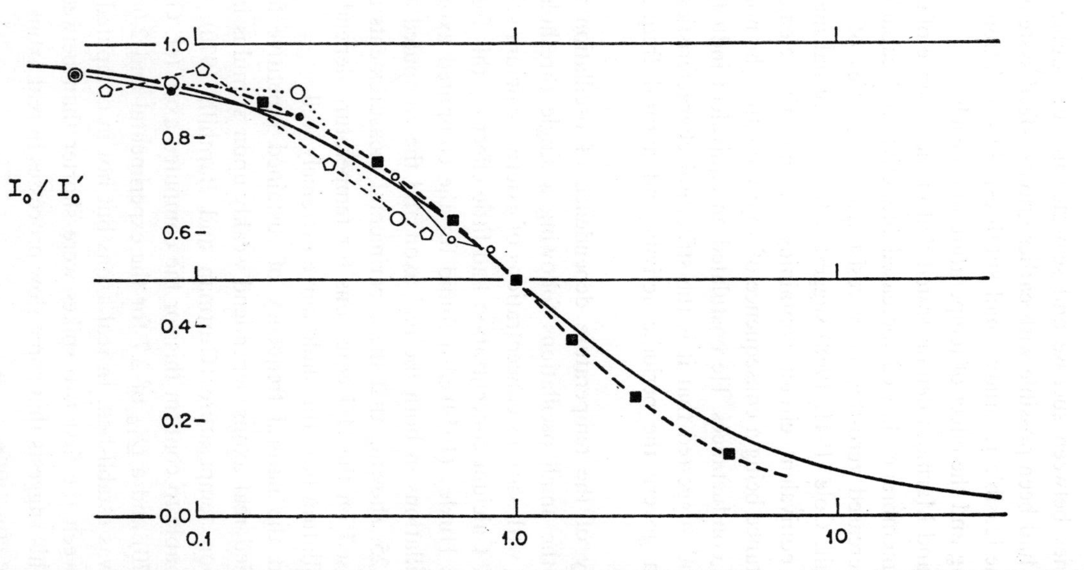

Figure 6 Dosage response to TTX. Experimental data from four axons in which excitability (1/normalized rheobase) is compared with a simple adsorption isotherm (heavy solid line). 1/normalized rheobase on a linear scale vs. relative concentration of TTX on a logarithmic scale. The data are slid horizontally for best fit. The concentrations for 50% excitability (or doubling of the rheobase) were 0.23, 0.23, 0.19, and 0.12 μg/ml TTX. Computed curve (dashed line, solid squares), based on assumption that TTX affects sodium conductance, normalized by sliding horizontally 0.7 unit to the left. See text for details

Robert Barnhill also developed a dual metering pump system that proved most helpful. The pumps were Sage model 220 unlimited-volume syringe pumps, modified and calibrated to pump at rates determined by external control signals. This sensitive control of solution flow rates provided much steadier shear lines between sucrose and seawater in our double sucrose gap system than had been possible with earlier gravity—feed systems.

Now it became feasible to study small subthreshold oscillations as well as repetitive firing and the effect of temperature upon each.

In 1942, Cole and Marmont demonstrated that in the absence of calcium, the squid axon membrane showed increased inductive and capacitive reactances and decreased zero-frequency resistance, properties of a rather good tuned circuit. Cole (1941, 1968) suggested for the normal membrane in seawater an equivalent circuit containing an inductive element, his equivalent inductance being a consequence of approximating the non-linear Hodgkin Huxley conductances. He postulated an equivalent inductance of 0.1 henry cm^2 and suggested that it is the structural characteristics of the membrane which govern the periodic activity and natural frequency of nerve.

The first study of the temperature dependence of oscillation was an investigation of the small oscillations following a single threshold spike during treatment with various concentrations of external calcium solutions, (Guttman, 1969). Calcium concentration had little effect on this frequency in real axons, as Huxley (1959) also found for the computed axons. The Q_{10} of these oscillations in both the real axon and the computed axon of FitzHugh was 2.25, showing that the experimental measurements and the computations based on the HH equations for temperature dependence of frequency of oscillation near threshold agree extremely well.

We found that the natural frequency of sustained repetitive firing in partially decalcified real axons dependend weakly upon stimulus intensity and strongly upon temperature (Guttman and Barnhill, 1970). Cooley and Dodge were able to confirm this for the computed axon (cf. Guttman and Barnhill, 1970) and a Q_{10} of 2.7 for the experimental and 2.6 for the computed axon was established. In real axons but not in computed axons, the intervals between the first two spikes were shorter than between subsequent spikes. This suggests that some slow processes in real axons are not represented in the equations.

Even more interestingly, subthreshold and superthreshold responses were sometimes intermixed in a train of responses from a real axon responding to a constant step of current, Figure 9. (This also is not predicted by the Hodgkin–Huxley equations). Whether these "skip runs" reflect random variation in threshold caused by fluctuation of voltage across the resting

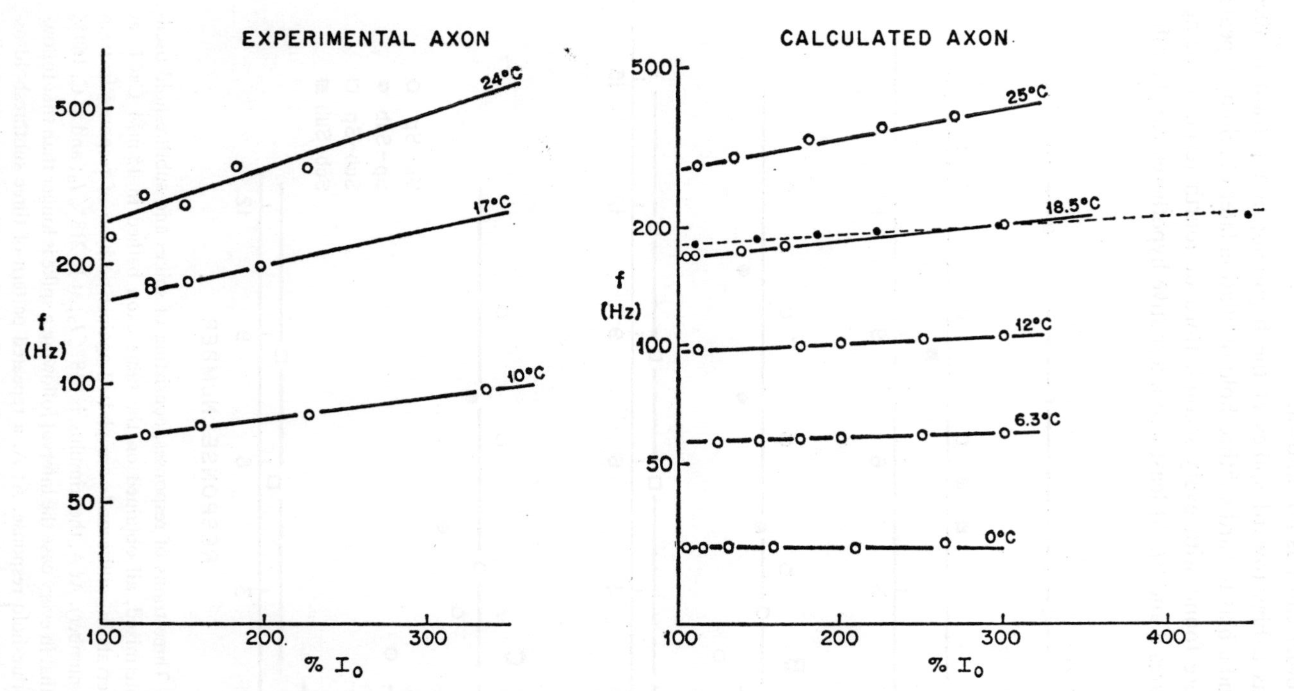

Figure 7 Effect of stimulus intensity, expressed as percentage of rheobasic value (% I_0) on frequency (Hertz) at various temperature in an experimental axon (left) and computed axon (right) in low calcium. The broken line is the result when all conductances were increased by a factor of 1.5 to take into account the change of conductance with temperature

membrane (cf. Verveen and Derksen, 1969; Lecar and Nossal, 1968) or have some other cause is not known.

There was a bias toward spikes at the beginning of the train in these skip runs and a bias toward subthreshold oscillation later on. Some repeated patterns were found with every second, third or fourth response being a spike, (Figures 8 and 9). At least two alternative hypotheses can be offered

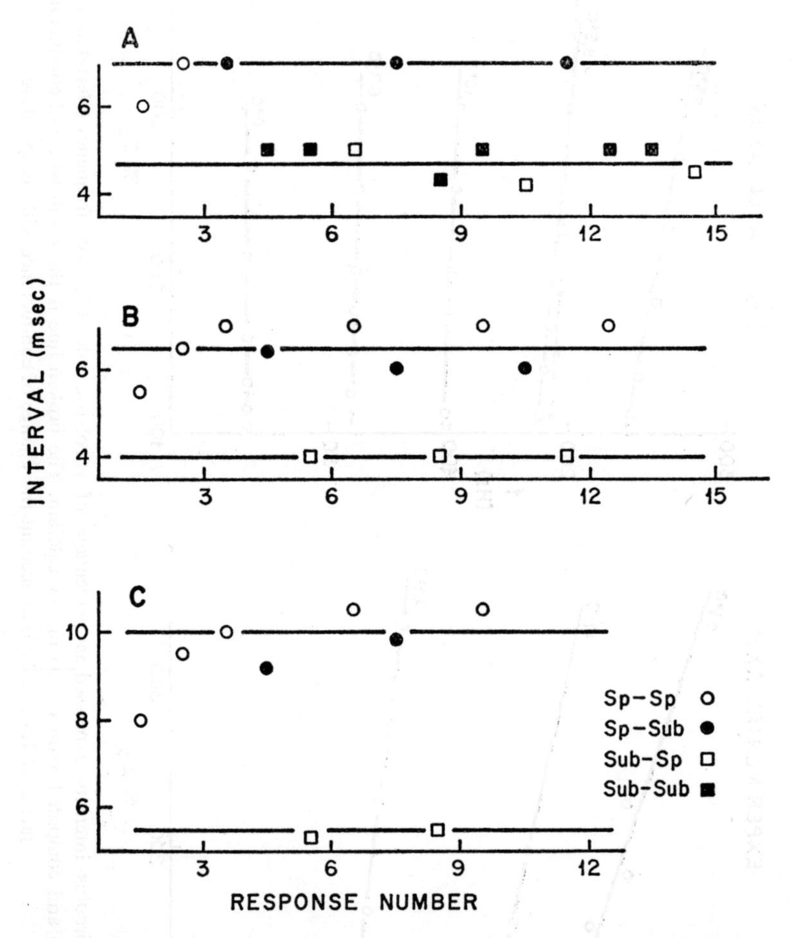

Figure 8 Three trains of responses, consisting of spikes and subthreshold oscillations intermixed, all obtained on the same axon bathed in 15 mM $CaCl_2$ at 15°C. Intervals in milliseconds are plotted against position of response in train (response number). At A, the stimulus, is 318% I_0; at B, 218% I_0; and at C, 169% I_0. Note that in every case the interval following a spike is longer than that following a subthreshold response. At A, a repeated pattern of three subthreshold oscillations and a spike is apparent. At B and C, the repeated pattern is two spikes followed by a subthreshold oscillation

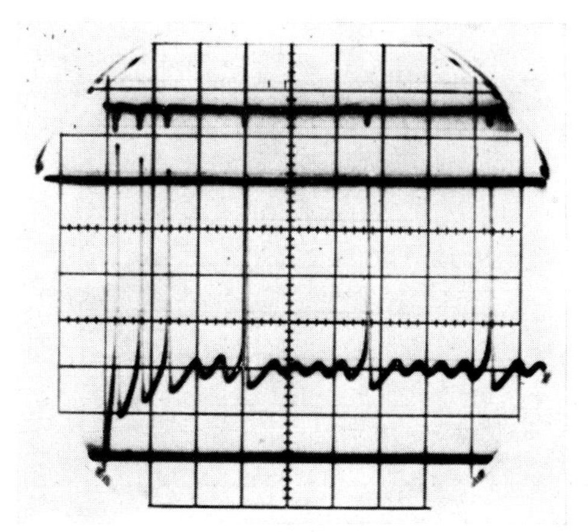

Figure 9 Photographic record of response trace (below) and stimulus trace (above) for an axon, stimulated by a uniform step of current, responding by a train of spikes and subthreshold ocsillations intermixed. The interval after the spike is longer than after the subthreshold response. There is bias toward spikes at the beginning of the train and toward subthreshold oscillation later on. Calibration: 1 μA, 20 mV, 5 msec. Temperature, 25°C. Typical record

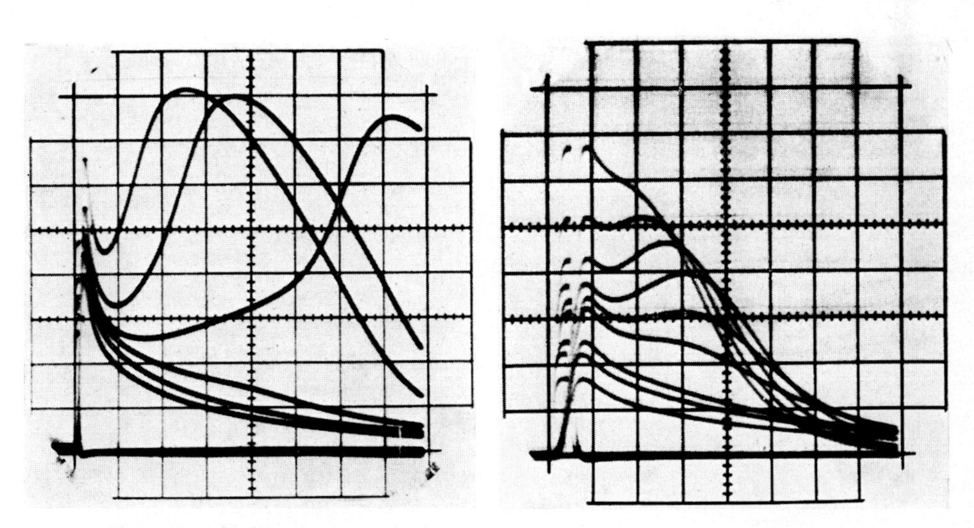

Figure 10 Oscilloscope record of membrane responses, ordinates (10 mV/div) vs time, abscissae, at 15°C, left, (0.2 msec/div) and 38°C right (0.05 msec/div)

for this rhythmicity. An accommodation may follow a spike, allowing only subthreshold responses to appear until it wears off. Or, alternately, a spike frequency and a subthreshold frequency may coexist and, depending upon the relative phases, one rhythm or another will be established.

An oscillating parameter with a period of about 15 msec at 25°C and 20 msec at 15°C could account for most of the "skip runs". The period of this oscillation is several times longer than the longest time constant of the HH equations and thus some parameter would have to be added to the HH equations to account for the skip run phenomenon.

In these skip runs, the time interval following a spike was always longer than that following subthreshold oscillation in slightly decalcified real axons, (Figures 8 and 9). Huxley (1959) and FitzHugh (cf. Guttman, 1969) also found this for computed axons.

Perhaps the most interesting temperature study of all was carried out recently in collaboration with Drs. Cole and Bezanilla (Cole, Guttman and Bezanilla, 1970). Here it was established that although the all or none law held for the propagating axon, in a space clamped axon, excitation occurred without threshold. This is true at all temperatures but more apparent at high temperatures, Figures 10 and 11. Calculations indicate that the maximum slope of the stimulus response curve decreases from 10^9 to 500 to 6 as the temperature is raised from 6.3°C to 20°C to 40°C. This is in agreement with the BVP analogue of FitzHugh (1969), which shows a fractional response. It was concluded that these measurements and calculations provide support for the conductances as important parameters of membrane structure and normal operation. They are continuous with continuous derivatives as functions of potential, time and temperature. It therefore seems unlikely that abrupt, threshold, all-or-none, or transition phenomena are major factors in the processes by which ions cross the normal axon membrane.

Indeed in all the experimental results that have been presented here near threshold and over a large temperature range, the data have given no evidence of a critical temperature or of significant differences between two distinct ranges of temperature. Also, in the HH equations the various parameters are rather gradual and smooth functions of temperature, with no indication of a sharp transition in any temperature range. Thus, though it has been suggested by Changeux *et al.* (1966) by Tasaki, (1968) and by Adam, (1968) that first order phase transitions in the membrane may be responsible for excitation, neither the HH equations nor our observations give any evidence in support of such a process.

In conclusion, all the work reported here has been done at or near threshold and with the axon in space clamped condition. These conditions confer certain advantages and certain disadvantages. The main value of threshold

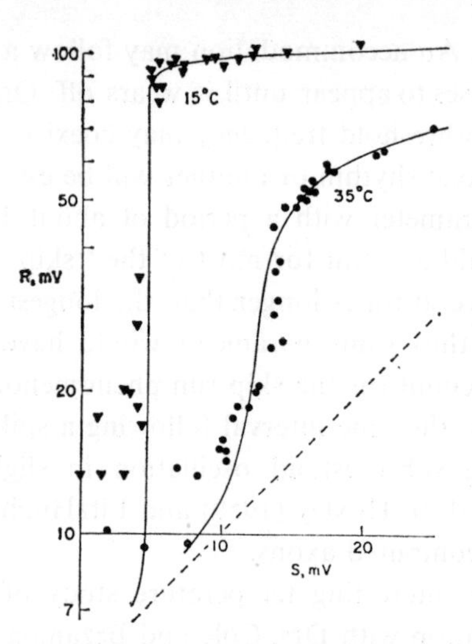

Figure 11 Comparisions of responses, *R*, vs. stimuli, *S*, for 15°C and 35°C. Curves are calculated, points are by experiment

information is that a great many factors affect it and it therefore may be a sensitive test of the HH equations. The space clamped condition is not a physiological condition for the impulse is made to stand still. The voltage clamp data give a more critical evaluation of criteria necessary for excitation than do space clamp data. However, the space clamped condition is an excellent test for the HH equations. The mathematics is simpler and it is a more direct test. It should be noted, however, that the voltage clamp and space clamp, are, of course, not mutually exclusive methods. (For a theoretical review of differences between thresholds for space clamped and propagated action potentials, see Noble and Stein, 1966.)

To summarize, most of the temperature and excitation data referred to are in fairly good agreement with the HH formulation. In fact, the work done in collaboration with Cooley and Dodge on the dependence on stimulus duration and the Cole, Guttman and Bezanilla (1970) study demonstrate some rather subtle threshold properties, which are predictable from the HH theory and probably represent triumphs of the formulation. The major areas of disagreement are in the realm of accommodation and oscillation phenomena. Lecar (personal communication) has pointed out that these phenomena depend upon the recovery parameters in the HH equations and suggests that this may be where the difficulty lies.

The work which I have outlined here could not have been done without the continued cooperation of many, particularly Drs. FitzHugh, Cooley and Dodge and most especially, Dr. Cole, himself. It is a tribute to one of

Dr. Cole's basic ideas: data without theory are useless as is theory without data. Only when people cooperate and try to link up theory and data will we, he believes, begin to understand. And so I wish to thank him and to thank Drs. FitzHugh, Cooley and Dodge, who have helped me so much in trying to link up these measurements on real squid axons with their calculations.

REFERENCES

Adam, G. (1968). *Zeit. für Naturforsch.* **23b**, 181.
Bonhoeffer, K. F. (1953). *Naturwiss.* **40**, 301.
Changeux, J. P., J. Thiery, Y. Tung, and C. Kittel (1966). *Proc. Nat. Acad. Sci.*, **57**, 335.
Cole, K. S. (1951). *J. Gen. Physiol.* **25**, 29.
Cole, K. S. (1955). Ions, potentials and the nerve impulse. In Electrochemistry in Biology and Medicine. T. Shedlovsky, editor. John Wiley and Sons, Inc. New York. 129.
Cole, K. S. (1968). Membranes, Ions and Impulses. Univ. of California Press. Berkeley and Los Angeles, Calif.
Cole, K. S., and G. Marmont (1942). *Fed. Proc.* **1**, 15.
Cole, K. S., R. Guttman, and F. Bezanilla (1970). *Proc. Nat. Acad. Sci.* **65**, 884.
FitzHugh, R. (1966). *J. Gen. Physiol.*, **49**, 989.
FitzHugh, R. (1969). Mathematical models of excitation and propagation in nerve. In Bioengineering. H. Schwan, editor, McGraw, Hill. New York.
Guttman, R. (1962). *J. Gen. Physiol.* **46**, 257.
Guttman, R. (1966). *J. Gen. Physiol.* **49**, 1007.
Guttman, R. (1968a). *J. Gen. Physiol.* **51**, 759.
Guttman, R. (1968b). *J. Gen. Physiol.* **51**, 621.
Guttman, R. (1969). *Biophys. J.* **9**, 269.
Guttman, R., and R. Barnhill (1970). *J. Gen. Physiol.*, **55**, 104.
Guttman, R., and L. Hachmeister (1972). Biophys. J. **12**, 552.
Hodgkin, A. L., and A. F. Huxley (1952). *J. Physiol.*, (London). **117**, 500.
Huxley, A. F. (1959). *Ann. N.Y. Acad. Sci.* **81**, 221.
Julian, F. J., J. W. Moore, and D. E. Goldman (1962). *J. Gen. Physiol.*, **45**, 1195.
Lecar, H. Personal communication.
Lecar, H., and R. Nossal (1968). *Biophys. Soc. T* **11**, **8**, p. A-132.
Noble, D., and R. Stein (1966). *J. Physiol.*, **187**, 129.
Sjodin, R. A., and L. J. Mullins (1958). *J. Gen. Physiol.* **42**, 39.
Stämpfli, R. (1954). *Experimentia* **10**, 508.
Tasaki, I. (1968). Nerve Excitation. Charles C. Thomas, Springfield, Illinois.
Verveen, A. A., and H. E. Derksen (1968). *Proc. IEEE.* **56**, 906.
Yamasaki, T., and T. Narahashi (1953). *Botyu-Kagaku* **19**, 39.

On the Constancy of the Membrane Capacity

R. D. KEYNES

Agricultural Research Council Institute of Animal Physiology, Babraham, Cambridge, England

---

In view of the seminal influence of the two classical papers by Cole and Curtis (1938, 1939) on the change in electrical impedance of excitable membranes during activity, it is rather surprising that one of the issues with which they were concerned has subsequently received no further attention from biophysicists. This is the question of the constancy of the membrane capacity during the impulse. It is, of course, well recognised now that the drop in impedance whose occurrence they demonstrated with such striking success arises from a sequence of changes in the ionic conductance of the membrane, affecting the highly selective channels through which sodium and potassium ions pass. In the conventional version of the equivalent circuit of the membrane, there are three conductive elements in parallel with the capacitative element, these being the sodium conductance g_{Na}, the potassium conductance g_K, and an unspecific leakage pathway g_1; excitation depends on the voltage and time dependent variations in g_{Na} and g_K described by Hodgkin and Huxley (1952). The membrane capacity has no role in the mechanism of propagation that could be described as dynamic, and has come to be regarded as more or less inert, and therefore of relatively little interest to the neurophysiologist.

Having recently been involved in studies on the one hand of the structural changes that can be detected optically (Cohen, Keynes and Hille, 1968), and on the other of the temperature changes during nervous activity (Howarth, Keynes and Ritchie, 1968), I have come to take a fresh interest in the supposed constancy of the membrane capacity. In the course of this work it became necessary to know both how the capacity of the nerve membrane varies with the electrical potential difference between its two sides, and how it varies with ambient temperature; but we soon realised that in neither case was there much experimental evidence in the literature. I had hoped, in this essay, to be able to remedy the situation by presenting the first results of some observations on the membrane capacity of voltage-clamped squid axons made with an impedance bridge. But when Professor Rojas and I set up the equipment for some preliminary measurements at Plymouth last winter, we quickly found that in order to obtain reliable values for the capacity we would have to pay much greater attention to detail, and to devote more time to the problem than we could spare. The work has therefore had to be postponed to a later occasion. I shall accordingly confine myself here to an explanation as to why the behaviour of the membrane capacity under different conditions is interesting, and to a brief account of the evidence that is already available.

Early attempts by Schmitt and Schmitt (1940) to detect a change in the birefringence of squid giant axons during the impulse were unsuccessful, and they were only able to conclude that any change must be smaller than 0.2%. However, the advent of signal averaging techniques has made it possible to obtain respectable records of responses which in a single sweep are hopelessly buried in noise. With the help of a CAT 400 C computer, Cohen, Hille and Keynes (1970) have shown that the birefringence of cleaned axons from *Loligo forbesi* decreases by about 1 part in 10^5 during a single impulse. The change is located in or very close to the membrane, and definitely has a radial optic axis. Almost all the resting birefringence of the axon, whose sign is positive referred to the longitudinal axis, arises from longitudinally arranged tubules and filaments in the axoplasm and in the Schwann cell layer, and from the form birefringence of the Schwann cells. By analogy with the work of Mitchison (1953) on erythrocyte ghosts, the contribution of the membrane itself is probably no more than 1% of the total resting birefringence, and is made up of a positive intrinsic birefringence and a negative form birefringence, each with a radial optic axis. Since the optic axes of the membrane and of the components responsible for the resting birefringence are mutually perpendicular, the observed decrease in the total birefringence must correspond, if it is referred—as seems legitimate—wholly to a change taking place in the membrane, either to an increase in its intrinsic birefringence or to a decrease in its form birefringence. The former might arise from a Kerr effect involving a reorientation of charged groups in or closely associated with the membrane, while the latter might result from an increase in membrane thickness due to a reduction during the depolarization of the compressional force exerted by the opposite charges on either side (Cohen, Hille and Keynes, 1969). If there is a variation in membrane thickness during the action potential there will be a parallel variation in membrane capacity, hence my interest in the observations of Cole and Curtis (1939).

A reorientation of dipoles or charged side groups in the membrane would also affect the membrane capacity, and might give rise to anomalies in the frequency impedance locus diagram at frequencies corresponding to the relaxation time constants of the birefringence change. The plural is used here because it turns out (Cohen, Hille, Keynes, Landowne and Rojas, 1971) that the birefringence change has both fast and slow components. Studies under voltageclamp conditions have shown that there is a fast change whose time constant is about 40 μsec at 13°C; this is a relatively stable phenomenon, always present. In freshly dissected axons, it appears to have superimposed on it a form of rebound which brings the birefringence back towards the resting level with a much longer time constant of around 5–40 msec. When, however, the axons are fatigued in various ways, or treated with agents such as tetrodotoxin or high-Ca solutions, the slow rebound disappears, and is replaced by an appreciable larger component which is additive with the fast change, and has a time constant of intermediate size, in the region of 0.5–3 msec. Possibly the prospects of detecting membrane capacity effects corresponding to the fastest and slowest birefringence changes are not good because of their small size. But the last-mentioned change can sometimes become so large that it is readily visible on a single sweep, without recourse to signal-averaging. The difference in membrane capacity displayed by a squid axon after prolonged exposure to tetrodotoxin would therefore be well worth examining.

In their first paper on the variation in membrane impedance during a propagated impulse, Cole and Curtis (1938) found that in *Nitella* the membrane capacity decreased by about 15%, their evidence being that when they plotted the impedance loci for cells at the height of activity, the experimental points consistently fell well to one side of the arc corresponding to a change of resistance alone. In the squid giant axon, however, the deviations from a pure resistance change were much less marked (Cole and Curtis, 1939); it was concluded that the capacity decreased by less than 2%, and that even so, phase shifts in the bridge, modulator and amplifier circuits might have been responsible for much of the apparent change. In the face of this evidence that the capacity of the nerve membrane is very nearly constant, it would probably not be profitable to reexamine the behaviour of the membrane capacity during the spike, which is too brief to offer much chance of success. But I would suggest that by using an internal electrode for the capacity measurements, and by looking for changes during voltage-clamp pulses lasting for several msec, it should be possible to detect smaller variations in capacitance than could have been measured reliably with external electrodes. This is the approach that I intend to adopt when the opportunity offers, but I cannot yet report any experimental results obtained on these lines.

It is also relevant to enquire whether membrane potential has any effect on artificial phospholipid bilayers. Babakov, Ermishkin and Liberman (1966), working with a membrane of unspecified composition, whose capacity was 0.37 $\mu F/cm^2$, reported that the application of 100 mV increased the capacity by about 1%, the change developing with a time constant of 0.3 sec. Observations on the intensity of light reflected from the membrane suggested that its thickness was not altering measurably, and the capacity change was attributed mainly to an increase in the effective area of the bilayer at the edge. Rosen and Sutton (1968) found appreciably larger capacity changes, and noted that 100 mV applied across a membrane of egg lecithin dissolved in n-heptane increased its capacity by as much as 14%; but the effect developed with a time constant of many seconds, and they too must have been observing an increase in total membrane area involving a slow retreat of the meniscus at the edge. The most thorough study yet reported is that of Ohki (1969), who again found that 100 mV gave a 1% increase in capacity for pure egg lecithin in n-decane at pH 7, but who measured the specific capacitance of the membrane, and made an allowance for area changes. As far as I am aware, membrane potential has not yet been shown to have any obvious effect on the optical properties of artificial bilayers, so that it is not known whether or not the capacity change is directly related to thickness. In any case, the bilayers that have been examined so far all contain appreciable amounts of solvent which may be squeezed out into islands when electrical pressure is applied. A direct comparison with nerve membranes that are probably less compressible is therefore dangerous.

My second reason for enquiring into the constancy of membrane capacity arises from a consideration of the change in entropy of the dielectric when a capacitance is charged and discharged. Howarth, Keynes and Ritchie (1968) measured the initial rise in temperature accompanying the impulse in the vagus nerve of the rabbit, and after allowing as carefully as possible for conduction time effects, estimated that it corresponded to a heat liberation of 24.5 μcal/g nerve. The most obvious source of this heat was a release as circulating electric currents of the energy ($= \frac{1}{2} CV^2$) stored in the membrane capacity (C) charged with the resting potential (V). Although neither C nor V could be measured directly, a reliable upper limit for the value of $\frac{1}{2}CV^2$ was estimated to be 11.5 μcal/g; over half the heat liberated was therefore not accounted for. After considering and dismissing other possibilities such as the heat of ionic mixing, it was concluded that the missing heat was most likely to arise from the occurrence of an appreciable decrease in the entropy of the membrane dielectric on reduction of the potential across it. In support of this conclusion, it was argued, following Gurney (1962), that when a capacitance C is charged or discharged

$$\frac{T.\Delta S}{\Delta F} = \frac{T}{C} \cdot \frac{dC}{dT},$$

where T is absolute temperature, ΔS is change in entropy, and ΔF is change in free energy (i.e. $-\frac{1}{2}CV^2$ when the condenser is discharged). An entropy change of the required sign implies that dC/dT must be positive, which is the reverse of expectation for a liquid dielectric, but is supported by the meagre evidence in the literature on the temperature coefficient of the capacity of biological membranes. Thus Taylor and Chandler (1962) found that for the squid giant axon the temperature coefficient, expressed as $\frac{I}{C} \cdot \frac{dC}{dT}$, of the appropriate component of the complex membrane capacity was 0.0073, whence the ratio of $T\Delta S$ to ΔF would be $+2.0$, and the total liberation of heat on discharge would be three times greater than $\frac{1}{2}CV^2$. Other figures cited by Howarth *et al.* (1968) similarly indicate that muscle membranes and artificial bilayers also have positive temperature coefficients of the same order of size for their capacities.

I have tried to show that a decrease in membrane capacity on depolarization, and an increase with rising temperature, are consistent both with the few direct measurements that have been reported and with my own thermal and optical observations. However, the existing evidence on the variation in membrane capacity under different experimental conditions is exiguous, to say the least, and needs to be greatly extended and refined in order to be properly correlated with the results of other types of study of membrane structure. Not only is there an urgent need for more precise quantitative data, but aspects of the subject such as the consequences that would follow from the presence of pre-existing stresses in the membrane should be further explored. It would not be unfair to say that capacity has become the most neglected property of nerve, though not perhaps of artificial membranes. But it is in a way a tribute to the pioneering work of Kenneth Cole that this should be so. The quality and thoroughness of his studies on membrane capacity made during the 1930's were such that it was impossible to improve on them. Even now, with considerably improved electronic equipment and techniques at our disposal, it will not be easy to make rapid advances in this field; and if progress is made, we shall always owe a great deal to the sure foundation provided by Cole on which we have to build.

REFERENCES

Babakov, A. V., L. N. Ermishkin, and E. A. Liberman (1966). Influence of electric field on the capacity of phospholipid membranes. *Nature, Lond.* **210**, 953–955.

Cohen, L. B., R. D. Keynes, and B. Hille (1968). Light scattering and birefringence changes during nerve activity. *Nature, Lond.* **218**, 438–441.

Cohen, L. B., B. Hille, and R. D. Keynes (1969). Light scattering and birefringence changes during activity in the electric organ of *Electrophorus electricus. J. Physiol., Lond.* **203**, 489–509.

Cohen, L. B., B. Hille, and R. D. Keynes (1970). Changes in axon birefringence during the action potential. *J. Physiol., Lond.* **211,** 495–515.

Cohen, L. B., B. Hille, R. D. Keynes, D. Landowne, and E. Rojas (1971). Analysis of the potential-dependent changes in optical retardation in the squid giant axon. *J. Physiol., Lond.* **218,** 205–237.

Cole, K. S., and H. J. Curtis (1938). Electric impedance of *Nitella* during activity. *J. gen. Physiol.* **22**, 37–64.

Cole, K. S., and H. J. Curtis (1939). Electric impedance of the squid giant axon during activity. *J. gen. Physiol.* **22**, 649–670.

Gurney, R. W. (1962). Ionic processes in solution. New York: Dover.

Hodgkin, A. L., and A. F. Huxley (1952). A quantitative description of membrane current and its application to conduction and excitation in nerve. *J. Physiol., Lond.* **117**, 500–544.

Howarth, J. V., R. D. Keynes, and J. M. Ritchie (1968). The origin of the initial heat associated with a single impulse in mammalian non-myelinated nerve fibres. *J. Physiol., Lond.* **194**, 745–793.

Mitchison, J. M. (1953). A polarized light analysis of the human red cell ghost. *J. exp. Biol.* **30**, 397–432.

Ohki, S. (1969). The electrical capacitance of phospholipid membranes. *Biophys, J.* **9**, 1195–1205.

Rosen, D. and A. M. Sutton (1968) The effects of a direct current potential bias on the electrical properties of bimolecular lipid membranes. *Biochim. biophys. Acta.* **163**: 226–233.

Schmitt, F. O., and O. H. Schmitt (1940), Partial excitation and variable conduction in the squid axon. *J. Physiol., Lond.* **98**, 28–46.

Taylor, R. E., and W. K. Chandler (1962). Effect of temperature on squid axon membrane capacity. *Biophys. Soc Abstracts,* TD 1.

PAPER 9

Voltage Clamp Data Processing

J.W.MOORE and E. M. HARRIS

Department of Physiology, Duke University Medical Center, Durham, North Carolina

---

INTRODUCTION (J. W. MOORE)

"Kacy" Cole and I spent many hours in developing the voltage clamp for the squid giant axon to the point of being a reliable as well as powerful tool. We sought to obtain as accurate data as possible at as high a rate as possible, in order to avoid long experiments during which the condition of an axon would deteriorate. The net result of successful efforts in this direction was the generation of another problem: how to handle efficiently the vast quantities of data which we were able to obtain with the voltage clamp.

During the early 1960's, I became completely frustrated with the delay and tedium involved in analyzing the large number of voltage clamp curves which had been recorded on photographic film (from a cathode ray oscilloscope screen) for later manual tracing from enlargement. There was such a long delay between experiment and analysis that frequently we never saw the results of a summer's work on squid fully plotted out until late in the fall or winter. Therefore, I looked into alternative methods of having the data available for "instant replay", so that decisions could be made about the direction of the experimental thrust while experiments were still in progress.

Magnetic tape recording was obviously the best candidate for rapid retrieval of data. Inspection of a number of current patterns under voltage clamp conditions indicated that a bandwidth approaching 10 kilocycles was necessary in order to record these signals faithfully. Furthermore, the bandwidth had to extend to zero frequency in order to record the steady current. A frequency-modulated (*FM*) tape recorder was required to satisfy these conditions. There were on the market several *FM* tape recorders capable of providing this bandwidth. However, the speed of the tape required was so high as to allow only a very few records on a tape of modest size. In other words, the packing of the data on the tape would be most inefficient because the event repetition rate was at most to be once per second and the period of the events of interest was only a few milliseconds long. Furthermore, the tape could not be started and stopped promptly when recording at these high speeds. However, the most serious difficulty was that the tape could not be slowed enough to provide a "hard copy" with pen and ink. Slowing the tape as much as one hundred-fold and passing the signal to a fast pen motor strip chart recorder was clearly not a solution. In order to have any resolution in the shape of the pattern, the paper had to be driven at a very high speed. From the duty cycle noted above, of only a few milliseconds of data per second, the paper waste was quite obviously inordinate.

Therefore, I looked to other methods of pre-processing the signal in order to slow it significantly before recording. At that time, small magnetic core memories were becoming commercially available as well as a few analog-to-digital (A/D) and digital-to-analog (D/A) converter components. I decided to convert the analog data to digital from, store it in a buffer memory during the few milliseconds during which data was taken and then the data would be read out from the buffer memory at a much slower rate between voltage clamp pulses, when there was no data of interest to be obtained. The readout speed could be adjusted to complete the reading of the buffer memory just in time for resetting the system for the next data burst. Because I wanted a family of voltage clamp currents drawn with pen and ink, preferably on an X–Y recorder, I chose to convert the digital signal stored in the memory back to analog form before recording on a *FM* tape recorder. The *FM* tape recorder thus could be run with a much slower speed and a greatly reduced bandwidth. I selected a recorder which had a tenfold speed change rather than a binary speed change. Thus, with a 100- to 200-fold slowdown between writing into and reading out of the buffer memory followed by an additional 10-fold slowdown between the speed of recording and playback of the magnetic tape, a net reduction speed of 1000 to 2000 was possible. This brought the signal bandwidth within the capabilities of X–Y plotters. That is to say, 10 milliseconds of data could be plotted in 10 or 20 seconds on the X–Y plotter. Synchronization signals on a separate channel would allow appropriate raising and lowering of the plotter pen and the generation of a sweeping voltage to correspond to time. Additional channels could carry the test potential step, voice, time signals, temperature, etc.

This system was designed and built in the early 1960's and has been used almost continuously since 1966, proving to be most reliable and convenient. It was very helpful to be able to "play back" the results of an experiment within minutes after it was completed so that one could ascertain whether the desired records had actually been obtained or not.

On several occasions Kacy Cole has expressed his keen desire to have the voltage clamp data immediately analyzed into conductance parameters such as those used by Hodgkin and Huxley[1,2]. This system was constructed as one very necessary step in this direction, namely the high speed acquisition of the data. Kacy has expressed his interest in this system several times and requested that it be written up. This chapter is in response to Kacy's request and is dedicated to him for his many efforts in my behalf in the past.

SYSTEM DESCRIPTION

Voltage control and current measurement

The system used is essentially the same as has been previously published on several occasions with Cole and others[3,4,5,6]. Subsequent to the designing of the new data acquisition system, it was clear that the digital logic elements used therein were quite appropriate for controlling the timing and pulse generation in the voltage control circuit. Thus the control and measurement system is simply a modernized completely solid state system, using second generation digital computer logic elements and miniature solid state operational amplifiers.

The only real inovation is a circuit which measures the membrane capacitance during the period between voltage clamp pulses and provides an output voltage directly proportional to the capacitance. One important use of this voltage signal is in compensating for fluctuations in the patterns of solution flow.

The new method of obtaining the capacitance measure signal is a simple and straightforward application of modern instrumentation shown schematically in Figure 1. The equations

$$I_R = \frac{V}{R} \tag{1A}$$

$$I_C = C\frac{dV}{dt} \tag{1B}$$

give the resistive and capacitive components of membrane current. For a sinusoidal voltage across the membrane, the resistive component is in phase with the voltage while that through the capacitor is 90° out of phase. Therefore we simply added a high frequency (about 15 KC) sine wave to the voltage command signals when voltage step response data was not being taken. This represented more than 90% of the time. The observed membrane potential in the voltage clamp was examined by a comparator circuit which tested for a zero crossing and was adjusted to deliver a very short pulse (about 1 microsecond) just at the time of transition. This pulse was then used to track, very briefly, the membrane current.

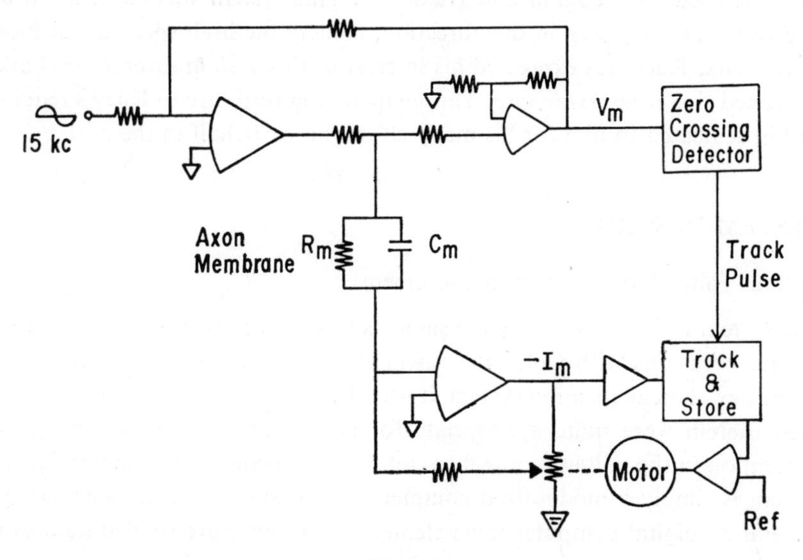

Figure 1

When the sinusoidally driven membrane potentials goes through zero, the resistive component of the current should be zero, and the only current flow should be capacitive, if the phase shifts in the voltage and current circuits have been held to a minimum or properly adjusted. Notive that dV/dt will be a maximum when V is zero; that is to say that the capacitive current flow will be maximum at this time. The value of the membrane current sampled with the 1 μs pulse at the zero corssing time is held for a fraction of a millisecond until the next transition of the membrane potential through zero causes the next tracking pulse to sample the membrane current. Thus one can obtain a voltage signal which is directly proportional to pure membrane capacitive current uncontaminated with ionic components. This is updated more than 10 times per millisecond.

Thus we were able to obtain a signal which rapidly followed small fluctuations in the measured membrane capacitance caused by fluctuations in the exposed artificial nodal area. In order to make compensating adjustments in the gain of the current channel to provide a current density signal independent of the perturbations in the area, this signal voltage was compared with a fixed reference voltage. Any difference was amplified to drive a servo motor which automatically adjusted the current gain by moving a potentiometer wiper making contact at a point proportional to the exposed area of the artificial node in the sucrose gap.

This system performed quite satisfactorily in response to a number of tests in which the exposed area of the artificial node was manually varied over a larger range of values than those which usually occurred during experiments. The observed ionic currents were directly proportional to the exposed area without the servo but, with the servo compensation turned on, were independent of the area.

A/D converter

We considered two types of analog-to-digital (*A/D*) converters from amongst several designs which existed in the early 1960's. The successive approximation and continuous converter types were considered in some detail.

The successive approximation type of *A/D* conversion starts *de novo* after each conversion. It first determines whether the input is greater or less than one-half of full scale, then in sequence, whether a quarter of full scale needs to be added, then an eighth, etc. The total conversion time is the product of the number of binary digits (bits) desired multiplied by the clock time interval (which has to exceed the total time for decision, switching, and settling the feedback resistors). In 1965 the minimum clock interval was about 4 microseconds. This meant that a conversion of analog data to 8 bits (1 part in 256) took 32 microseconds or a conversion to 10 bits took 40 microseconds (resolution of 1 part in 1024).

Furthermore, if the input signal changes significantly during the period of full conversion, there will be a considerable ambiguity and error in the resultant digital output unless the analog voltage has been stored in a "track-and-store" circuit just prior to initiation of conversion.

Such a converter was built and tested for its ability to respond faithfully to squid axon membrane currents. It was found, as expected, that the points were far too sparsely spaced in the rising phase of the sodium inward current to give any fidelity ot the recording.

A continuous *A/D* converter simply adds or subtracts the least significant bit at each decision time (or clock pulse), depending on whether the digital output is higher or lower than the analog input. Because it simply starts from the previous value of the analog input signal, such a converter appeared to have a significant advantage because it did not have to be reset after each conversion.

The successive approximation converter seems to be most appropriate for a system in which different input signals are being looked at successively and there is no prior information as to where any one might happen to be.

In contrast, the continuous converter type would seem to be most appropriate for following a single variable The only problem with the continuous converter is its lack of ability to follow a very rapidly changing input. It normally takes such a converter a long time to go from negative full scale to positive full scale. This equals the time per conversion of each bit multiplied by the total number of increments for full scale. For example, in a full-scale eight-bit converter which we were considering, with a four microconsed bit conversion time, there are some 256 levels and it would require about 1 millisecond for the continuous counter converter to go from negatve full scale to positive full scale. In contrast, a successive approximation converter of similar precision would take only 32 microseconds. Computer simulated responses of these converters to steady and step function inputs are shown in Figure 2. The resolution is very low (1 part in 32) (or 5 bit) but the difference between the logic systems is quite evident. The initial value of the input indicated by dots is small and steady. Just previous to the first simulated output, each type of converter is assumed to have found the correct value of output to match the input. The continuous converter type "hunts" about the input level with each succeeding clock pulse (dot), Figure 2B. The successive approximation converter (in Figure 2A) takes 5 clock periods to arrive at an equal resolution and then starts the whole cycle of search over gain. Although it cannot "settle-down" about a steady value, after a step input change, it only takes 5 clock periods to find the new level in contrast to the sluggish following of the continuous converter.

Thus it appeared that, if we could improve the continuous converter system by making it capable of following rapid transients, we would have both a fast and faithful converter. Therefore, we set out to design and build an additional "hurry-up" circuit which would command the counter to "slew" at a higher rate. It was obvious that this could be done by telling it to increment by bits more significant than the least. It only remained then to determine what sort of criterion was appropriate to initiate this faster response capability. We postulated that a very simple separate short-term memory circuit which kept a record of the counter's recent past performance would be a useful device. Therefore, we inserted into a short (4 bit) shift-register the status of the direction of the count command, "1" for a "count-up" or "0" for a "count-down", as shown in Figure 3. As each new count command was made, the previous commands were shifted on by one step and the latest was entered at the input end of the shift register. Thus the shift register always held the direction of the count for the latest four commands. If the converter was properly balancing the analog input signal, this register would hold an alternating pattern of "1"s and "0"s (representing the previous count-up and count-down commands). On the other hand,

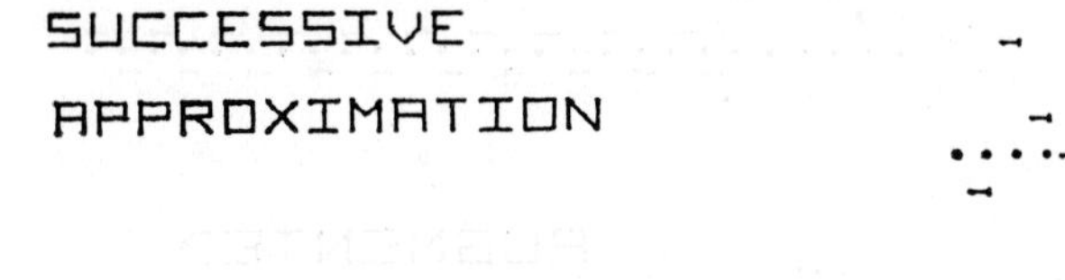

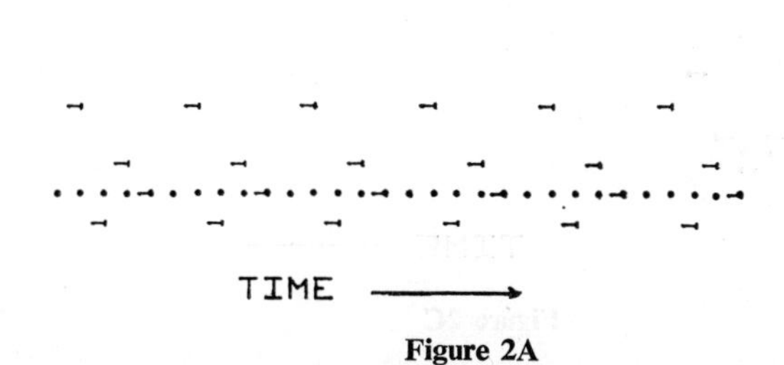

Figure 2A

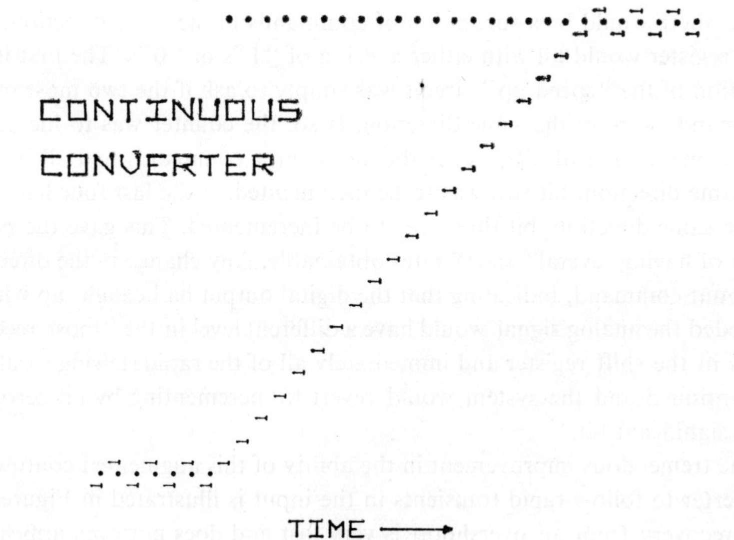

Figure 2B

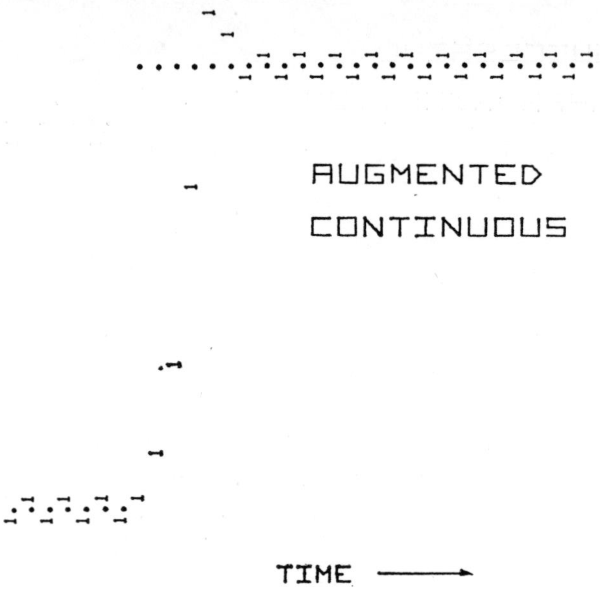

AUGMENTED
CONTINUOUS

TIME →

Figure 2C

if the converter had been sluggish in following a rapidly changing analog input, there would be a succession of commands in the same direction. The shift register would fill with either a string of "1"s or "0"s. The first interrogation of the "speed-up" circuit was simply to ask if the two most recent commands were in the same direction. If so, the counter was to increment by bit one instead of bit zero. If the latest three commands had all been in the same direction, bit two was to be incremented. If the last four had been in the same direction, bit three was to be incremented. This gave the possibility of having several "slew" rates obtainable. Any change in the direction of count command, indicating that the digital output had caught up with or exceeded the analog signal would have a different level in the "most recent" locus in the shift register and immediately all of the rapid slewing would be discontinued and the system would revert to incrementing by bit zero, the least significant bit.

The tremendous improvement in the ability of this augmented continuous converter to follow rapid transients in the input is illustrated in Figure 2C. The recovery from an overshooti is very fast and does not even appear on experimental records (taken at slower sweep rates).

The full analytic expression for frequency response of this converter is complex. However, the fact that it can rapidly adjust to a maximum slewing

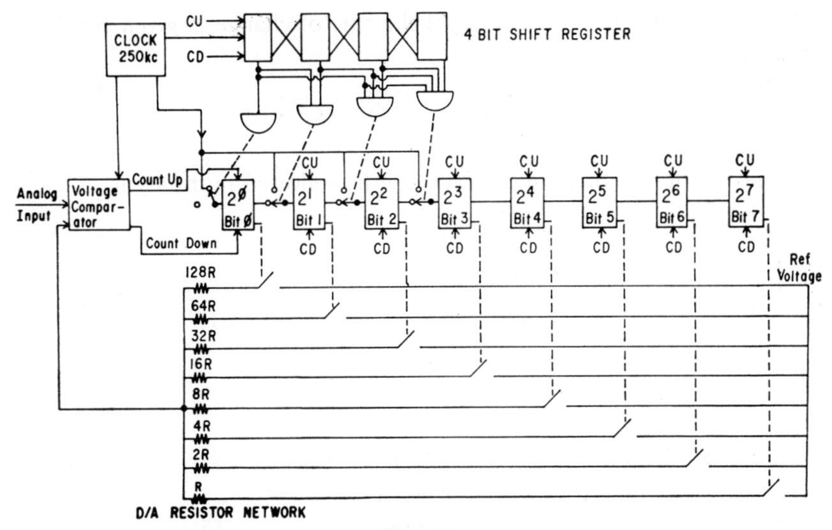

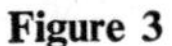

Figure 3

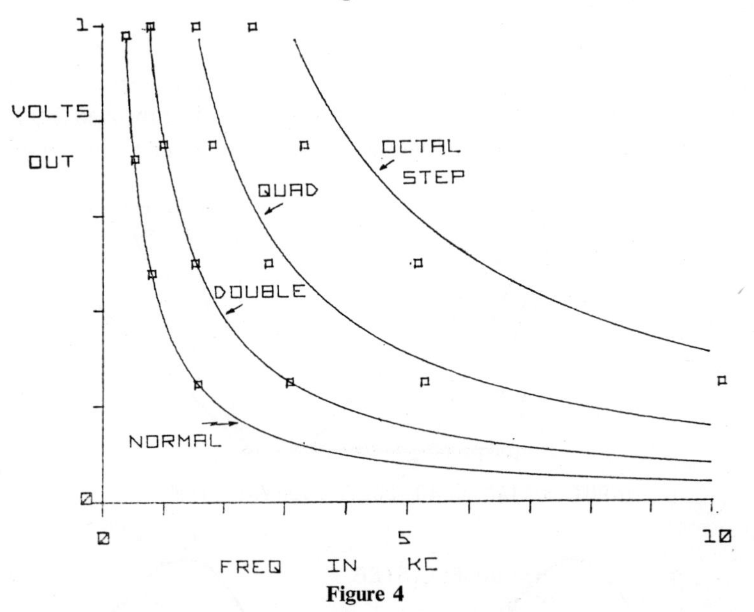

Figure 4

rate, allows us to approximate the cut-off characteristics in a simple way. The basic converter takes 1.024 ms to move the output from minimum to maximum. We used an additional offset signal of one-half full scale so as to provide a measure of both polarities, resulting in a slewing time of 0.512 ms from zero to + or − full scale. At the signal selector switch (Figure 6), full scale was ± 1.25 V, giving an effective input slewing rate (SLR) of $\simeq 2441$ V/sec.

For a sine wave ($Y = A \sin 2\pi ft$) the rate of change is given by the expression

$$\frac{dY}{dt} = A\, 2\pi f \cos 2\pi\, ft. \tag{2}$$

The maximum rate of change which can be followed faithfully is

$$\frac{dY}{dt_{\max}} = \text{SLR} = 2441\ \text{V/S}. \tag{3}$$

Thus the maximum amplitude sine wave which can be followed is

$$A_{\max} = \frac{\text{SLR}}{2\pi f} = \frac{2441\ \text{V/S}}{6.28\, f} \tag{4}$$

Speeding of the response by addition of various levels of "hurry-up" increase the slewing rate by 2, 4, or 8-fold.

Graphs of these frequency response are given in Figure 4 along with experimentally observed points.

The performance of this converter is illustrated in Figure 5 where its response to a short pulse and a sine wave are shown. The speed of catching up to an input jump (smoothest curve) in Figure 5A is improved 2-fold with each additional stage shift of logic to cause the counter to increment at a

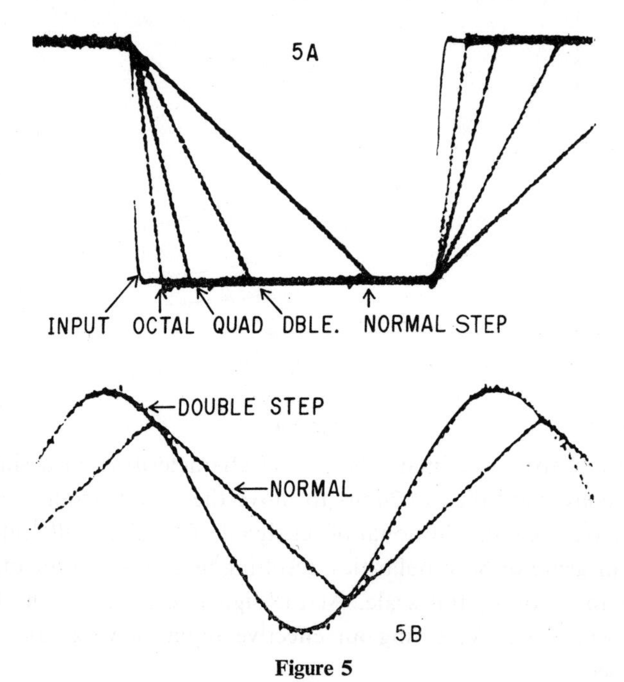

Figure 5

more significant bit. In Figure 5B, it can be seen that although the normal continuous converter cannot follow (it gives a triangular try) a 2 V p–p 1500 cps sine wave, the tracking becomes perfect when the "hurry-up" logic is switched in.

This augmentation or "hurry-up" scheme for the continuous converter was implemented with the addition of a relatively few inexpensive logic elements. It would seem that an extension of this concept to a longer shift register for a larger slewing rates and perhaps different algorithms to select which slewing rate to which to jump would make a very versatile *A/D* converter, not only for following a single rapidly changing signal, but also for multiplexing several signals.

Memory

We selected a magnetic core memory with 4096 words of 10 bits each. This gave the possibility of storing more than enough data points, each with one-part-in-a-thousand resolution. This possible precision was more than we could expect to have in the way of accuracy in original data, and would likely be degraded by the 35 db signal to noise ratio in the tape recorder which was to follow. We wanted to convert the signals from analog-to-digital form at the maximum speed capability of our converter, one conversion every 4 microseconds.* The core memory selected had a five microsecond read and write time, about as fast as any readily available. The write-only cycle took three microseconds; this was adequate for the 4 μs conversion rate because we did not plan to read-out until after all the data was entered.

The rate at which the converter samples were entered into the memory could be selected by a panel switch. It could either be sampled after each conversion or, for more slowly rising currents under some experimental conditions, the rate could be reduced by sampling each eight or sixteen microseconds. The core memory was reset to address zero just prior to the initiation of each data burst. The memory address was incremented by 1 as each data value was entered. In order that a selected number of words of data could be filled, a panel switch was connected to levels which indicated when the address 1024, 2048, or 4096 had been reached. Achievement of this condition set a logic level in order to inhibit further input and initiate readout.

The readout speed was also adjustable by a panel switch. The address could be advanced and read at 800 microsecond, 1.6 or 3.2 millisecond intervals. During the process of readout, a "memory busy" signal was available to

+ The *A/D* converter was interrogated by a pulse from a crystal clock each 4 microseconds.

inhibit any attempt at writing in, if the master repetition rate signal had inadvertantly been set at too high a value.

A table of data times and rates is given in Table 1.

Table 1

Read data in			Write data out			
Sample rate (μs)	Sample length (words)*	Data time (ms)	Write rate (ms)	Write time (sec)	Slow down ratio	Repetition max per sec
4	1000	4	0.8	0.8	200/1	1
4	2000	8	0.8	1.6	200/1	0.5
4	4000	16	0.8	3.2	200/1	0.25
4	1000	4	1.6	1.6	400/1	0.5
4	2000	8	1.6	3.2	800/1	0.25
8	4000	32	0.8	3.2	100/1	0.25

* The 1024, 2048, or 4096 memory addresses are rounded for convenience.

FM tape

The output of the memory was converted back from digital to an analog form with a resistor network similar to that used in the original *A*/*D* conversion. The large value of the input-to-output slowdown (usually 4 μs-to-800 μs), meant that the bandwidth requirements for recording had been reduced by 200-fold. Thus a 10 kilocycle component in the original signal could be recorded faithfully on a *FM* tape with a bandwidth capability of only 50 cycles per second.

We selected a Precision Instrument Type 6200 recorder because it had 3 speeds in decade speed ratios rather than the more usual 2-fold ratios. The intermediate speed of 3.75 inches per second was most convenient for recording allowing a further 10-fold slowdown in playback for *X*–*Y* plotting at 0.375 inches per second. On recording, the bandwidth was restricted to 100 cycles per second instead of the possible 1 kilocycle per second for the recording speed. Thus we were able to im provesignal-to-noise in recording, we did not have to change the bandwidth for playback, and we were even able to insert enough filtering of the signal before plotting to reduce the noise-to-signal ratio considerably before any apparent loss of signal or change in the signal pattern.

Alternatively, experiments could be scanned rapidly by playing back at 37.5 inches per second, speeding up the experimental process by 10-fold and giving the same subjective impression felt in time lapse photography.

A block diagram of the high speed data acquisition system is shown in Figure 6.

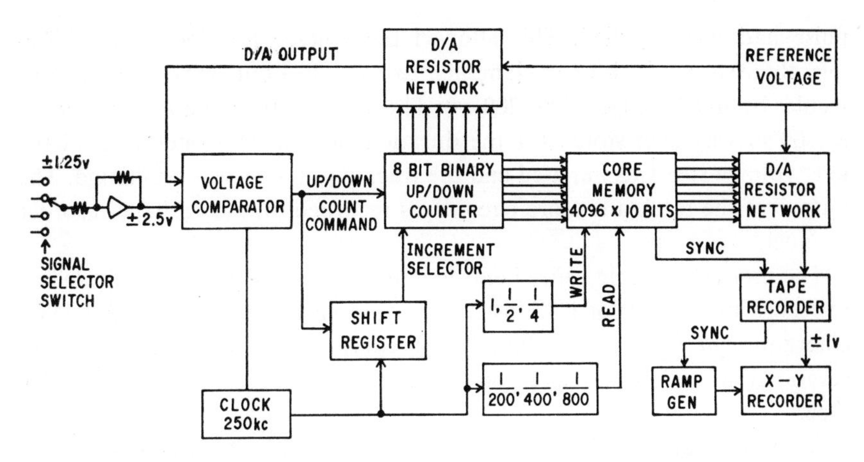

Figure 6

A second channel for synchronization was driven by the "memory unload" pulse so that whenever data was coming to the tape from the memory, the level on the synchronizing channel was driven away from zero. This was very convenient for triggering an oscilloscope sweep circuit, driing an integrator to generate a ramp for the time axis on the *X–Y* recorder, and dropping the pen.

The running time of the experiment was made available on a digital clock which was advanced one digit each ten seconds. With each transferral of data out at the memory, these clock time digits were "jammed" into a shift register in parallel and then the shift register contents were rapidly (within a few milliseconds) transferred serially into another channel (3) of the tape. This period was so short that the time signals and voice comments could share the same channel; the microphone signal was simply inhibited while the time pulse code was being transferred from the shift register. There was no apparent loss of voice information with this time-sharing scheme.

It was also necessary to record on another channel the level or voltage step being applied to the membrane. There were some problems in arranging to record the appropriate value of this voltage:

a) in a way which could be independent of the length of the voltage pulse;
b) independent of any early transient ringing at the beginning of the step;
c) available over a long period of time to activate a slow readout device or to be available for sampling by a computer after the period of data on the current channel had been completed.

These objectives were achieved by inserting a track-and-store circuit which "tracked" the membrane voltage for 200 microseconds after the master pulse had been applied. This allowed the voltage transient ringing, frequently associated with fast clamping, to have died out (most voltage steps would be much longer than 200 μs). The value of the membrane potential at 200 μs was then stored for the duration of that data interval until the next voltage step was initiated and the tracking process was reinitiated. The output of the track-and-store circuit was recorded on the 4th tape recorder channel.

The modest recording speed and high density packing of data allowed us to record on 7″ reels of 1/4″ instrumentation magnetic tape or high fidelity entertainment tape. Many families of curves or the usual equivalent of several hours of experimentation could be recorded on a 1200′ or 1800′ tape.

As soon as an experiment was completed, the tape could be rewound to any desired position, the speed reduced to 0.375″ per second and started. The *X–Y* plotter pen then would be dropped when the synchronizing level was on the tape. This level also initiated the generation of the ramp to drive the plotter horizontally at a rate related to the rate of stepping through the memory addresses. The *Y* axis would be driven by the data signal. This was usually the membrane current density, but other functions, such as the membrane potential and test points in the voltage clamp (appropriately scaled to fit into the *A–D* converter's full scale limits) could be selected for the data channel by push-button switches.

In order to avoid ambiguities and possible manual errors in writing down sensitivity changes, we decided to build a fixed gain into the system as a whole, adjusting the gain to the full scale range of the *A–D* converter. This in turn was attenuated to match that of the tape recorder. The tape recorder had the standard ± 1 volt for full scale input.

The availability of inexpensive logic elements for counters and pulsers also made it simple and feasible to have, in addition to the ordinary manual push buttons for various potential levels, an automatic programmed set of potential levels which could be initiated by pushing a single button. This allowed a family of currents to be taken at a rate of a new value of potential for each pulse of another 1 second clock driven from the 60 cycle power line*. Thus we had some 16 curves recorded within 16, 32 or 64 seconds, depending on the number of data points taken. For a situation in which it was desired to follow the effect of a drug with time, we simply set the pulse

* Synchronization of data taking with the 60 cycle line, causes any noise introduced from the power line to have a fixed phase with respect to the beginning of the data burst. It is easily measured (by insertion of no signal) and thus can be subtracted out readily.

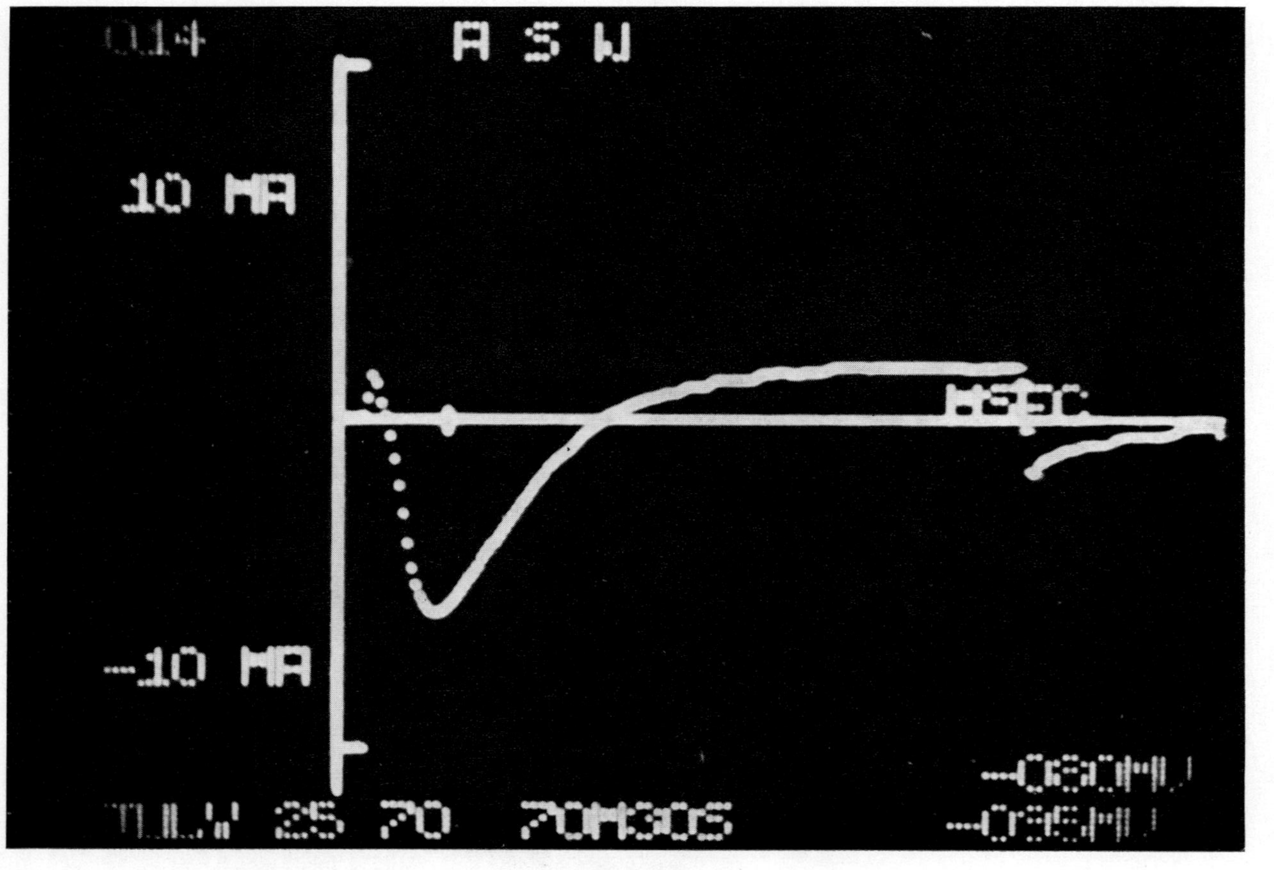

Figure 7

amplitude and duration at fixed values and continued to record the signal with each pulse (usually at a rate of 1 per second). This allowed us to follow rather rapid changes in the membrane characteristics quite conveniently.

COMPUTER ANALYSIS

Not only is it desirable to have a rapid way to view the experimental results but it is also necessary to be able to correct for the leakage current, and to be able to subtract a record under one experimental condition from one under another in order to find the difference current. The most frequent representation of the data is a plot of the peak transient (sodium) and steady state (potassium) currents as a function of voltage. These operations are tedious and very time consuming where a large number of curves is involved.

Automation was obviously the answer. Therefore we chose a LINC-8 computer which had all the necessary features to meet these requirements. Analog tape signals were played back at 3.75 inches per second and monitored by programs in the LINC-8. The LINC-8 sampled the data channel, usually current, at intervals which were specified by the operator and converted these data back again into digital form and stored them on its own small digital tape. At the same time it sampled the voltage record and the time signals. It also requested the date and experimental conditions for each family of experiments; these data were entered by keyboard. It then displayed on the cathode ray screen all the pertinent data that had been read from the analog tape and entered by keyboard. A photograph of the screen is shown in Figure 7. Each record was stored with the ancillary date in a 256 word block on the digital tape. Programs were provided to subtract the leakage current from a given curve, subtract one curve from another, or subtract a family of curves from another family, which had been obtained under different experimental circumstances. The only manual aspect was the selection of the peak height and steady-state values of the current. We decided that because there were some unusual situations, it would be better for the investigator to monitor the values of the peak and steady-state currents that the computer was storing. These values were very rapidly modified by simply rotating two knobs, one for the peak current and one for the steady-state current and were stored as soon as the operator verified that the selected values were appropriate. These values could then be recalled by another program to plot out the peak and steady state currents as a function of voltage. Other programs were written to allow a "hard copy" to be made of any chosen display on the cathode-ray screen by driving either an analog or a digital incremental plotter. A sample is shown in Figure 8.

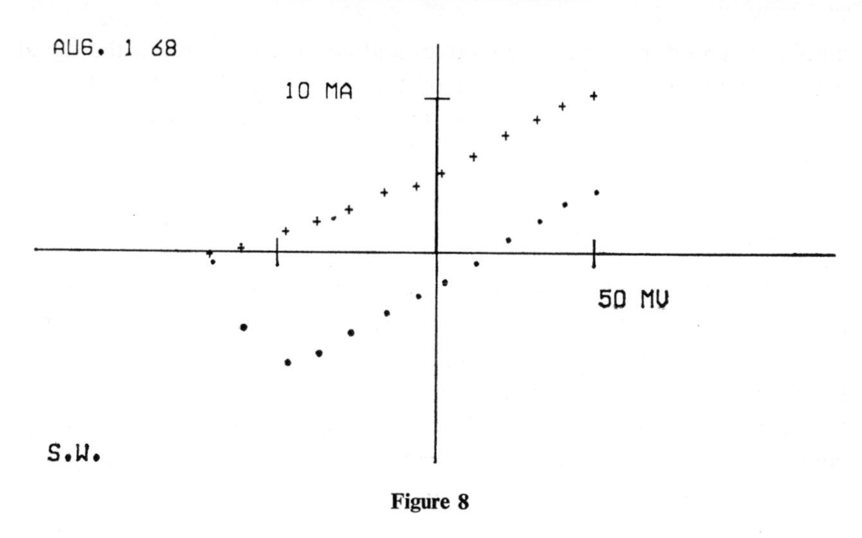

Figure 8

DISCUSSION

The feasibility of implementing such a system in the early 1960's was largely dependent on the vastly improved reliability of the solid state circuit elements over conventional tube type circuits. We used Computer Control Corporation* cards which were reinforced with heavy metal frames which were restrained in metal guide slots with a retaining bar to hold each bay of cards in place. These cards were all discreet component circuitry. The solid state contsruction, heavy-duty card sockets, and hardware made the system virtually impervious to vibration. The double relay rack packed nearly full of these elements has been transported by truck between Durham, N. C. and Woods Hole, Massachusetts for several summers without any problems whatsoever. Adjustment of a few trimming potentiometers for recalibration was all that was needed to have the system completely operational again.

Having such a reliable system with a capability of taking data at a very high rate has allowed us to take and analyze vast quantities of data in a short period each summer. This subsequently revealed the one weak link in this process; namely, that it takes ten times as long to plot a curve on the *X–Y* plotter as it does to generate it in the experiment. Thus, if one takes data for two or three hours during a day from the axon, it takes only 20 to 30 hours to plot this data out on a *X–Y* plotter at night!

Because of this bottleneck in the system, we pushed the LINC-8 to its limit and arranged for it to sample the analog *FM* tape at the normal recording speed of 3.75 inches per second. With this accommodation, it

* Now the Computer Control Div. of Honeywell, Inc., Framingham, Mass.

takes just a bit more than the experimental time to enter the data into the computer. (Some additional time is taken in locating the desired records on the *FM* tape and starting the tape at a point just preceding the data desired.)

The cost and size of miniature computers have now been reduced to such a magnitude that, were we to start to plan a data acquisition system today, we would certainly consider building in a "mini" computer which, in combination with *A/D* and *D/A* converters and a digitally formated magnetic tape, could perform functions of both amplitude and timing control on the generation of pulses for the clamp. It would also have a buffer area in memory into which data would be entered in a burst and then withdrawn at a rate appropriate to go on to the digital tape in a fashion similar to that described for an analog tape system. Such a process would also eliminate the additional time of playing back the analog tape to obtain digital tapes because data would be entered into the digital computer for analysis as the experiment was in progress.* One would probably want to consider having an analog tape recorder with inexpensive magnetic tape functionally in parallel with the digital tape so as to provide rapid and convenient viewing of data in the familiar analog form.

ACKNOWLEDGEMENT

The financial support of NIH in the form of Grant NB 03437 is gratefully acknowledged. We are most appreciative for the superb programs written for the LINC-8 by Mr. Ronald W. Joyner.

REFERENCES

1. Hodgkin, A. L., and A. F. Huxley (1952d). A quantative description of membrane current and its application to conduction and excitation in nerve. *J. Physiol.* **117**, 500–544.
2. Hodgkin, A. L., A. F. Huxley, and B. Katz (1952). Measurejent of current-voltage relations in the membrane of the giant axon of *Loligo*. *J. Physiol.* **116**, 424–448.
3. Cole, K. S., and J. W. Moore (1960). Ionic current measurements in the squid giant axon membrane. *J. Gen. Physiol.* **44**, 123–167.
4. Cole, K. S., and J. W. Moore (1960). Potassium ion current in the squid giant axon: Dynamic Characteristic. *Biophys. J.* **1**, 1–14.
5. Moore, J. W. (1963). Operational Amplifiers. *Physical Techniques in Biological Research* **6**, 77–97. Academic Press.
6. Moore, J. W., and K. S. Cole (1963). Voltage Clamp Techniques. *Physical Techniques in Biological Research* **6**, 263–321. Academic Press.

* The digital tape recorder would also circumvent the problem of drift in the *FM* tape recorder. Although small relative amplitude differences are readily discernible on analog tape, absolute values to within a few millivolts are questionable because of output drifts associated with stopping and starting the analog tape between families of curves.

Comments on the Theory of Ion Transport Across the Nerve Membrane[(1)]

TERRELL L. HILL

Division of Natural Sciences,
University of California, Santa Cruz, California

It is indeed a pleasure to contribute to a volume in honor of Kacy Cole at his seventieth birthday. Kacy is a scientist's scientist; his intelligence, skill, and enthusiasm—in huge doses—are an inspiration to all of us who know him.

Kacy was my boss (Research Director) at the Naval Medical Research Institute in Bethesda from 1949 to 1954. Off and on during that period he tried to interest me in the theoretical aspects of the nerve membrane problem, but my background was too limited to take the idea very seriously.

More recently, when his splendid book *Membranes, Ions, and Impulses*[1] came out, he sent me a copy inscribed with: "I'm expecting you to help with all the problems we've made for you." It has in fact turned that I am now trying to "help", in a necessarily limited sort of way, but it remains to be seen whether any positive contribution will come of the effort.

Despite the fact that I had earlier talks on the subject with Tasaki[2], Goldman[3], and Adam[4], nothing really registered until I saw a preprint (thanks to Max Delbrück) of the paper by Blumenthal, Changeux, and Lefever[5] (BCL). Since I had had previous interests in cooperative and steady-state transport problems, this paper provided a not-too-strange entry into the new area. Then, in a paper[6] which at least served to provide a baptism of fire for us, Chen and I extended the BCL approach to a very simple model for cooperative steady-state K^+ transport when $K_e = K_i$. We were concerned entirely with the K^+ steady-state negative conductance[(2)].

However, having recently acquired a somewhat improved understanding of the Hodgkin-Huxley[7] results (with much help from Kacy and Robert

Blumenthal), it is already clear to us that we will need to modify our model[6] considerably.

As of this writing we are in fact at the "back to the drawingboard" stage. My main purpose in the remainder of these comments will be to give a largely qualitative outline of the model we intend to try next. There is no reason to expect it to survive very long: the odds are very much against any particular modelbuilder making the right guesses, especially at the present stage of ignorance about membrane structure on the molecular level. The eventual successful theory will almost certainly contain a mixture of ideas from various sources.

My main objective with this model is to try to introduce the idea of cooperativity into a modified HH analysis, rather than to abandon completely[2,4,6] HH as a starting point for a molecular theory. There is so much elegance in the HH analysis, it is a little hard to believe that it is entirely emprirical and completely divorced from the molecular realities.

I gratefully accept here the great theoretical simplification which follows from the HH additivity of Na^+ and K^+ currents and from the linear form of the instantaneous currents. However, I introduce cooperativity as a likely aid in explaining the rather sharp potential dependency of the HH parameters n_∞, m_∞, and h_∞. This is, of course, essentially the same reason that Chen and I[6], following BCL[5], introduced cooperativity. That is, the steep K^+ steady-state negative conductance (when $K_e = K_i$) is no doubt closely related to the HH $n_\infty(V)$ function (even though $K_e \ll K_i$ in the latter case).

More or less related ideas on cooperativity have been used or discussed by Adam[4], Blumenthal[5,8], Changeux[5,9], Starzak[10], Tasaki[2], and Wyman[11]. Relevant treatments of other theoretical aspects of the problem are in works by Agin[12], Bass and Moore[13], Cole[1], Goldman[3], Hoyt[14], Mullins[15], and Tasaki[2]. Other articles which I have found especially helpful are by Lehninger[16], Mueller and Rudin[17], Singer[18], and Triggle[19].

I shall confine the present argument almost entirely to the HH normal axon. Variations in inside and outside concentrations, and other complications, will be dealt with later if the model seems to warrant it. Generally speaking, I shall try singlemindedly to discuss what appears to me at the moment to be the "main chance" and not be distracted by controversies and alternatives.

THE MODEL (ESPECIALLY K^+)

The HH additivity of I_K and I_{Na} suggests separate and independent channels for the two ions. The differing mechanisms (Na inactivation) and time constants reinforce this suggestion. I assume further that the channels (or gates) are made of protein molecules or subunits, possibly related in structure to the much-discussed macrocylic compounds[17,20]. These protein channels are presumed to be dispersed[18] in the membrane and possibly to extend from one side of the membrane to the other.

Cooperativity is invoked for the reason already mentioned, but is assumed to be confined *within* each protein channel—which might consist of, say, four to twelve subunits (hemoglobin and some polymeric enzymes serve as models here[21]). That is, there are interactions between subunits of a single channel (or gate) but not between channels. This is "small-system"[22] cooperativity, then, rather than cooperativity in an extended two-dimensional lattice[2,4,5,6,9]. The cooperativity modulates a potential dependent conformational change in the protein subunits of a channel. In turn, conformational change alters the ion permeability values per channel, for simplicity[(3)]).

In order to make these ideas more explicit, let us turn first to the K^+ channels. If we combine the two-state assumption just mentioned[(3)] with the implications of the experimentally observed[1,7] linear instantaneous K^+ current, we are led immediately to the following formulation for I_K (using Cole's notation[1] as a guide):

$$I_K(V, t) = g_K(V, t)\,(V - E_K), \tag{1}$$

$$g_K(V, t) = [1 - p_K(V, t)]\,g'_K + p_K(V, t)\,g''_K, \tag{2}$$

where p_K is the fraction of K^+ channels in the high permeability state (i.e., $g''_K \gg g'_K$), which we shall still[6] refer to as state II (p_K corresponds to our earlier[6] Pc). The low permeability state is I. The essential points are that (a) g'_K and g''_K are steady-state values which have been reached in tens of microseconds (whereas the time scale we are really concerned with here is milliseconds) and (b) that g'_K and g''_K are independent of V as well as of t.

The adjustment of the electrostatic potential and of ion distributions and fluxes within a channel must be very fast compared to the rate determining process implicit in $p_K(V, t)$—which I take to be some kind of cooperative macromolecular conformational change in the channel.

In general we would except g'_K, g''_K, and $p_K(V, t)$ to depend on inside and outside ion concentrations, though the dependence may not be strong in some cases (E_K of course depends, via the Nernst relation, on K_e and K_i). There is also the question of the extent of ion concentration variation over which linear instantaneous currents are still observed[23,24]. Linearity *is* found in the $K_e \cong K_i$ case[23].

On combining Equations (1) and (2), we have

$$I_K(V, t) = [1 - p_K(V, t)]\,g'_K(V - E_K) + p_K(V, t)\,g''_K(V - E_K). \tag{3}$$

The pure state I flux is $g'_K(V - E_K)$ and the pure state II flux is $g''_K(V - E_K)$. These are "instantaneously" achieved steady-state *linear* relations. An unsatisfactory feature of the model used in our earlier paper[6] is that it did *not* give linear "pure" fluxes.[(2)] With linear fluxes we have the very considerable simplification that, as seen in Equation (2), experimental values of g_K can be translated almost immediately into values of p_K.

Of course, $I_K(V, t)$ in Equation (3) is in general not linear in V (e.g., at steady-state), because p_K is a function of V.

The theoretical problem (for K^+; Na^+ is similar) thus breaks up into two subproblems, though they are unfortunately interrelated: one has to do with conformational kinetics (the relation between p_K and HH); the other is the question of the origin of "pure" steady-state fluxes which are sometimes linear over as much as 200 mV. The same molecular model of a K^+ channel has to do both jobs. The second subproblem is not necessarily trivial: in *molecular* models of ion flux, V tends to appear ubiquitously[(2)] in exponentials—which cannot be linearized over such a large range in V. In the rest of this paper I shall discuss the first subproblem only, except for the following remark: continuum treatments of electrostatic potential and ion flow through a slab (representing the membrane) are seemingly inapplicable to the present model because the transport is confined here to isolated protein channels imbedded in a membrane matrix (with presumably quite different properties).

In a typical instantaneous (K^+) current experiment, the membrane is held at, say, $V = V_0$ long enough so that only K^+ ions contribute to the current. Then I_K is measured after a sudden switch (at, say, $t = t_0$) from V_0 to V, but before there is time for conformational relaxation. This "instantaneous" current is then linear in V,

$$I_K(V) = \{[1 - p_K(V_0, t_0)]\, g'_K + p_K(V_0, t_0)\, g''_K\}\, (V - E_K), \tag{4}$$

because the conductance $\{\,\}$ does not depend on V: it is a property of the initial state (V_0, t_0). But, as time passes, $\{\,\}$ becomes a function of V and $t < t_0$, and evolves its toward its steady-state value

$$\{[1 - p_K(V, \infty)]\, g'_K + p_K(V, \infty)\, g''_K\}. \tag{5}$$

The explicit connection with the HH formulation may be seen from

$$I_K(V, t = g'_K(V - E_K) + p_K(V, t)\,(g''_K - g'_K)\,(V - E_K). \tag{6}$$

In the normal HH axon, $g'_K \cong 0$ and g''_K is the same as the HH $\bar{g}_K$. The probability p_K is then to be associated with the HH n^4.

ROLE OF COOPERATIVITY[4]

It might be helpful to begin by simply reinterpreting the HH expression $g_K = \bar{g}_K n^4$ in the present context, as follows. The K^+ channel contains four identical and non-interacting protein subunits, each of which can be in one of two conformations, *i* and *ii*. The probability that a subunit is in conformation *ii* is n. There is room for a K^+ ion to pass through the four subunits (arranged in a square or tetrahedron?) only if *all* are in conformation *ii*, with probability $p_K = n^4$ (i.e., any one subunit in conformation *i* blocks the channel). This is "state II" of the channel; all other subunit arrangements are "state I". In addition, the conformational change follows first-order kinetics and n_∞ turns out to be (in order to match the experimental results) strongly dependent on V.

However, independence of the subunits seems unlikely (compare hemoglobin). Furthermore, the steepness in $n_\infty(V)$ and in the $K_e = K_i$ staedy-state negative conductance region of $I_K(V)$ suggest cooperativity. Thus it would appear worthwhile to explore generalizatio nsof the (reinterpreted) HH formulation in which there are, say, k_n ($4 \leqq k_n \leqq 12$?) subunits which interact cooperatively. Chen and I have just started to do this.[4] The interactions will influence the kinetics as well as the steady-state. In particular, cooperativity will tend to extend the "induction" period in g_K (before the rise) follwoing a sudden depolarization—this being the "small-system" equivalent of the nucleation time of a new phase (I → II).[5]

As a simple explicit *illustration*, let us consider $k_n = 4$ with tetrahedral symmetry. Let α_n be the "intrinsic" (i.e., for isolated subunits) rate constant for $i \infty ii$ and β_n for $ii \infty$ i. Let $[j_n]$ represent a K^+ channel with j_n subunits in state *ii* ($0 \leqq j_n \leqq k_n$), and let p_{j_n} be the probability of $[j_n]$. Then if we niclude interactions in the kinetics in the simplest (symmetrical) way possible, the kinetic scheme is

$$[0] \underset{\beta_n y^{-3}}{\overset{4\alpha_n y^3}{\rightleftarrows}} [1] \underset{2\beta_n y^{-1}}{\overset{3\alpha_n y}{\rightleftarrows}} [2] \underset{3\beta_n y}{\overset{2\alpha_n y^{-1}}{\rightleftarrows}} [3] \underset{4\beta_n y^3}{\overset{\beta_n y^{-3}}{\rightleftarrows}} [4], \tag{7}$$

where

$$y = e^{w/4kT}, \; w = w_{11} + w_{22} - 2w_{12}, \tag{8}$$

and w_{11} is the interaction energy between two subunits in state *i*, etc. We are especially interested in the case where $w < 0$ and $y < 1$. We assume the only channel state that conducts is [4]; hence $p_K = p_4$. That is, II = [4], I = [0] + ··· + [3]. This is precisely the HH scheme if $w = 0$ (and $y = 1$). The intrinsic constants α_n and β_n are functions of V. We assume (for simplicity) that y is independent of V, as would be the case, for example, of the predominant interaction is simply steric repulsion (poor fit) between a subunit in state i and another in state ii ($w_{12} < 0$).

From an empirical point of view, in this particular scheme with $k_n = 4$, we are simply adding a new parameter y to the HH K^+ treatment, with HH corresponding to a special case, $y = 1$. If we adjust y for best fit of the kinetic and steady-state data, we might expect improvement (for example, in the induction period[7]).

The conformational equilibrium distribution ($t = \infty$), corresponding to (7), is easily found to be

$$p_0^e = Q_n^4/\xi, \quad p_1^e = 4y^6 Q_n^3/\xi, p_2^e = 6x^8 Q_n^2/\xi$$
$$p_3^e = 4y^6 Q_n/\xi, pe = p_e^4(V, \infty) = 1/\xi \tag{9}$$

where

$$\xi = Q_n^4 + 4y^6 Q_n^3 + 6y^8 Q_n^2 + 4y^6 Q_n + 1$$
$$Q_n = \beta_n/\alpha_n. \tag{10}$$

The quantity ξ is a kind of statistical mechanical grand partition function. $Q_n(V)$ corresponds to y^{-1} (*not* the same y) in Equation (12) of our earlier paper[6]. Thus $Q_n(V)$ is the (intrinsic) i/ii partition function ratio at V—*not* including interactions between subunits (but induding binding of H^+, K^+, Ca^{++}, etc.).

When y is small, channel states [0] and/or [4] predominate at equilibrium. When V is such that $Q_n \gg 1$ (e.g., at the resting potential), [0], is the main state; when V is such that $Q_n \ll 1$ (e.g., at $V = 0$), [4] is the important state. In a sudden depolarization at $t = 0$ from, say, $V = -65$ mV to $V = 0$, the system would evolve from a $t = 0$ distribution given by Equation (9) with $Q_n(-65) \gg 1$ to a $t = \infty$ distribution determined by the same equations with $Q_n(0) \ll 1$. The kinetics for $t \geqq 0$ would be governed by (7) with $\alpha_n(0)$ and $\beta_n(0)$, where $\beta_n(0)/\alpha_n(0) = Q_n(0)$.

One can indicate schematically a more general formulation ($k_n > 4$) as follows:

$$[0] \rightleftarrows [1] \rightleftarrows \cdots \rightleftarrows [k_n - 1] \rightleftarrows [k_n]. \tag{11}$$

The high permeability state II might contain more than one channel state, for example, II $= [k_n - 1] + [k_n]$. Or, one might use, say, a three-state model (I, II, III)—with III $= [k_n]$ conducting better than II $= [k_n - 1]$; etc.[(3)] In the extreme, a linear gradation in conductance from [0] to $[k_n]$ would bring us back to a model formally indistinguishable from one in which individual subunits (in patches of size k_n) conduct[6]. Actually, all of the possibilities mentioned in this paragraph would seem to be ruled out by the HH observation of K^+ kinetic asymmetry: there is an induction period following depolarization, but a first-order process occurs after repolarization. This strongly suggests[(3)] that II is a single end-of-the-line state, II $= [k_n]$. The two-dimensional model (infinite, or in small patches) used by Hill and Chen[6] has the fault of predicting essentially symmetrical kinetics following depolarization and repolarization.

The lengthened induction period when depolarization follows hyperpolarization, observed by Cole and Moore[25], might be accounted for by (11), that is, by $k_n > 4$ combined with cooperativity. Suppose, say, $k_n = 8$, II $= [8]$, and at the resting potential there is an appreciable population of, say, [0], [1], [2], and [3], while $p_K = p_8 \cong 0$. Progressive initial hyperpolarization would successively eliminate [3], [2], and [1] from the $t = 0$ population—thus steadily lengthening the depolarization induction period.[(5)] An alternative but similar kind of possibility is that, say, $k_n = 4$ but each subunit has three conformations, 0, *i*, and *ii* (rather than only two), with 0 most stable at large hyperpolarizations, *i* most stable near the resting potential, and *ii* most stable near $V = 0$. Only the all-*ii* state conducts. Under normal HH conditions, only *i* and *ii* are involved. But initial hyperpolarization would interpose, after depolarization, prior (probably cooperative) steps involving $0 \rightarrow$ i. I am not too confident these schemes, with suitable parameters, could lead to a simple time delay with no appreciable effect on the shape of the subsequent $I_K(t)$ curve[25], but they seem worth trying out.[(6)]

STEADY-STATE NUMERICAL EXAMPLES

I would like to present in this subsection a few numerical results, for staedy-state, to illustrate some of the points made above. The tetrahedral $k_n = 4$ case will be used, but this particular choice should not be taken very seriously.

The function $n_\infty(V)$ is defined empirically from experimental $g_K(V, \infty)$ values by the HH relation $g_K = \bar{g}_K n_\infty^4$. The V-dependence in $n_\infty(V)$ is fairly strong (Figure 1). We might note in passing (Figure 1) that this property is not altered much by a switch to $g_K = \bar{g}_K N_\infty^6$ (i.e., $N_\infty - n_{2/3}^\infty$). If we assume that there are no interactions between subunits ($y = 1$), as implied by the HH formulation, then the steepness in the $n_\infty(V)$ function has to be accounted for entirely by a molecular model for α_n and β_n, since $n_\infty = \alpha_n/(\alpha_n + \beta_n)$. If, on the other hand, we assume that there *are* cooperative interactions ($y < 1$), then—as we shall see below—the "steepess" required of $\alpha_n/(\alpha_n + \beta_n)$, in order to account for the experimental $g_K(V, \infty)/\bar{g}_K$, is reduced. One hopes that this will present a less serious problem for a molecular model.

To see this effect of cooperativity, we define, for any y,

$$n'_{\infty}(V) = \frac{\alpha_n(V)}{\alpha_n(V) + \beta_n(V)} = \frac{1}{1 + Q_n(V)}\,, \tag{12}$$

where $Q_n(V)$ is related to the HH $n_{\infty}(V)$ by [see Equations (9) and (10)]

$$g_K(V, \infty)/\bar{g}_K = n_{\infty}^4 = (Q_n^4 + 4y^6Q_n^3 + 6y^8Q_n^2 + 4y^6Q_n + 1)^{-1}. \tag{13}$$

For a specified y, this equation determines Q_n as a function of n_{∞}.

The physical significance of n_{∞} is that this would be the steady-state fraction of subunits of type *ii* were it not for interactions. That is, $n_{\infty} = \alpha_n/(\alpha_n + \beta_n)$ is an "intrinsic" conformational equilibrium property—with interaction effects separated out.

Equations (12) and (13) determine n_{∞} as a function of n_{∞}, for any specified y (Figure 2). In the HH case (no cooperativity), $y = 1$ and $n_{\infty} = n_{\infty}$. Extreme cooperativity is represented by the limit $y = 0$ ("all or none" channels: all subunits in conformation i or all in ii). Cooperativity leads to $n_{\infty} < n_{\infty}$ (Figure 2). The significance of this is more obvious from Figure 1: when cooperativity is introduced (compare $y^2 < 1$ with $y^2 = 1$), less steepness is demanded of the $n_{\infty} = \alpha_n/(\alpha_n + \beta_n)$ curve in order to reproduce the experimental function $g_K(V, \infty)/\bar{g}_K$. Some of the burden, so to speak, of the observed steepness of $g_K(V, \infty)$ has been removed from $\alpha_n/(\alpha_n + \beta_n)$ by the introduction of cooperativity between subunits. In the HH treatment ($y = 1$), the entire burden rests on $\alpha_n/(\alpha_n + \beta_n)$.

I have included, in Figures 1 and 2, curves for $y^2 = \frac{1}{2}$ in the case of square symmetry, for which

$$\xi = Q_n^4 + 4y^4Q_n^3 + (4y^4 + 2y^8)\,Q_n^2 + 4y^4Q_n + 1. \tag{14}$$

Also, it should be noted that the $y^2 = 0$ and $y^2 = 1$ curves are the same for square and tetrahedron.

Along the lines of "explaining" $n'_{\infty}(V)$ curves by molecular theory, let us pursue this rather arbitrary numerical example ($k_n = 4$, tetrahedron) one step further: we attempt to fit the $n'_{\infty}(V) = [1 + Q_n(V)]^{-1}$ curves in Figure 1 using for Q_n an essentially empirical function of the form[6] $Q_n(x) = Qe^{\delta x}\, e^{8x^2}$ (see also the Discussion, below), where $x = \text{g}V/kT$ and $\varepsilon =$ unit positive charge (e.g., on K^+). Since there are three constants (Q, δ, γ) in this function, we may adjust their values, for each y^2, so that $[1 + Q_n(x)]^{-1}$ agrees exactly with $n'_{\infty}(V)$ at three values of V, chosen to be $V = -65, -15$, and $+50$ mV. Calculated values of $[1 + Q_n(x)]^{-1}$, obtained in this way, are compared in Figure 3 with $n_{\infty}(V)$ curves taken from Figure 1. The points match the curves fairly well for $0.7 \leqq y^2 \leqq 1.0$. The agreement, however, is less satisfactory if one compares corresponding values of $p_{\frac{e}{4}}(x)$ [see Equation (9)],

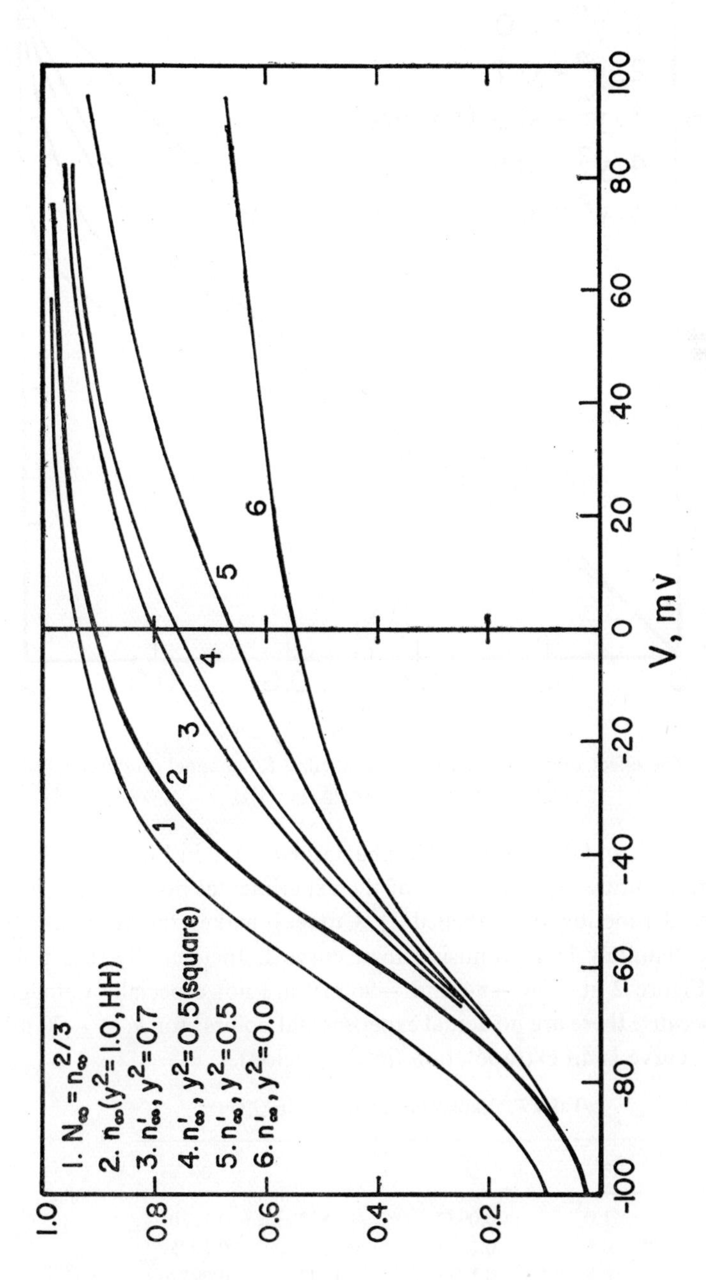

Figure 1 The effect of cooperativity in a tetrahedral K^+ channel on $n=\alpha_n/(\alpha'_{\infty n}+\beta_n)$

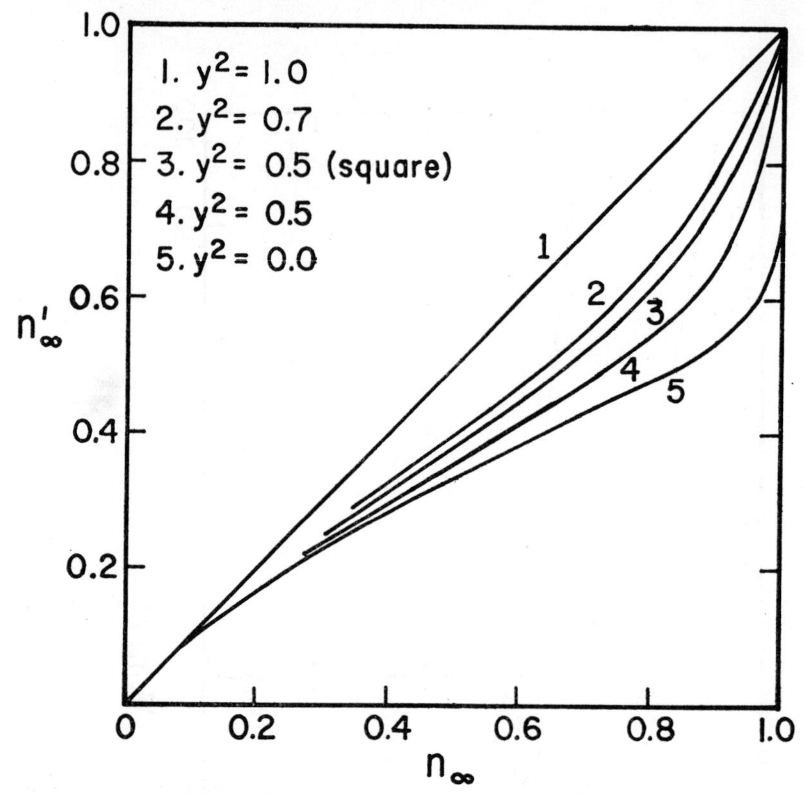

Figure 2 The effect of cooperativity in a tetrahedral K^+ channel on the function $n'_{\infty}(n_{\infty})$, where $n_{\infty} = (g_K/\bar{g}_K)g$ (HH)

for $y^2 = 1.0$, 0.7, and 0.5, with $n^4_{\infty}(V)$. The dashed curve in Figure 3, labelled 1, is a more accurate representation of the experimental points[7] than is the HH empirical function n_{∞} (labelled 2). Curve 1 makes the trial function $Q_n(x)$ look definitely less promising than curve 2. Incidentally, the points shown in Figure 2 at $V = -84$ and -96 mV are not especially damaging for $Q_n(x)$ because there are no actual experimental points[7] for $V < -70$ mV; the HH n_{∞} curve is an extrapolation (in this region).

Table I Values of parameters in $Q_n(x)$

y^2	Q	δ	γ
1.0	0.0933	−0.827	0.1220
0.9	0.1254	−0.792	0.1075
0.8	0.1723	−0.747	0.0912
0.7	0.242	−0.692	0.0746
0.5	0.465	−0.532	0.0567

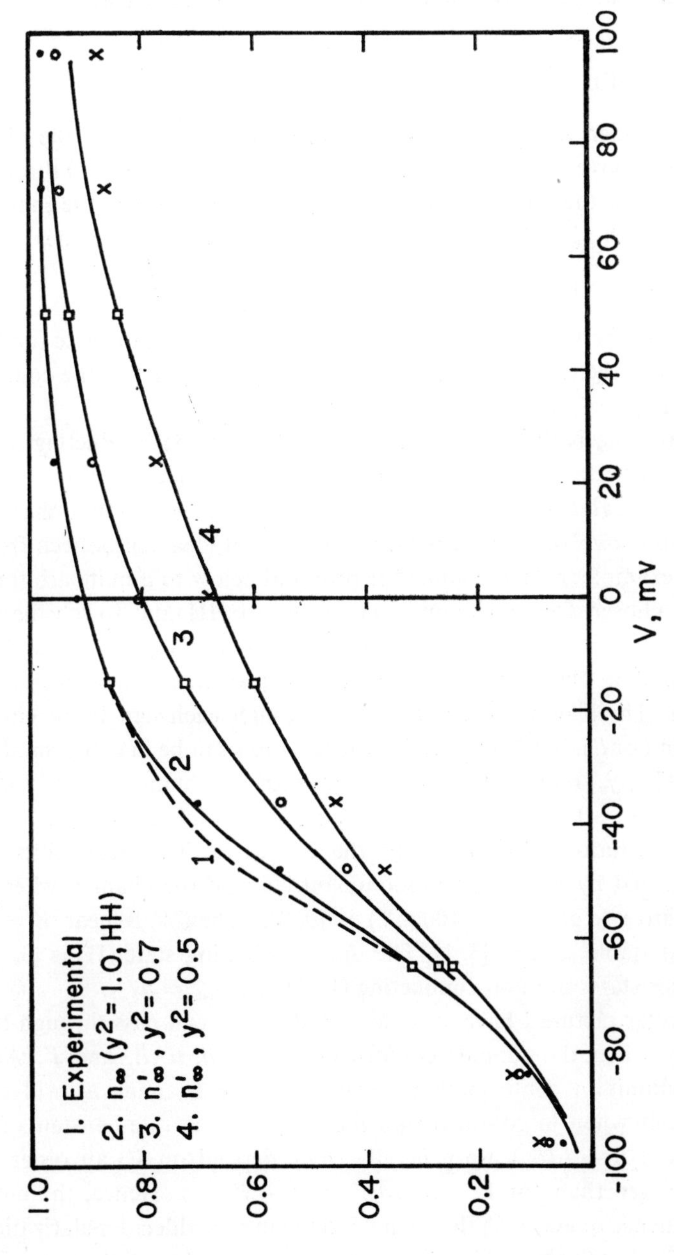

Figure 3 Attempts to fit $n'_{\infty}(V)$ curves from Figure 1 with the trial function $Q_n(x) = Qe^{\delta x}\, e^{\gamma x^2}$. Points are from the trial function (which is forced to agree with n'_{∞} at the squares)

Table I contains the values of Q, δ, γ referred to above. Their magnitudes are similar to those used earlier by Chen and me[6] but the meanings here are somewhat different [see above and Equation (12) of reference 6].

THE Na^+ MODEL

My remarks under this heading will be very brief because of the close relation to the above discussion of K^+. In particular, Equations (1) through (6), and comments thereon, apply to the Na^+ channel with very little change. For example, with $g'_{Na} \cong 0$, as in the HH case, Equation (6) becomes

$$I_{Na}(V, t) = p_{Na}(V, t)\, \bar{g}_{Na}(V - E_{Na}). \tag{15}$$

The difference between K^+ and Na^+ transport resides not in these formal equations but in the function $p_{Na}(V, t)$, that is, in the kinetics of the conformational change.

The quantity p_{Na} is the fraction of Na^+ channels in the conducting state II, and is to be identified with the HH empirical function m^3h. Again, I reinterpret the HH kinetic scheme in terms of conformational changes which probably involve cooperativity.[7] The general idea can be seen from a particular example: I shall confine the proposal below to a quite arbitrary special case, chosen for maximum resemblance to HH and to scheme (7) for K^+.

Suppose that a Na^+ channel contains four protein subunits arranged tetrahedrally. Three of these are identical (though each can be in either conformation *i* or *ii*), but the fourth is different (and can be in conformation *i'* or *ii'*). Let $[j_m, j_h]$ represent a Na^+ channel with j_m subunits in state *ii* and j_h in state *ii'*, where $0 \leqq j_m \leqq k_m$, $0 \leqq j_h \leqq k_h$, and, in this example, $k_m = 3$, $k_h = 1$ (this notation conforms with the K^+ case). The probability of $[j_m, j_h]$ is denoted by $p_{j_m \gamma_h}$. The predominant state of the channel when V is quite negative (e.g., $V = -100$ mV) is [0, 0]. When V is near $V = 0$, say, the most stable state is [3, 1]. The only conducting state (II) is [3, 0]; all seven other states are non-conducting (I). Hence $p_{Na} = p_{3,0}$.

The molecular picture I have in mind is that a Na^+ can pass through the channel only when the subunit conformations are *ii*, *ii*, *ii*, and *i'*. Any unprimed subunit in conformation *i* will block the channel, as will the primed subunit when in conformation *ii'*. The intrinsic rate constants for $i \rightarrow ii$ (α_m) and $ii \rightarrow i$ (β_m) are generally (they depend on V) an order of magnitude larger than for $i' \rightarrow ii'$ (β_h) or $ii' \rightarrow i'$ (α_h). Hence, the most common sequence of events in the channel following a sudden depolarization from, say, $V = -65$ mV to $V = 0$, is conversion of all $i \rightarrow ii$ (thus opening the channel) *before* $i' \rightarrow ii'$ (blocking or "inactivating" it again).

The above is identical with the HH picture, except for the reinterpretation in terms of conformational kinetics. However, I anticipate, as in the K^+ model, that cooperativity may be involved.[7] For example, if we assume that optimal fit of subunits occurs in the two extreme states ("phases") [0, 0] and [3, 1]—as in the K^+ model—then the analogue (and generalization) of (7) is

$$
\begin{array}{ccccccc}
[0,0] & \underset{\beta_m z_1^{-2} z_2^{-1}}{\overset{3\alpha_m z_1^2 z_2}{\rightleftarrows}} & [1,0] & \underset{2\beta_m z_2^{-1}}{\overset{2\alpha_m z_2}{\rightleftarrows}} & [2,0] & \underset{3\beta_m z_1^2 z_2^{-1}}{\overset{\alpha_m z_1^{-2} z_2}{\rightleftarrows}} & [3,0] \\
\alpha_h z_2^{-3} \uparrow\downarrow \beta_h z_2^3 & & \alpha_h z_2^{-1} \uparrow\downarrow \beta_h z_2 & & \alpha_h z_2 \uparrow\downarrow \beta_h z_2^{-1} & & \alpha_h z_2^3 \uparrow\downarrow \beta_h z_2^{-3} \\
[0,1] & \underset{\beta_m z_1^{-2} z_2}{\overset{3\alpha_m z_1^2 z_2^{-1}}{\rightleftarrows}} & [1,1] & \underset{2\beta_m z_2}{\overset{2\alpha_m z_1^{-1}}{\rightleftarrows}} & [2,1] & \underset{3\beta_m z_1^2 z_2}{\overset{\alpha_m z_1^{-2} z_2^{-1}}{\rightleftarrows}} & [3,1]
\end{array}
\tag{16}
$$

where $z_1 \leqq 1$ and $z_2 \leqq 1$ correspond to y (z_1 relates to interactions between pairs of unprimed subunits and z_2 to interactions between the primed subunit and the three unprimed ones). The HH case is $z_1 = z_2 = 1$.

A typical kinetic problem which arises in relation to (16) is the following: one starts with the equilibrium distribution among the eight states corresponding to, say, $V = V'$ (the four rate constants are functions of V) and then determines the time dependence of $p_{Na} = p_{3,0}$ following a sudden change in V to $V = V''$. One would often be able to use the approximation that the dominant states in the kinetic process are [0, 0], ..., [3, 0], [3, 1].

The equilibrium distribution, corresponding to Equation (9) and (10), is

$$
\begin{gathered}
p_{0,0}^e = Q_m^3 Q_h/\xi, \quad p_e^{1,0} = 3z_1^4 z_r^2 Q_w^3 Q_h/\xi, \\
p_{0,2}^e = 3z_1^4 z_2^4 Q_m Q_h/\xi, \quad p_{3,0}^e = p_{Na}(V, \infty) = z_2^6 Q_h/\xi, \\
p_{0,1}^e = z_2^6 Q_m^3/\xi, \quad p_{1,1}^e = 3z_1^4 z_2^4 Q_m^2/\xi, \; p_{2,1}^e = 3z_1^4 z_2^2 Q_m/\xi, \\
p_{0,3}^e = 1/\xi,
\end{gathered}
\tag{17}
$$

where ξ is the sum of the eight numerators in Equation (17) and

$$
Q_m = \beta_m/\alpha_m, \quad Q_h = \alpha_h/\beta_h. \tag{18}
$$

In the HH case ($z_1 = z_2 = 1$),

$$
p_{3,0}^e = p_{Na}(V, \infty) = \left(\frac{\alpha_m}{\alpha_m + \beta_m}\right)^3 \left(\frac{\alpha_h}{\alpha_h + \beta_h}\right) = m_\infty^3 h_\infty. \tag{19}
$$

DISCUSSION

I shall close with brief commets on several topics.

(1) The essential point in this paper is the suggestion that the HH school and the "cooperativity" school of theoreticians may have a model in

common, after all. I have proposed here the outlines of such a hybrid model, but it is very incomplete. Even if an analysis of Na^+ and K^+ kinetic and steady-state data lends support to the idea of cooperativity within channels, the model is still confronted with two key problems: (a) it must explain *linear* steady-state flux-potential curves for a K^+ or Na^+ channel in a "pure" state; and (b) it must explain the V-dependence of the intrinsic rate constants[(8)] (modified—as in the K^+ discussion—by taking cooperativity into account). Then, of course, one has to face—if the model still survives—the extensive experimental results obtained under non-HH conditions.

(2) An apparently common point of view is that excitability is a secondary property which is a consequence of the primary property of negative conductance (observed with Na^+ under normal conditions and with K^+ when $K_e \cong K_i$]. My view is that excitability and negative conductance are both secondary properties; the primary property is the rather sharp V-dependence ("two-states" transition) of m_∞, n_∞, and h_∞ (or their "cooperative" counterparts). The origin of either the Na^+ or K^+ ($K_e = K_i$) negative conductance as a secondary property, is easily seen from Cole's[1] Figure 3:36.

(3) As of this writing I am completely open-minded (that is to say, in the dark) about the nature of the assumed conformational changes in the subunits of the K^+ and Na^+ channels as well as about the membrane-potential triggering mechanism[(8)] (i.e., how and why the conformational equilibria—expressed, say, by Q_n, Q_m, ynd Q_h—depend on V). In this connection, it is essential to remember that one needs a *differential* effect of membrane potential or electric field. That is, the free energy *difference* between the two conformations must vary with V. For example,

$$Q_n(x) = i/ii = e^{-\Delta G(x)/kT}$$

$$\Delta G(x) = G_1 - G_2 = \Delta G(0) + [\Delta G(x) - \Delta G(0)], \tag{20}$$

where G_1 is the free energy of one subunit in conformation i, etc., and $x = \varepsilon V/kT$. In the case $Q_n = Qe^{\delta x} e^{\gamma x^2}$, discussed above,

$$Q = e^{-\Delta G(0)/kT}, \quad \Delta G(x) - \Delta G(0) = -(\delta x + \gamma x^2). \tag{21}$$

Some possible triggering effects depend on the electric field strength and others on the electrostatic potential in the channel. In either case knowledge of the membrane potential V is not enough: the way the potential varies through the channel is needed.

With the above reservation in mind, let us assume here for simplicity that the electric field strength is roughly proportional to V (or x). I also assume, as a reasonable guess, that a subunit conformational change involves a shift in the tertiary structure of the subunit and/or a small rotation or movement of the entire subunit. But, in a given conformation, the subunit is considered essentially fixed in structure, location, and orientation—except for possible local responses of charged groups to the field (as in (a), below).

The following, then, is at least a partial list of possible triggering mechanisms:

(a) *Net charge* This relates to fixed charges and to charges dependent on proton binding (e.g., —NH_2^+ and —COO^-). Chen and I have discussed this subject elsewhere[6]. If the conformational change moves these charges into appreciable different electrostatic potentials, or alters significantly the amount of proton binding, the approximate effect will be a contribution to the free energy difference of the form $-\delta x$, as in Equation (21). If, because of the charges, the field distorts the local structure of the conformations (with restoring forces), a term $-\gamma x^2$ would arise here.

(b) *Polarizability* The two conformations may have different polarizabilities (or may be located at different values of E) and hence different free energies in an electric field. Proton mobility[6,26] would seem to be the most likely source for this with the difference in polarizability arising either from somewhat different rotational orientations of an asymmetric subunit in the electric field or from different amounts of proton binding. This effect would lead, approximately, to a term of the form $-\gamma x^2$ in Equation (21).

Chen and I[6] were attracted to a term of this type because of definite indications, found by Gilbert and Ehrenstein[27], of a negative conductance region in steady-state $I_K(x)$ curves at positive x as well as negative x, when $K_e = K_e$. In fact, curve IV in their Figure 2 shows corresponding behavior with *normal* Kt (g_K reaches a maximum at about $V = 80$ mV and then decreases for $V > 80$ mV; the HH data extend only to about $V = 50$ mV). Some kind of contribution to $\Delta G(x)$ which is more or less symmetrical about $x = 0$ is required of any two-state transition theory in order to take these results into account.

(c) *Second Wien Effect*[13,28] This is the effect of an electric field on the degree of dissociation of a weak electrolyte. It the conformational change moves proton (or K^+, etc.) binding sites of the subunit to positions where the electric field is different, or if the change alters the proton dissociation constants, then $\Delta G(x)$ would have a contribution from the second Wien effect which, to a first approximation, would be a function of $|x|$. This is another possible source of a term symmetrical about $x = 0$.

(d) *Rotation of Permanent Moment* If the subunit has a permanent dipole moment and if the conformational change rotates the moment somewhat in the electric field, there will be a contribution to $\Delta G(x)$ which is linear in x (as in $-\delta x$).

(e) *Binding of Ca^{++}, K^+, Na^+* Suppose that say, Ca^{++} binds to a subunit and that the strength of binding is different in the two conformations and/or the electrostatic potential at the sites is different (the degree of binding depends on the local potential). The conformational equilibrium ratio will then have a V-dependent contribution from this source. This is basically the same kind of effect as for proton binding, but in the proton case we have, as an approximation[6], used an expansion about $x = 0$ to obtain the first two terms. See Equation (12) of reference 6 for an example.

A severe restraint must be kept in mind here as well as in (a), (b), and (c) above: the model must be so constructed that V-dependent changes in the ionic properties of the channel have no effect on $\bar{g}_K$ and $\bar{g}_{Na}$ (which are observed to be independent of V).

This work was supported in part by research grants from the National Science Foundation and the General Medical Sciences Institute of the United States Public Health Service.

I am indebted to Dr. Yi-der Chen for carrying out the computing for the figures, and to Drs. Kenneth Cole and Robert Blumenthal fortheir patient contributions to my education.

NOTES

(1) This paper is in the nature of a research proposal. It has turned out to be Part I of a series. In the time since it was submitted (August 1970), my colleague Yi-der Chen and I have been able to test some of the suggestions made here. These brief notes, added as the manuscript goes to press, are intended to bring the reader up to date.

(2) The treatment in reference 6 was extended considerably in Hill and Chen, *Biophys. J.* **11**, 685 (1971).

(3) A much more satisfactory generalization of HH (who take $f_K' = 0$) than Equation (2) or Equation (11) is contained in Appendix 3 of Part IV of this series, Hill and Chen, *Biophys. J.* (in press). Actually, there is no evidence we are aware of that any such generalization is needed.

(4) In Parts II and III of this series, Hill and Chen, *Proc. Nat. Acad. U.S.A.* **68**: 1711, 2488 (1971), we found that the early induction and the Cole-Moore[25] superposition of $I_K(t)$ experimental curves seem to rule out any very significant amount of cooperativity within a channel. That is, $\omega = 0$ below. We also found, in the models studied, that co-operativity *between* channels is in even more conflict with the experimental $I_K(t)$ curves. Thus, kinetic analysis forces us back to the original HH implicit assumption ($\omega = 0$).

(5) It turns out [see the references in note (4)] that in general the *entire* $gg(t)$ curves is "extended" so that, on a readjusted time scale, the induction period is not lengthened.

(6) In Part V (submitted), Chen and I present two illustrative models which do account for the Cole-Moore hyperpolarization delay. In one model, each of, say, four subunits requires about six (or more) substates of conformation i in order to achieve sufficient delay. The substates are assumed to involve binding of phospholipid molecules.

(7) We have not studied the Na^+ case further. We have no information about cooperativity in the Na^+ channel. On the basis of K^+ results [see note (4)], one might anticipate a lack of cooperativity in the Na^+ channel as well.

(8) Our present work (Part VI?) is on this subject. Our working hypothesis is that a K^+ channel or gate is a complex of, say, four independent (as in HH) protein subunits. Each subunit undergoes a V-dependent $i - ii$ conformational change. Similarly, the Na^+ channel is assumed to have $3 + 1$ independent protein subunits (two types: m and h). An incidental bonus already encountered is a simple explanation of the approximate coincidence of the maximum in $\tau_i(V)$ with the rising part of the $i_\infty(V)$ curve, where $i = n, m, h$.

REFERENCES

1. Cole, K. S., *Membranes, Ions, and Impulses* (University of California Press, Berkeley, 1968).
2. Tasaki, I., *Nerve Excitation* (Charles C. Thomas, Sprinfield, Illinois, 1968).
3. Goldman, D. E., *Biophysical. J.* **4**, 167 (1964).
4. Adam, G. in *Physical Principles of Biological Membranes* (Gordon and Breach, New York, 1970), eds., F. Shell, J. Wolken, G. Iverson, and J. Lam.
5. Blumenthal, R., J.-P. Changeux, and R. Lefever, *J. Membrane Biol.* **2**, 351 (1970).
6. Hill, T. L., and Y. Chen, *Proc. Nat. Acad. Sci.* **66**, 607 (1970).
7. Hodgkin, A., L. and A. F. Huxley, *J. Physiol.* **117**, 500 (1952).
8. Blumenthal, R., private communication.
9. Changeux, J.-P. in *Symmetry and Function of Biological Systems* (John Wiley and Sons, New York, 1969), eds., A. Engström and B. Strandberg.
10. Starzak, M., to be published.
11. Wyman, J., in *Symmetry and Function of Biological Systems* (John Wiley and Sons, New York, 1969), eds., A. Engström and B. Strandberg.
12. Agin, D., *Biophysical. J.* **9**, 209 (1969).
13. Bass, L., and W. J. Moore in *Structural Chemistry and Molecular Biology* (W. H. Freeman, San Francisco, 1968), eds., A. Rich and N. Davidson.
14. Hoyt, R. C., *Biophysical. J.* **3**, 399 (1963).
15. Mullins, L. J., *J. Gen. Physiol.* **42**, 1013 (1959).
16. Lehninger, A. L., *Proc. Nat. Acad. Sci.* **60**, 1069 (1968).
17. Mueller, P., and D. O. Rudin, *Nature* **217**, 713 (1968).
18. Singer, S. J., in *Membrane Structure and Function* (Academic Press, New York, 1971), ed. L. I. Rothfield.
19. Triggle, D. J. *Neurotransmitter-Receptor Interactions* (Academic Press, New York, 1971), Chapter 2.
20. Pressman, B., and D. H. Haynes in *Molecular Basis of Membrane Function* (Prentice Hall, Englewood Cliffs, New Jersey, 1969), ed. D. C. Tosteon.
21. Engström, A., and B. Strandberg, eds., *Symmetry and Function of Biological Systems* (John Wiley and Sons, New York, 1969).
22. Hill, T. L. *Thermodynamics of Small Systems, Parts I and II* (W. A. Benjamin, New York, 1963 and 1964), pp. 115–117, Part II, and various other examples.
23. Adelman, W. J. Jr., F. M. Dyro, and J. Senft, *L. Gen Physiol.* **48**, (52), 1 (1965).
24. Hodgkin, A. L., and A. F. Huxley, *J. Physiol.* **117**, 473 (1952).
25. Cole, K. S., and J. W. Moore, *Biophysical J.* **1**, 1 (1960).
26. Kirkwood, J. G., and J. B. Shumaker, *Proc. Nat. Acad. Sci*, **38**, 855 (1952).
27. Gilbert, D. L., and G. Ehrenstein, *Biophysical J.* **9**, 447 (1969).
28. Onsager, L., *J. Chem. Phys.* **2**, 599 (1934).

PAPER 11

Membranes and Ionic Double Layers

D. E. GOLDMAN

Medical College of PENNSYLVANIA
Philadelphia, Penns.

The ionic double layers present at the interfaces of a membrane with its environmental media are capable of exerting a marked influence on its electrical behavior. In biological systems the complexity of membranes, their extreme thinness, the asymmetry of the media, etc., make it very difficult to estimate a priori just how the role of these double layers should be evaluated. Quantitative treatments of ion transfer processes have generally had to be based on highly simplified structural assumptions and have dealt primarily with the membrane alone (see Cole, 1968; Lakshminarayanaiah, 1969); the adjacent solutions function essentially as ionic reservoirs. Limited attention to have been given to double layers a they may relate to membranes although the influence of Donnan potentials at the interfaces has been considered (see Teorell, 1953). A complete analysis of ion flow in biological membranes would require more detailed information on both membrane structure and on the relevant physico-chemical rules of operation than is now available.

What will be attempted here is a general treatment of an electrolyte-membrane-electrolyte system, taking explicit notice of ionic distributions as they modify the electrical properties of the membrane. This will be based on the concepts and approach developed by Gouy (1910) for diffuse layers and by Stern (1924) for compact layers (see also Grahame, 1947). In the system considered there are interactions not only between layers within a membrane, but between layers on opposite sides of an interface. Verwey and Overbeck (1948) have discussed certain aspects of double layer interaction at equilibrium. Here we shall describe a more general approach and will include time dependent behavior and steady states. To the extent that such an analysis may be successful, it should be particularly useful in studies of biological systems, but should be applicable to artificial systems as well, often in simpler form.

In dealing with this problem one of the realities that must be faced is the impossibility, certainly at present, of making direct measurements in systems where both the membranes and the double layers are very thin. If one wishes to study ion current flow, one can perhaps control to some extent the composition of the system but is obliged to make one's measurements at a respectful distance from the regions of major interest. This fact, as will appear, makes it necessary to consider the system as a whole; piece-wise treatment can only be carried out in special cases.

As a first step in the analysis, let us consider a segment of cell membrane to have been flattened out and fixed between two aqueous media, which are in turn limited by plane parallel electrodes. This avoids consideration of many characteristics of the cell interior. If, also, the membrane can be considered as (laterally) uniform, at least on an average, one has only a one-dimensional system to deal with. In practice the separation of the electrodes need not be great and there is rarely any need to consider extremely rapid transients. Thus, wave propagation can be neglected and this, together with the one-dimensional character of the system, means that electromagnetic phenomena can be ignored. The flow of electric current in the membrane and in the environmental media under convection-free conditions can then be described rather concisely be the classical relations:

$$\frac{\partial D}{\partial x} = -\varrho, \tag{1}$$

$$E = \frac{-\partial \phi}{\partial x}, \tag{2}$$

$$\varrho = e \sum z_r n_r, \tag{3}$$

$$\frac{\partial j_r}{\partial x} = -z_r e \frac{\partial n_r}{\partial t} \quad \text{(plus source terms if present)}, \tag{4}$$

$$J = \sum j_r \tag{5}$$

In addition, we have:

$$D = f(E\ldots), \tag{6}$$

$$j_r = f(E, n_r, \ldots). \tag{7}$$

Here,

D is electric displacement,
E is electric field,
ϕ is electric potential,

ϱ is charge density,
J is total electric current density,
j_r is current density of rth ion,
n_r is concentration of rth ion,
z_r is valence of rth ion,
e is electronic charge.

Conventionally, $D = \varepsilon E$ although ε, the permittivity, may vary with field strength, local ion concentrations, conformational factors, etc. Hopefully, these dependences are not strong enough to invalidate the usefulness of the simple relation.

In aqueous media j_r can be expressed directly in terms of the relevant forces and parameters. If one is concerned with electrical phenomena it is usually sufficient to invoke the Nernst-Planck equations (see Planck, 1890). Within the membrane, a more complicated expression may be needed although the Nernst-Planck relations are often used as an approximation. Also, it is very rarely necessary to deal with other than isothermal conditions.

A complete specification of boundary conditions requires, among other things, precise knowledge of electrode phenomena. The difficulties arising from this can, however, be circumvented by placing the electrodes far enough away from the membrane so that any diffuse double layers in the solution are well separated and by having the measuring times short enough to avoid contamination of the solutions near the membrane by possible electrode products.

The boundary conditions at the membrane surfaces require close inspection. These surfaces are not ideal planes. They are likely to be uneven and to be quite heterogeneous at the molecular level, especially in biological membranes. They may contain both polar and nonpolar groups. The question then arises as to when it may be appropriate to represent these surfaces by uniform average structures. There may also be charged elements near the surface which, if fixed, could participate in chemical reactions with some of the environmental ions or, if mobile within the membrane, could participate in formation of carrier particles. Here also one must consider under what circumstances averaging procedures are justifiable (see Cole, 1969). If, in fact, the membrane boundary is uneven as distinguished, say, from a clean mercury surface, and if the molecular arangements permit, one may expect fixed charge elements present at the surface either to be combined with ions of opposite sign or have such ions as part of their ion atmospheres. There may also be difficulties in identifying a compact double layer. However, the boundary relations should be stated in as much detail as knowledge permits.

Here, for example, we may write simply:

$$\bar{\bar{D}} = \bar{D} \tag{8}$$

where the bars and double bars refer to values on opposite sides of an interface. Further, there are likely to be significant energy barriers to ion penetration of the interface, due to differences in such things as permittivity, ion hydration, potential, and molecular packing of membrane constituents. Thus,

$$j_r = z_r e \lambda_r (\bar{n}_r \exp(-\bar{w}_r) - \bar{\bar{n}}_r \exp(-\bar{\bar{w}}_r)) \tag{9}$$

where λ_r is a "barrier permeability" parameter and w_r is the potential energy barrier (divided by kT) as seen from one side or the other. W_r is somewhat arbitrarily separated into a "chemical" and an electrical term, i.e.:

$$W_r = W_r^* + z_r e \phi^* \tag{10}$$

with $w_r = W_r/kT$.

Conventionally, for a very thin barrier, ϕ^* is one-half the potential difference across the boundary and this convention is a substitute for a precise knowledge of barrier structure. It should be evident that the potential difference across the boundary is not in general an equilibrium phase boundary potential and cannot be determined independently. Equation [9] can be written in the more convenient form:

$$j_r = z_r e \lambda_r^* (\bar{n}_r \exp(-\tfrac{1}{2} z_r \theta^*) - \xi_r \bar{\bar{n}}_r \exp(+\tfrac{1}{2} z_r \theta^*)) \tag{11}$$

where

$$\lambda_r^* = \lambda_r e^{\bar{w}_r*}, \; \theta^* = \frac{e\phi^*}{kT}, \quad \xi_r = \exp(w_r^* - \bar{\bar{w}})$$

and ξ_r is, in effect, the distribution coefficient for the ion between the two media. If there is binding of ions, one must include an appropriate source term in the equation of continuity [equation (4)].

One can now, in principle, solve equations (1) through (7), subject to the boundary conditions and so reach the desired results. Because of the presence of the phase boundaries and particularly because the potential, field and concentrations are given only at the external boundaries, an iterative procedure is required and, indeed, is to be expected in a system in which controls are applied at the ends and the interior is left to adjust itself.

To illustrate this, let us assume that there are no fixed charges and no chemical reactions, and that, knowing the appropriate values of the parameters, we wish to determine the steady state behavior of the system. We drop equation (4), noting that each current component must be the same everywhere. We begin at some point, $x = -a$ in the left hand solution where the ion concentrations are known constants and where the potential, as measured by an idealized probe electrode, has a reference value of zero. We assume values for the current components, j_r, calculate the total current, and multiply be the resistivity of the (bulk) solution. This yields the electric field at that point. Next, we solve the field equations for n_r, E, and ϕ and calculate their values at the membrane boundary. We assume a value for ϕ_1^* at this boundary and calculate n_r, E, and ϕ on the membrane side. We then solve the equations for the membrane phase, jump the barrier in the other side in the same way, assuming a ϕ_2^* at that boundary, and so calculate n_r, E, and ϕ at a point $x = +b$ in the second bulk solution. These calculated values are then compared with the known values of n_r, E and ϕ at that point. If there is a discrepancy, the values of j_r, ϕ_1^*, and ϕ_2^* are corrected and the entire process is iterated to convergence. Since the problem is real and physical, one can safely assume that if a steady state exists convergence must occur. However, convergence problems sometimes arise in the computer solution process and have to be guarded against. The procedure for solving the time dependent case is considerably more complex since there are partial differential equations to be dealt with; however, the principle is the same. If the system is allowed to reach equilibrium, there are, of course, no current flows; all j_r's are zero. If some of the ionic components cannot penetrate the membrane, a Donnan equilibrium state results.

It should be observed that the double layers have failed to make an explicit appearance in this analysis. They are identifiable as regions close to a boundary where the concentrations, potentials, etc. deviate fairly rapidly from their values in the bulk phase. If the membrane is thin enough, the distributions inside it may be quite complex, and no double layers will be recognizable as such. The concept of a double layer, fairly clear when one is dealing with a simple boundary of an infinite medium, may be less so in cases where strong interactions occur.

An important calculation which can be made is the total net charge between the two reference points in the bulk solutions. If the system is asymmetrical we have no a priori assurance that the membrane and its adjacent layers will have no net charge, even though the entire system including the electrodes must be electrically neutral. Should a charge be present between $x = -a$ and $x = +b$ there would be a force tending to drive the membrane toward one of the electrodes and this would result in a mechanical pressure difference between the two solutions. If, in addition, the membrane is permeable to solvent molecules or to other uncharged particles present, osmotic effects may be substantial. If fixed charges are also present, electro-osmosis may occur. Interaction between electrical, mechanical and diffusional forces then offers the possibility of oscillatory phenomena (Teorell, 1957), provided that the system is not rigidly sealed.

One can also include other factors in the analysis. For example, there may be mechanical forces of electrical origin tending to distort the membrane. The occurence of very high electric fields may affect the structure of the membrane and so make some of the parameters field dependent; or they may be concentration dependent. Indeed, some of these factors may be extremely important in relation to the behavior of excitable membranes.

The problem and its analysis have been presented here in a fairly general way. Specific cases can be handled whereever one can write the appropriate equations, determine parameter values and carry out the necessary computations. As a rule an iterative mix of assumption, calculation and measurement is required. The complexity and nonlinearity of the system is such that it is generally very difficult to tell under what circumstances one can rely on approximations designed to linearize it and thus permit the use of direct analytical methods. Interesting and useful examples of a simplified treatment applied to nerve include those of Chandler, Hodgkin and Meves (1965) on the effects of internal ionic strength on measured membrane potentials and those of Gilbert and Ehrenstein (1969) on the interaction of calcium with charges sites on the external side of the axom membrane. It is perhaps characteristic of biological systems that their effectiveness often appears to depend on the presence of many interacting phenomena, as well as on a complicated arrangement of many different molecular elements. In this connection, the interests of the physical chemist and of the biologist may be quite different.

REFERENCES

1. Chandler, W. K., A. L. Hodgkin, and H. Meves, *J. Physiol.* (London), **180**: 821–836, (1965).
2. Cole, K. S. "Membranes, Ions and Impulses". Univ. of California Press, Berkeley, (1968).
3. Cole, K. S. *Biophys. J.* **9**, 465–469, (1969).
4. Gilbert, D. L., and G. Ehrenstein *Biophys. J.* **9**, 447–463, (1969).
5. Gouy, M. *J. de Physique* (4), **9**, 457–468, (1910).
6. Grahame, D. C. *Chem. Rev.* **41**, 441–501, (1947).
7. Lakshimarayanaiah, N. "Transport Phenomena in Membranes". Academic Press, New York, (1969).
8. Planck, M. *An.. Phys. u. Chem.* (neue Folge) **39**, 161–186, (1890).
9. Stern, O. *Z. Elektrochem.* **30**: 508–516, 1924.
10. Teorell, T. *Prog. Biophys. and Biophys. Chem.* **3**, 305–369, (1953).
11. Teorell, T. *Acta Soc. Med. Upsal.*, **62**, 60–66, (1957).
12. Verwey, E. J. W., and J. Th. G. Overbeek, "Theory of the Stability of Lyophobic Colloids", Elsevier, Amsterdam, (1948).

The Use of the Flux Ratio Equation under Non-Steady State Conditions

HANS H. USSING

Institute of Biological Chemistry, University of Copenhagen
2 A, ϕster Farimagsgade 1353 Copenhagen K, Denmark

INTRODUCTION

Let me begin with a confession: When I was a student I was sure of one thing: Electrophysiology was not worth touching. It was mainly the reading of Kenneth Cole's papers which converted me and convinced me that the study of electrical phenomena in biological membranes could be a science and not just a collection of incoherent observations. Thus I am greatly indebted to Dr. Cole and I am most thankful for this occasion to contribute this little paper to the volume to honour him on his 70th birthday.

The flux ratio equation (Ussing, 1949, 1952, see also Teorell, 1949) sometimes called the independence equation, states that for an ionic species passing through a membrane without interacting with moving particles and without being subject to active transport, the flux ratio is independent of conditions in the membrane phase and is determined solely by the electrochemical potential difference across the membrane for the ion in question. The general case where the interaction with the flow of solvent is also taken into account has been treated by Ussing (1952), Koefoed-Johnsen and Ussing (1953) Meares and Ussing (1959), Hoshiko and Lindley (1963), and Kedem and Essig (1965).

The equation has been used widely in studies of biological membranes because it may disclose the contribution of active transport or other interactions in the movement of both electrolytes and non-electrolytes.

It has been tacitly assumed in all previous treatments that the fluxes should be measured under steady state conditions. This requirement is readily met in many cases. The amount of isotope passing in unit time usually becomes constant within a very short time in most systems. Thus equilibration is always rapid for cell membranes. Even composite systems like frog skin and toad bladder exhibit a constant sodium flux after a building up period of say, 10 to 15 minutes. This is because the transport rate is high and the sodium pool in the tissue is low. In other cases, however, the permeability is low and the pool is large so that it may take hours to get a steady state flux. Another related problem occurs when the the tissue is slowly changing its properties because of general deterioration or because of the action of poisons or hormones. Is it possible to determine the presence or absence of active transport or other interactions under such circumstances? It will be demonstrated in the following that, provided certain precautions are taken, meaningful information can be obtained from flux ratio determinations even under non-steady state conditions. In the following derivation the word "membrane" is used in a wider sense, viz: a unstrirred sheet consisting of any number of parallel layers with arbitrary permeability properties. The membrane is supposed to separate two well mixed bathing solutions.

DERIVATION OF FLUX RATIO EQUATION

The rate of passage of a given substance through a plane within the membrane parallel to its surface is given by

$$\frac{dn}{dt} = -\frac{A}{g} \cdot C \cdot \frac{d\tilde{\mu}}{dx}, \tag{1}$$

when dn is the amount of the substance passing through during the time dt, A is the area, g, a generalized frictional coefficient, c the concentration of the ion, $\tilde{\mu}$ its electrochemical potential, and x the distance in the membrane from the surface bathed with solution 1.

In the simple case of no interaction with other moving particles and absence of active transport, we have

$$\frac{d\tilde{\mu}}{dx} = RT\frac{d \ln c}{dx} + RT\frac{d \ln f}{dx} + zF\frac{d\psi}{dx}, \tag{2}$$

where R, T, z, and F have their usual meanings, f is the activity coefficient of the ion, and ψ the electrical potential at x.

We now define a quantity $\tilde{a}$ (the electrochemical activity) by the following equation

$$RT \ln \tilde{a} = RT \ln c + RT \ln f + zF\psi = \tilde{\mu} - \tilde{\mu}_{\text{standard}}. \tag{3}$$

From (3) we obtain

$$C = \frac{\tilde{a}}{f} \cdot \exp\left(-zF\psi/RT\right). \tag{4}$$

Combining (1) and (4) we then get

$$\frac{dn}{dt} = -\frac{ART}{g} \cdot \frac{\tilde{a}}{f \cdot \exp\left(-\frac{zF\psi}{RT}\right)} \cdot \frac{d \ln \tilde{a}}{dx},$$

or (5)

$$\frac{dn}{dt} = \left[-\frac{ART}{g \cdot f \cdot \exp\left(-\frac{zF\psi}{RT}\right)}\right] \cdot \frac{d\tilde{a}}{dx}.$$

Let us consider the movement of two different ideal tracers for the substance in question through the membrane. Subscripts 1 and 2 will be used to designate the symbols describing the behaviour of tracer 1 and 2, respectively.

We want to obtain an expression for the ratio of the fluxes determined by the movement of the two isotopes across the plane considered. In equation (5) the variables within the brackets describe the essentially inaccessible properties of the interior of the membrane. However, since the variables do not change with time, they are identical for the two isotopes and cancel in the expression of the flux ratio [equation (6)]

$$\frac{\frac{dn_1}{dt}}{\frac{dn_2}{dt}} = \frac{\frac{d\tilde{a}_1}{dx}}{\frac{d\tilde{a}_2}{dx}}. \tag{6}$$

The condition for the elimination of the expression in brackets, viz., that all its variables be non-variant with time, requires some consideration. R, T, z and F are constants but ψ, g, A and f may be functions of both x and t. The most important variables by far are ψ and g. As long as they are only functions of x they present no problem. The potential profile and the "resistance profile" may vary with x in any conceivable way, because they will always be the same for an ion of species 1 and one of species 2. However, if ψ and g are also functions of time, the situation is different. On an average the plane within the membrane which we are considering will be reached at different times depending on whether the ion comes from bathing solution 1 or 2. If we call the expression in brackets $B_1(x, t)$ for isotope 1 and $B_2(x, t')$ for isotope 2 they will in general not be identical for a given value of x (except when the membrane has a symmetry plane) if the B-function varies with time. For the following treatment we shall therefore consider a period which is so short that the B-function may be considered only a function of x.

By the procedure outlined above we have gotten rid of all the unmanageable variables describing the interior of the membrane and we have obtained a separation of the remaining variables so that the equation can be integrated. The left hand expressions are integrated from time zero when the two isotopes are added to the bathing solutions 1 and 2 to the time t. We shall assume that the isotopes immediately acquire constant concentrations in the solutions to which they are added and that the concentrations remain constant throughout the experiment. Since the tracers are considered ideal, we shall make the usual assumption that one tracer represents all ions originating in solution 1 and the other one those originating in solution 2. We shall also assume that the bathing solutions are of so large a volume that we can disregard back diffusion of isotope which has passed the membrane once. These considerations lead us to the boundary conditions: for the right hand expression:

$$\text{for } x = 0: \tilde{a}_1 = \tilde{a}_{1(1)} \text{ and } \tilde{a}_2 = 0$$

$$\text{for } x = x_0: \tilde{a}_1 = C \quad \text{and } \tilde{a}_2 = \tilde{a}_{2(2)}$$

where $\tilde{a}_{1(1)}$ and $\tilde{a}_{2(2)}$ are the (constant) values of the electrochemical activity of the ion in solutions 1 and 2 and x_0 is the total thickness of the membrane. The result is given in equation (7)

$$-\frac{n_{1(t)}}{m_{2(t)}} = \frac{\tilde{a}_{1(1)}}{\tilde{a}_{2(2)}} = \frac{C_{1(1)} \cdot f_{1(1)}}{C_{2(2)} \cdot f_{2(2)}} \cdot e^{\frac{zF(\psi_2 - \psi_1)}{RT}} \tag{7}$$

This expression is formally identical to the flux ratio equation for steady state fluxes, but the interpretation is different. It now means that at any time after the first appearance of the isotopes on the opposite sides of the membrane, the flux ratio is constant and is determined solely by the (constant) ratio between the electrochemical activities of the ion in the two bathing solutions.

It is important to realize that the flux ratio equation is only valid if we consider fluxes which have passed the total thickness of the membrane. The concentrations of tracers in the membrane phase may build up in a very asymmetrical way. Thus early values for the ratio between the amount of tracer leaving solution 1 and that leaving solution 2 in unit time may differ markedly from the steady state values. The equation is valid for "appearing fluxes" and not for "disappearing fluxes". The reason obviously is that the two appearing fluxes have to pass all barriers in the membrane, albeit in a different sequence, whereas the disappearing fluxes do not pass all barriers.

It is implicit in the above result that one can obtain correct figures for the flux ratio, even when the properties of the tissue are changing with time, provided that the isotope fluxes are measured for a period which is short in relation to the rate of change in the resistance and potential profiles.

So far we have assumed that all ions follow a single path through the membrane. In composite membranes like, for instance, epithelia, one can easily envisage a cellular path and an intercellular bypass. Even in apparently homogenous membranes the ultrastructure may create more than one type of pathway. The treatment of a composite membrane will lead to a sum of expressions of the same type as equation (7). Let us assume, for the moment, that the properties of the different pathways are not changing with time, but the pathways differ with respect to the rate of passage through them. It is then clear that if the flux ratio is determined at different times after the addition of the isotopes, it will first be dominated by the fast pathway and later by the slower ones. Thus, if one pathway is active and another is passive, the flux ratio will be changing with time and it may be possible to separate the two components by proper numeral calculations.

In order to appreciate the justification for such a procedure, one must realize that the determination of the flux ratio from "early" non-steady state fluxes is also permissible in cases of active transport and other types of deviation from ideal behaviour. Thus, in the case of active transport one may introduce an active transport potential, π, in the expression for the electrochemical potential gradient. Similarly, the effect of solvent drag can be included in the electrochemical potential gradient as described by Ussing (1952), see also Koefoed-Johnsen and Ussing (1953) Meares and Ussing (1959).

Instead of letting the radioactive tracers "represent" the non-active ion whose transport characteristics we are interested in, it is equally permissible to consider the tracers as independent ionic species which just happen to have the same transport characteristics as the inactive ion under study.

If this interpretation is used, the concentration of tracer ion may be defined, for instance, as counts x min^{-1} x ml^{-1}. This means that in order to performe the flux ratio test according to equation (7) one does not even have to know the chemical concentrations of the inactive ion under study in the two bathing solutions. What is needed is a measurement of the potential difference and a determination of the amounts of the two isotopes passing through in a given period of time. A certain ambiguity comes from the fact that the activity coefficients for the ion in the two bathing solutions must be estimated, but for monovalent ions, at least, the ratio between the activity coefficients can often be neglected.

So far we have assumed that the concentrations of tracer 1 was kept constant in solution 1 and tracer 2 in solution 2 throughout the flux determination. This is, in fact, not always necessary. If the membrane has a symmetry plane or if the establishment of a steady state within the membrane phase is rapid, the functions $B_1(x, t)$ and $B_2(x, t)$ will always cancel. In such cases it is a necessary and sufficient condition for the flux equation to hold for passive ions that the ratio $a_{1(1)}/a_{2(2)}$ be kept constant. Granted that the potential does not change, this means that one should keep $c_{1(1)}/c_{2(2)}$ constant. Obviously, in most experimental situations the only sensible thing is to keep both concentrations constant. There may be cases, however, where the experimentalist is able to control only the concentration of tracer 1 on one side of the membrane whereas the concentration of tracer 2 on the other side is changing with time. In such cases it may be possible to program the concentration of the tracer on the controlable side of the membrane so that the concentration-time functions are identical for the two tracers. The isotope fluxes should then be measured for exactly equal periods of time, but if the tissues are reasonably stable, it should be permissible to introduce a time-lag between the monitoring of the changes of tracer 2 on side 2 and the simulation of these changes with respect to tracer 1 on side 1.

SUMMARY AND CONCLUSIONS

1. For a non-transported ionic species which passes through a membrane without interacting with other moving particles and without being subject to active transport, the flux ratio is independent of conditions in the membrane phase and is determined solely by the electrochemical potential difference across the membrane for the ion in question.

2. If the flux ratio for an independently diffusing ion is calculated from the amounts of two ideal tracers which pass through the membrane simultaneously in opposite directions, the value of the flux ratio is the same from the moment of the first appearance to the isotopes on the opposite sides of the membrane until the time of steady state flux.

3. In a system which is changing because of treatment with a drug, a hormone or other chemical agent, or simply because of general deterioration, the flux ratio may still give correct information with respect to the presence or absence of active transport or other interactions, provided that the fluxes are measured for a period which is short compared to the rate of change of the resistance and potential profiles inside the membrane.

4. If the ion in question can follow more than one pathway through the membrane the faster pathway may dominate the flux ratio initially whereas

the slower ones may dominate later. A change in the flux ratio during the run, thus may help disclose the simultaneous existence of a passive and an active transport path.

REFERENCES

Hoshiko, T., and B. D. Lindley. The relationship of Ussing's flux-ratio equation to the thermodynamic description of membrane permeability. *Biochim. Biophys. Acta*, **79**, 301–317 (1964).

Kedem, O., and A. Essig. Isotope flows and flux ratios in biological membranes. *J. Gen. Physiol.*, **48**, 1047–1070 (1965).

Koefoed-Johnsen, V., and H. H. Ussing. The contributions of diffusion and flow to the passage of O_2D through living membranes. *Acta physiol. scand.*, **28**, 60–76 (1953).

Meares, P., and H. H. Ussing. The fluxes of sodium and chloride ions across a cation-exchange resin membrane. *Trans. Faraday Soc.*, **433**, 142–155 (1959).

Teorell, T. Membrane electrophoresis in relation to bioelectric polarization effects. *Arch. Sci. Physiol.*, **3**, 205–219 (1949).

Ussing, H. H. The distinction by means of tracers between active transport and diffusion. *Acta physiol. scand.*, **19**: 43–56 (1949).

Ussing, H. H. Some aspects of the application of tracers in permeability studies. *Adv. in Ezymology*, **13**, 21–65 (1952).

Specific Ionic Conductances at Synapses

J. C. ECCLES

Department of Physiology, School of Medicine, State University of New York
Buffalo, New York 14214

A fundamental contribution to the understanding of the nerve membrane during activity was made by Cole and Curtis (1939). High frequency alternating current was passed transversely across the giant squid axon. Their analysis of the experiment results revealed that during a propagated impulse there was a very large increase in the membrane conductance, but virtually no change in its capacity or in the axoplasm conductance. Subsequent investigations employing more sophisticated recording with intracellular electrodes and voltage clamp analysis have confirmed these original observations (Cole, 1949; Marmont, 1949; Hodgkin and Huxley, 1952a). Furthermore it was shown that the great increase in membrane conductance during the nerve impulse is due to an initial rapid increase in sodium conductance overlapping with a later developing potassium conductance (Hodgkin and Katz, 1949; Hodgkin and Huxley, 1952a, 1952b). The sum of these two conductances accounts for the large conductance increase discovered by Cole and Curtis. There is a comprehensive and authoritative review of the whole of this era (Cole, 1968).

A conceptual model of the events occurring in the membrane represents the changes in membrane conductance as being due to specific carrier-like mechanisms for transporting ions entirely in accord with respective electrochemical gradients. The sodium and potassium ions have almost independent (cf. Chandler and Meeves, 1965) carrier mechanisms (Hodgkin, 1964; Katz, 1966) operating in transmembrane channels that can be envisaged as having a gate-like control. Voltage clamp experiments (Hodgkin and Huxley, 1952a) reveal that the degree of opening of both sodium and potassium ionic gates depends on the level of membrane depolarization.

Quite early in the investigations on the excitatory postsynaptic potentials at neuromuscular junctions and at many other synaptic junctions it was discovered that they also had the properties to be expected for membrane potentials generated by ions moving down their electrochemical gradients. In parenthesis it should be noted that these ionic currents differed from those associated with impulse propagation in that they were turned on, not by membrane depolarization, but by the specific action of synaptic transmitter molecules. In the earlier investigations the depolarizing action of the synaptic transmitter was found to be diminished when the membrane potential was reduced, and even to be reversed when this change exceeded a critical level, which was thus identified as the equilibrium potential for the ionic currents generating the postsynaptic potential.

With frog skeletal muscle fibers it was not possible to apply through the intracellular electrode currents powerful enough to reverse the endplate potential (EPP) (Fatt and Katz, 1951). However the reversed action of the endplate currents was revealed by the very effective reduction in the summit of a muscle action potential arising from an EPP. For example in Figure 1A, trace 5, the potential is greatly reduced relative to spike potentials recorded progressively more distant from the endplate zone (traces 3 to 2 to 1 to 8). Correspondingly, for impulses propagating along the muscle fiber there is reduction of the action potential at the activated endplate zone (Fatt and Katz, 1951). Evidently the currents at the activated endplate are very powerfully shunting the action potential. Figure 1B shows diagrammatically by the directions and lengths of the arrows the action of the activated endplate on the muscle action potential. The cross hatched zone represents the approximate equilibrium potential for the ionic currents generating the endplate potential, the E_{EPP} being about -10 to -20 mV. Since that is approximately the liquid junction potential that would be expected between the myoplasm and the fluid surround of the muscle fiber (Nastuk and Hodgkin, 1950), it was initially proposed by Fatt and Katz (1951) that the synaptic transmitter substance (acetylcholine) caused a large non-selective increase in ion permeability, which acted virtually as a short circuit of the endplate zone of the postsynaptic membrane, there being reduction to a leak resistance of about 20,000 Ω.

This simple hypothesis proved satisfactory for many years (cf. Katz, 1958) and was adopted for other excitatory synaptic actions, where it was likewise observed that reversal occurred beyond a critical level of membrane depolarization (cf. Eccles, 1953, 1957). Thus the excitatory postsynaptic potentials (EPSPs) of motoneurones were reversed when the membrane potential was reversed so that it was internally positive (Figure 1E; Coombs, Eccles, and Fatt, 1955b). However it was found that, in contrast to the EPP (Figure 1A), the EPSP did not appreciably reduce the peak voltage of the action potential that was superimposed on it, as may be seen by comparing the spike potentials set up antidromically Figure 1C and by synaptic excitation Figure 1D (Coombs, Curtis, and Eccles, 1957; Fatt, 1957). A very impressive demonstration of reversal of the EPSP at about -10 mV was made by Nishi and Koketsu (1960) for the frog sympathetic ganglion cell (Figure 1G). It can be stated that these investigations raise important problems concerned with the identification of the ion species whose increased conductances result in the transmembrane current flows responsible for generating EPSPs. Figure 1F diagrammatically summarizes the simple hypothesis for excitatory synapses where E_E and R_E represent respectively the equilibrium potential and the resistance for the activated synapses. With motoneurones E_E is about 0 mV and R_E the shunt resistance. R_E may be as low as $5 \times 10^5\ \Omega$.

By employing the voltage clamp technique on the frog neuromuscular synapse Takeuchi and Takeuchi (1959, 1960) have obtained direct recordings of the current that flows across the motor endplate (the EPC) and that otherwise would generate an endplate potential (EPP). Figure 2A gives specimen records of these currents at the indicated membrane potentials, and in the plotted open circles in B and C it is seen that normally there is a linear relationship of the EPC to the membrane potential, much as occurs with the EPSPs in Figures 1E and G. Unfortunately, it was not possible to pass currents through the impaling microelectrode that were large enough to displace the membrane potential sufficiently for inverting the EPC. However, this reversal potential of -12 to -15 mV can be determined by extrapolation on the assumption of a linear relationship, which is justified by the findings on preparations such as those in Figures 1E and G.

As Katz (1962) has pointed out, the reaction between the excitatory transmitter and the membrane receptors has no regenerative link; the local conductance change produced by the ACh is independent of the level of membrane potential (Fatt and Katz, 1951; Takeuchi and Takeuchi, 1960). The excitatory transmitter substance must greatly increase the ionic conductance of the subsynaptic membrane and the only ion species in sufficient abundance to participate appreciably in this conductance are sodium, potassium and chloride, which originally were assumed to participate non-selectively in the ionic conductance. The fraction of the total conductance due to each one of these ion species was assessed for the first time by Takeuchi and Takeuchi (1960), who observed the effects that changes in the concentration of each have on the E_{EPC}.

In Figure 2B the reduction of the external sodium from 111.36 to 33.6 mM causes the EPC to be decreased at any particular membrane potential and the E_{EPC} is displaced by about -17 mV for a decrease of 3.4 times.

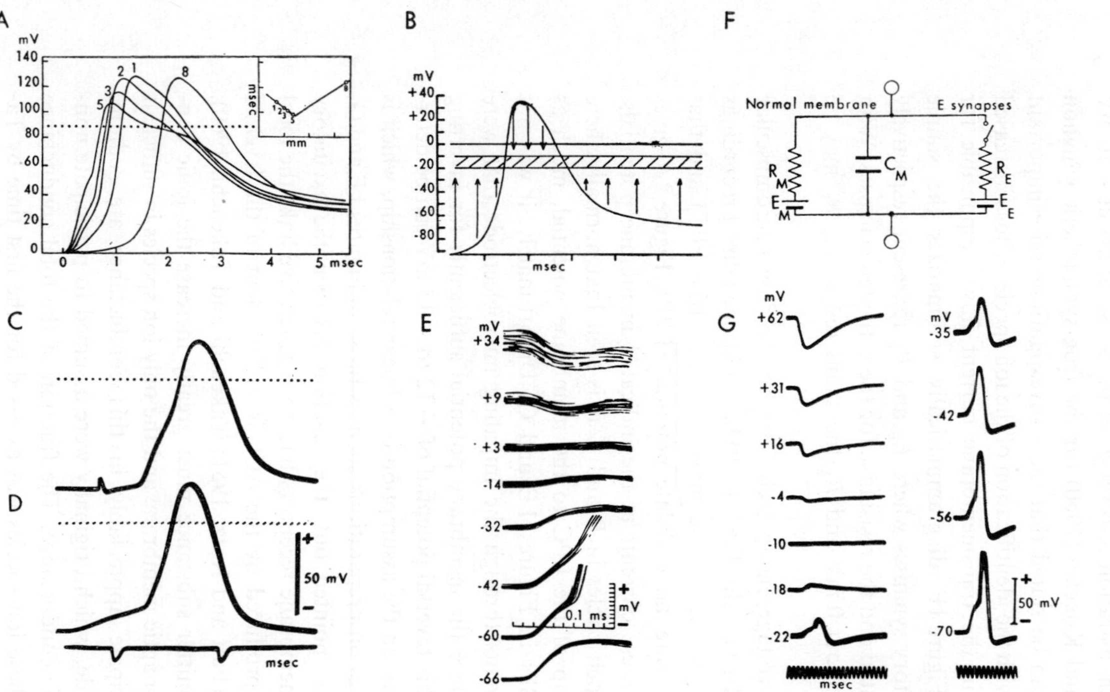

Figure 1 A. Action potentials generated in a muscle fiber by a single nerve impulse and recorded intracellularly at the junctional region (5) and at various distances therefrom as shown in the inset diagram where the time of the spike peak was plotted against the place of insertion of the microelectrode along the fiber. The broken line signals zero membrane potential. B shows a muscle action potential, the arrows indicating the directions in which it is affected by endplate currents (Fatt and Katz, 1951). C and D show intracellular recording from a motoneurone of action potentials set up by an antidromic impulse and by synaptic excitation (Coombs, Curtis and Eccles, 1957). E. EPSP's of a cat motoneurone at various levels of membrane potential, as indicated, −66 mV being the resting potential. The membrane potential was displaced to the indicated levels by transmembrane currents applied through the microelectrode (Coombs, Eccles and Fatt, 1955b). F. Formal electrical diagram of postsynaptic membrane with areas of excitatory synapses shown in component to the right. G. A series of EPSP's set up in a frog ganglion cell and with membrane potential displaced as in E to the indicated levels (Nishi and Koketsu, 1960)

Increase of the intracellular sodium by electrophoretic injection into the muscle fiber through the microelectrode has the same effect, decreasing the EPC for any particular membrane potential. These changes indicate that sodium ions carry a considerable part of the EPC. Similarly in Figure 2C increasing the extracellular potassium from 0.5 mM to 4.5 mM causes a large increase in the EPC, which is reversible on return to 0.5 mM. In Figure 2C there is a displacement of the E_{EPC} of about -28 mV for a nine-fold decrease of extracellular potassium. Thus the trans-membrane flux of potassium ions must also contribute to the EPC. The transmitter substance (ACh) must greatly increase the conductance to both sodium and potassium, and at any particular membrane potential the component of the EPC due to each species of ion may be assumed to be proportional to the product of the ionic conductance and the respective electrochemical gradient across the subsynaptic membrane. From curves such as those of Figures 2B and C it is calculated that the increase in conductance for sodium ions is about 30 per cent greater than that for potassium (Takeuchi and Takeuchi, 1960).

In contrast to these experiments with cations, variation of the external chloride concentration produces no detectable change in the E_{EPC}. In Figure 3A the open circles and the crosses are obtained after replacement of the chloride in the external solution by the presumably impermeable glutamate ion. The change in slopes of the lines is due to change in the tubocurarine concentration. The significant finding is that, despite the different slopes, both lines cross the base line at the same point. This shows that virtually no change occurs in the E_{EPC}; hence, increased chloride conductance can make little if any contribution to the EPC.

This elimination of chloride from an effective participation in the EPC is strongly supported by the finding that, under very wide ranges of extracellular and intracellular chloride concentration, there is no appreciable change in the equilibrium potential for the current generated by the electrophoretic application of ACh to the endplate zone (Takeuchi, 1963a). This technique of electrophoretic application of ACh has also corroborated the conclusion from observations such as those of Figures 2B and C that normally the EPC is due to a conductance increase that is larger for sodium than for potassium ions (Takeuchi, 1963a). However, if the external concentration of potassium is greatly increased, the conductance ratio (sodium/potassium) is decreased from the normal value of about 1.3 to 1.6 in 8 mM potassium and to as low as 0.44 in 100 mM potassium. It was also found (Takeuchi, 1963b) that an increase in extracellular calcium from 2 to 30 mM reduces the sodium conductance to about 70 per cent with little or no effect on the potassium conductance.

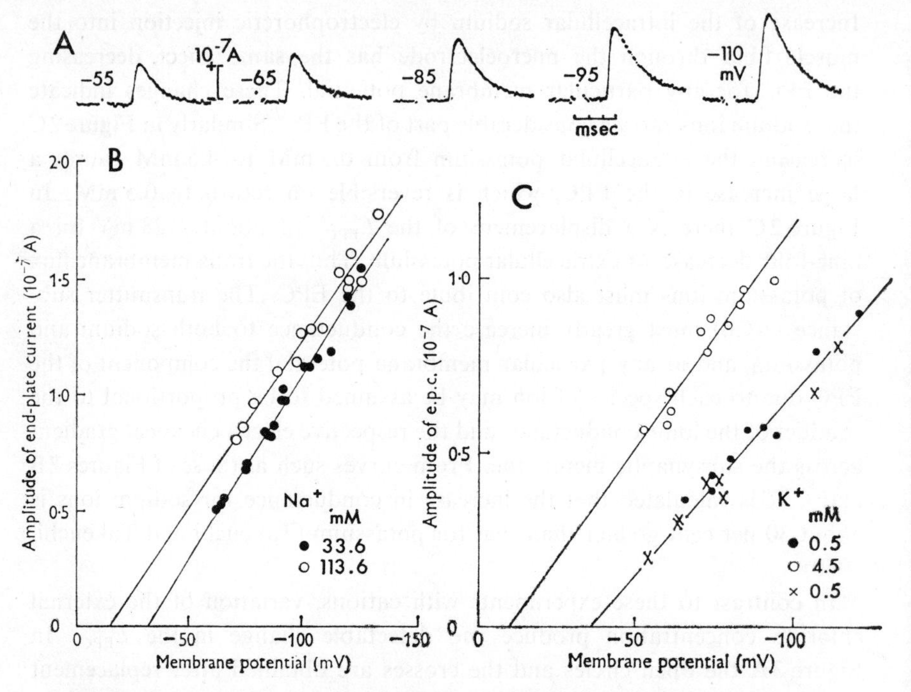

Figure 2 A. End plate currents (EPCs) recorded from a frog neuromuscular synapse under voltage clamp conditions in reponse to a single nerve impulse. The membrane potentials are indicated for each record (Takeuchi and Takeuchi, 1959). B. Relationship between amplitude of EPC and membrane potential as in A, but obtained from an end-plate in two different concentrations of external sodium as indicated. C. As in B but for an end-plate in Ringer with low then high then low potassium as indicated (Takeuchi and Takeuchi, 1960)

The impermeability of the ACh-activated subsynaptic membrane to chloride can be accounted for if there are fixed negative charges on the ionic channels through the membrane. However, the simple hypothesis of a non-selective cationic channel cannot account for the variations in the sodium/potassium conductance ratios that arise as a consequence of an increase in extracellular potassium or calcium. If the ionic channels opened up in the subsynaptic membrane by ACh are available for all cations that are sufficiently small, there is no way of accounting for the normal value of 1.3 for the sodium/potassium conductance, because hydrated sodium ions are larger than hydrated potassium ions and so should have a lower relative conductance. On account of these two findings it was postulated (Eccles, 1966) that there are separate channels for sodium and potassium ions (Takeuchi, 1963a and 1963b), and that the diagram of Figure 3B can be taken literally as indicated, there being two separate conductances that have a variable linkage in this switching, as is symbolized by the broken line. In view of this conceptual development, the chloride impermeability can be attributed to the absence of a specific chloride channel and not to fixed negative charges on the pores.

Maeno (1966) has presented further evidence supporting the postulate of independence of the potassium and sodium conductance channels in the endplate membrane. Figure 4B shows the change in the EPP when the frog nerve muscle preparation is soaked in 10^{-4} procaine. The EPP was

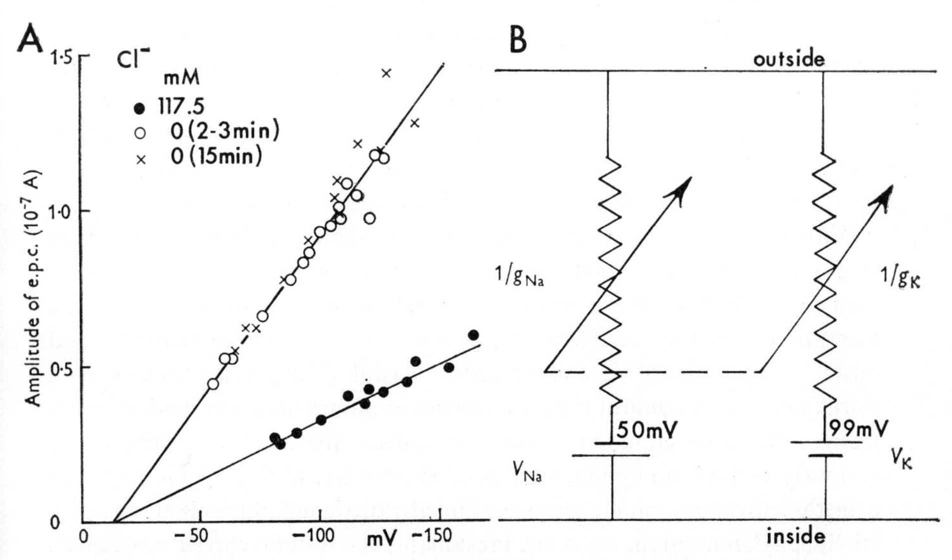

Figure 3 A. Plot of EPCs for an end-plate under voltage clamp conditions as in Figure 2A and plotted as in Figures 2B and C, but for normal Ringer and then in Ringer with replacement of chloride by glutamate as indicated, and also with a reduction in the tubocurarine concentration. B. Schematic diagram of the assumed changes that occur in sodium (gNa) and potassium (gK) conductances during an EPC (Takeuchi and Takeuchi, 1960)

reduced to 0.37 times the control response in A and the decaying phase was greatly prolonged. These changes were observed also in miniature EPPs. It appears that the change in the EPP was due to the action of procaine in the sodium conductance channel. For example, when an extracellularly recording electrode was located accurately on an endplate zone at normal resting potential the recorded EPPs are regarded as being due to endplate currents flowing through the sodium conductance channels. Figure 4D shows that procaine reduced the sodium conductance, as so measured, to 0.78 of the control (6) and there was also as in B a prolonged residual current flow.

This investigation on endplate currents has been continued by Gage and Armstrong (1968) using the voltage-clamp technique to set the membrane potential of the muscle fiber to any desired level. In E the lower trace gives the miniature EPC when the membrane potential was clamped at the equilibrium potential for potassium ions (upper trace). It therefore is a record of the time course of sodium conductance for a min. EPC. Clamping of the muscle fiber at depolarized levels, even to the equilibrium potential for sodium (+50 mV), was possible after glycerol treatment (Figure 4F). When clamped at −50 mV (Figure 4G) it was at the equilibrium potential for K^+ (in 15 mM KCl). Normally there was little difference in the min. EPC's for the sodium and potassium channels. However, in the presence of procaine, the sodium min. EPC (G) developed the slow phase already illustrated in B and D, whereas the potassium min. EPC (F) exhibited no such prolongation. It can be concluded that there is now strong experimental support for the independence of the sodium and potassium conductance channels.

With the frog neuromuscular junction Furukawa, Takagi and Sugihara (1956) showed that ammonium ions can substitute for sodium ions, being even more effective in the process of depolarization produced by synaptic transmission and by acetylcholine, as also can hydrazinium (Koketsu and Nishi, 1959). In more systematic studies Nastuk (1959) and Furukawa and Furukawa (1959) confirm the effectiveness of ammonium ions and find that many substituted ammonium ions can replace sodium. For example, the variously methyl ammoniums are good substitutes, while larger ions such as trimethylethylammonium, choline, dimethyldiethanolammonium and triethylphenylammonium are poor. Presumably the sodium carrier mechanism has not an exclusive specificity, but somewhat related molecules can also travel through the sodium conductance channel, though usually less effectively. In a very comprehensive review of the ionic movements at excitatory synapses Ginsborg (1967) has discussed in detail the biophysical problems arising from specific ionic movements across the postsynaptic membrane.

The other special ionic channel concerned in excitatory synaptic transmission is for Ca^{++} ions and is located in the presynaptic membrane (Katz, 1969). Just like the Na^+ and K^+ gates on the ionic channels involved in impulse propagation, the Ca^{++} gates are opened by membrane depolarization of 30 mV or more, but they differ in being restricted to the presynaptic membrane at the synaptic region (Katz and Miledi, 1965c; 1967a; 1967c; 1969b). Just as with the Na^+ ionic channels, the Ca^{++} ions move down their electrochemical gradient into the presynaptic terminal. This gradient is even steeper than for Na^+ ions, the $E_{Ca^{++}}$ being about 200 mV from the resting potential – from −70 mV to +130 mV (Katz and Miledi, 1969a).

This influx of Ca^{++} into the presynaptic terminal is essential for triggering

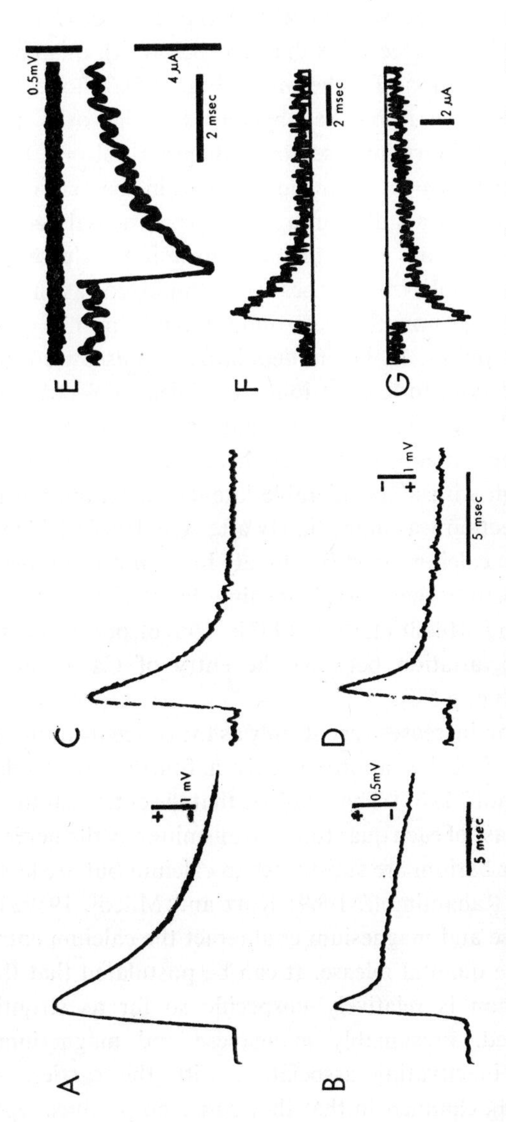

Figure 4 A and B are endplate potentials of a frog muscle fiber, A being control and B after soaking in 10^{-4} procaine. C and D are extracellularly recorded EPPS, which represent approximately the conductance change in the sodium channels C being before and D after soaking in 10^{-4} procaine (Maeno, 1966). E. Lower trace is miniature EPC recorded under voltage clamp conditions at E_K. Upper trace is the voltage trace. F and G are miniature EPC's at E_{Na} and E_K respectively and after soaking in 10^{-5} procaine (Gage and Armstrong, 1968)

the release of quanta of transmitter by a nerve impulse. On the other hand the release of transmitter occurs even at zero level of external Na^{+} ions (Katz and Miledi, 1969b). In the uppermost traces of Figure 5A, B there is an absence of external Ca^{++}, and a nerve impulse fails to evoke any release of quanta, but in the lower traces, when calcium is injected electrophoretically close to the nerve terminal (see inset diagram below A), quantal release is restored during the injection. Furthermore, Figure 5C shows that, when nerve impulse transmission is blocked by tetrodotoxin, brief applied depolarization to the nerve terminal (see inset diagram below C) causes the release of quanta only when Ca^{++} ions are present in the external medium. Figures 5D, E, F illustrates a still more refined experimental demonstration of the relationship of inward Ca^{++} flux to transmitter release (Katz and Miledi, 1967c). Brief depolarizing pulses were employed, as in Figure 5C, to release transmitter, but now the electrophoretic injection of calcium was also by brief current pulses. In D the depolarizing pulses were ineffective due to the lack of calcium. In E Ca^{++} ions were injected electrophoretically by a brief current (1 msec) in close proximity to the endplate, and only 1 msec before the depolarizing pulse; yet there was now always a release of transmitter, though with a considerable latency variation. On the other hand, if the Ca^{++} injection was immediately after the depolarizing pulse (F), it was ineffective. The calcium must be already in position to move through the calcium gates before their brief opening by the depolarizing pulse. However, as Katz and Miledi (1965b, 1967b, 1967c) point out, there is a considerable latency variation between the entry of Ca^{++} ions and the quantal release (cf. Figure 5E).

The quantal emission increases very steeply as the concentration of external calcium is raised, there being approximately a fourth power relationship (Dodge and Rahamimoff, 1967). This suggests that the entry of four Ca^{++} ions is required for the release of each quantum of transmitter by the nerve impulse. External strontium or barium are substitutes to calcium but are less effective (Dodge, Miledi and Rahamimoff, 1969; Katz and Miledi, 1969a). On the other hand manganese and magnesium counteract the calcium entry and so diminish or block the quantal release. It can be postulated that the carrier mechanism for calcium is relatively unspecific so far as strontium and barium are concerned. Presumably manganese and magnesium act by entering into some inactivating association with the carrier. The Ca_D channels resemble Na_D channels in that they can even produce regenerative responses when K_D channels are blocked by TEA (Katz and Miledi, 1969a, 1969b).

Figure 6 gives a diagrammatic display of the various ionic channels across the postsynaptic membrane of the neuromuscular junction, there

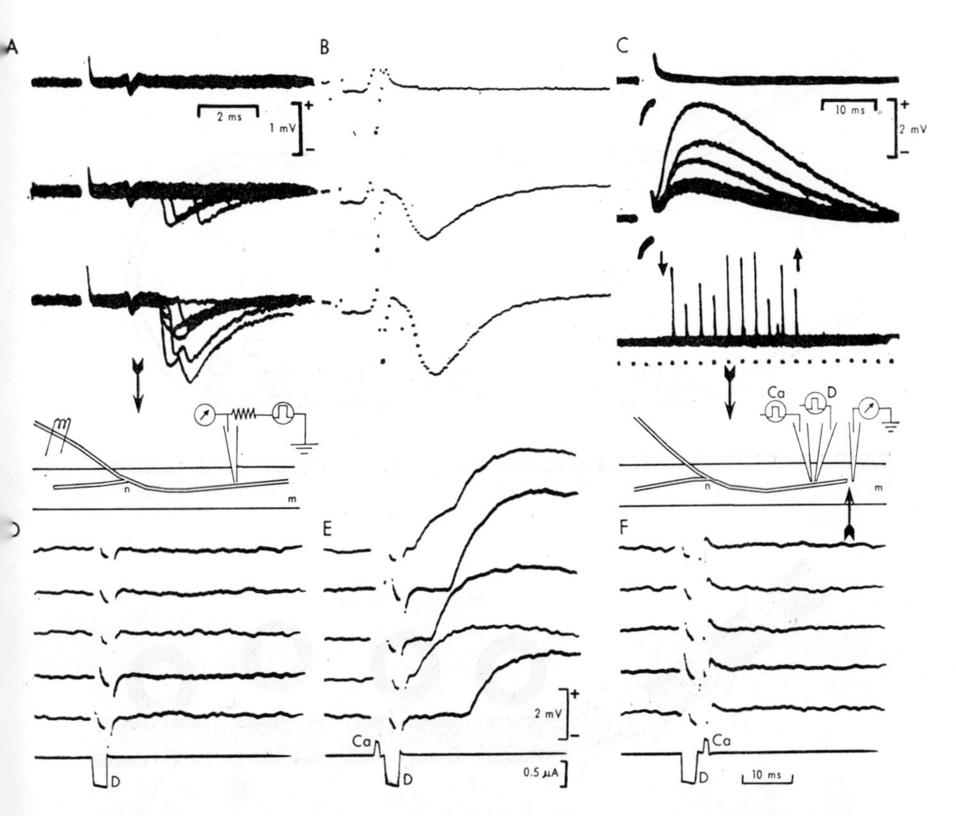

Figure 5 Effect of calcium on transmitter liberation. A. The frog sartorius was immersed in a Ca^{++}-free solution containing 0.84 mM Mg. A micropipette containing 0.5 M $CaCl_2$ was recording extracellularly from a junctional spot. In A Ca^{++}-efflux was stopped by negative biassing of the pipette. This bias was reduced in 2 steps, so restoring transmitter liberation as shown by the EPCs. B. Averages of 600 traces of equivalent responses of A (Katz and Miledi, 1965c). The inset diagram below A shows the recording and the Ca^{++} injection micro-pipette. C. Intracellular recording close to the neuromuscular junction in which TTX has blocked nerve transmission, and in a Ca^{++}-free solution containing 1.7 mM magnesium. There was focal application of depolarizing pulses at 2/sec. A separate micropipette containing 1 M $CaCl_2$ was closely adjacent, as shown in the inset diagram below C. The upper trace shows the ineffectiveness of the depolarizing pulses, but in the middle trace EPPs were evoked because Ca^{++} was being continuously injected ionophoretically. In the lowest trace at a very slow sweep speed the dots indicate depolarizing pulses at 0.5 sec intervals and Ca^{++} injection was given between the arrows (Katz and Miledi, 1967a). D, E, F. Same arrangement as for C, but Ca^{++} injections were given in E and F by brief (1 msec) pulses. D is control for depolarizing pulses only. In E the Ca^{++} injection is 1 msec before the depolarizing pulse and in F 1 msec after (Katz and Miledi, 1967c)

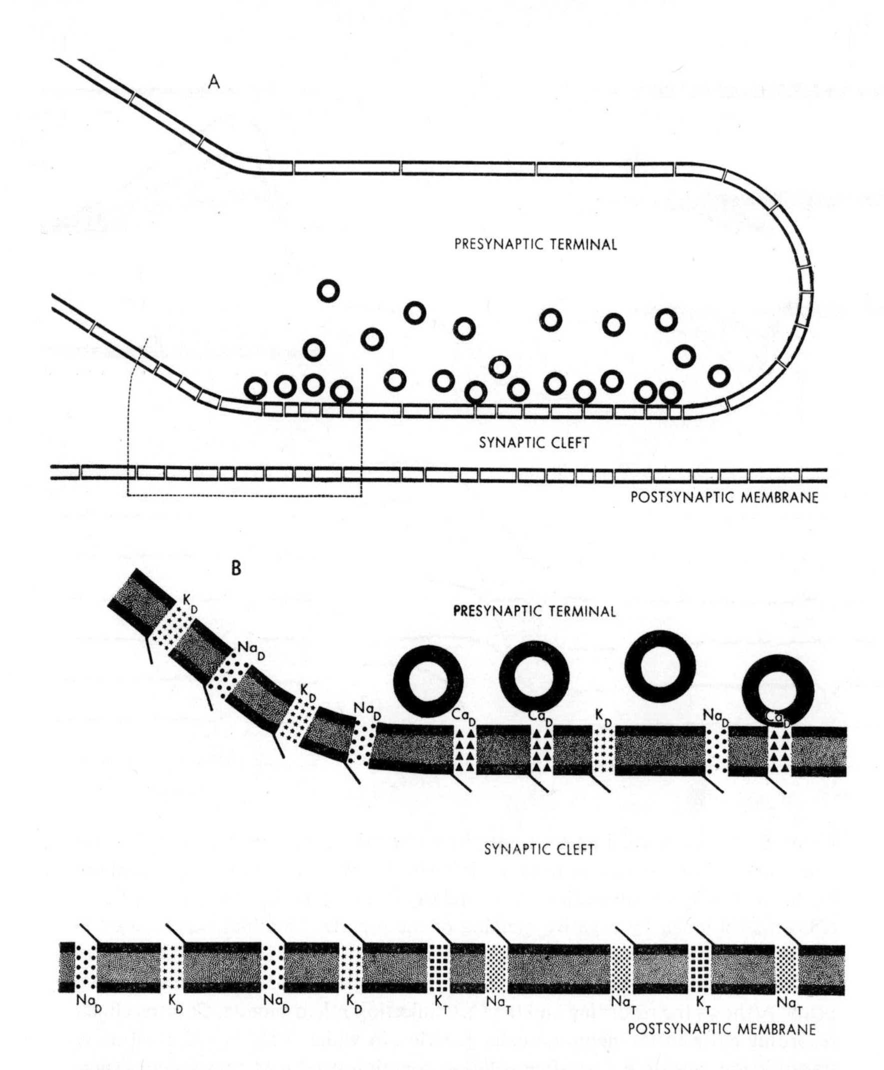

Figure 6 Diagrams of neuro-muscular synapse of the frog. A shows the presynaptic terminal separated by the synaptic cleft from the postsynaptic membrane, and containing many spherical synaptic vesicles, some being close to the membrane fronting the synaptic cleft. Note the various channels through the membrane. In B the segment outlined by the broken line in A is shown enlarged to give fine detail. Note that all channels have a gate control, and that symbols in the channels indicate five different kinds of chemical constitution of the carrier molecules. The subscripts D and T indicate opening by depolarization, and by the transmitter substance, respectively

being separate Na^+ and K^+ channels for the EPC's. Also shown in the presynaptic nerve fiber are the Na^+ and K^+ channels with gates opened by depolarization, and that generate the action potential. They are labelled Na_D and K_D. Though selectively permeable to the same ions these channels are quite distinct from those operated by the chemical transmitter acetylcholine. For example the Na_D and K_D channels are blocked by extracellular TTX and intracellular TEA respectively, whereas the Na_T and K_T channels opened by ACh action are unaffected. For this reason the channels in Figure 6B are shown filled by different carrier mechanisms, as indicated be the different symbols, and by the labels Na_T and K_T.

Figure 6B further illustrates a special feature of the presynaptic membrane, namely the presence of specific calcium channels operated by membrane depolarization, Ca_D (Katz and Miledi, 1965c, 1967a, 1967c, 1969b). In accordance with the evidence of Katz and Miledi (1965a, 1968) the Na_D and K_D channels are shown along the nerve terminal in Figures 6A and B, but these channels are not shown in the subsynaptic component of the postsynaptic membrane where there are exclusively the Na_T and K_T channels. It is not known how far the Na_D and K_D channels are interspersed with the Na_T and K_T channels, but at least they are in close proximity (cf. Eccles, 1964, pp. 107–113).

Analytical investigations on excitatory synaptic transmission at the squid stellate ganglion now rival those on the neuromuscular junction in providing evidence on some of the ionic mechanisms. The giant synapse (cf. Figure 7A) offers the great advantage of having large presynaptic and postsynaptic components, so that intracellular electrodes can be inserted both for recording and for polarizing. In Figure 7B an action potential is set up by direct stimulation of the postsynaptic fiber, and the presynaptic fiber is excited at various times relative thereto (Gage and Moore, 1969). The superimposed traces show that the postsynaptic action potential had exactly the same size as the action potential generated synaptically, which is the thick record formed by several superimposed traces when the postsynaptic stimulus followed the presynaptic and consequently was ineffective. When the presynaptic stimulus followed the postsynaptic, synaptic transmission was occurring into a refractory postsynaptic fiber, hence it set up the superimposed EPSPs without spikes except for the one small late spike. Figure 7B thus resembles the observations on the motoneurone in Figures 1C and D.

In Figure 7C the size of the excitatory postsynaptic current (EPSC) is plotted against the membrane potential just as for the EPCs in Figures 2 and 3, but with a very different result. By extrapolation a value of almost +60 mV is derived for the E_{EPSP}. In a total of nine such experiments Gage and Moore (1969) find that the E_{EPSP} (mean, +50 mV) is almost identical

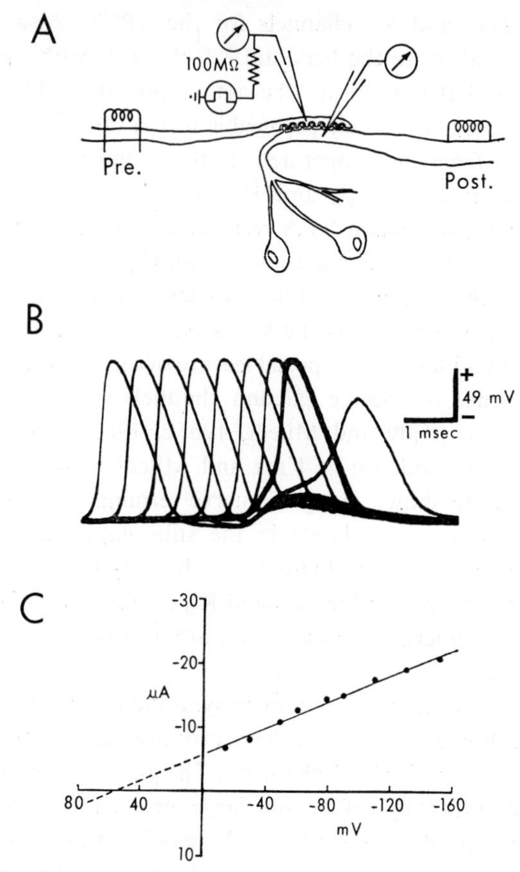

Figure 7 A is a drawing of the giant synapse in the stellate ganglion of the squid. Pre and Post indicate the presynaptic and postsynaptic components, and microelectrodes have been inserted in both for recording and for passing currents. Also both have external stimulating electrodes (Miledi and Slater, 1966). B shows effect of synaptic transmission on antidromic action potentials set up by stimulation of post-fiber (see A), recording being by the microelectrode in the post-fiber at the synaptic region. Superimposed traces with post-stimulus progressively later relative to the pre. Further description in text (Gage and Moore, 1969). C. Plotting of peak amplitude of postsynaptic currents against clamped membrane voltage. As in Figure 2 extrapolation of the curve gives the E_{EPSP} for the squid giant synapse (Gage and Moore, 1969)

with the E_{Na} (mean, +55 mV). It is concluded that at this synapse the excitatory transmitter substance (probably glutamate, Miledi, 1967; J. S. Kelly and P. W. Gage, unpublished observations, 1969) acts almost exclusively in opening gates (Na_T) for the Na^+ ionic channels.

It is the ionic conductance through the potassium channel (K_T) that gives the large reduction of the action potential at the endplate in Figures 1 A and B. The absence of any reduction in Figure 7B is therefore to be expected. However the absence of reduction by the EPSP of motoneurones (Figures 1C, D) has yet to be explained. Because the E_{EPSP} is at about 0 mV (Figure 1E) an appreciable shunting effect on the action potential would be expected. However the intensity of the postsynaptic currents is much less than the endplate currents, and the E_{EPSP} (0 mV) indicates that the potassium current is proportionately smaller than at the endplate, where the E_{EPP} is about −15 mV.

In Figure 8A (Miledi and Slater, 1966) synaptic transmission was almost completely blocked in a low Ca^{++} solution. It was restored in Figure 8B by the ionophoretic injection of a minute amount of calcium through a microelectrode close to the synapse (cf. Figures 5A, B, C). Under such conditions the release of 10^{-14} to 10^{-13} moles in one second may be sufficient to restore transmission. Figure 8C shows the time course of recovery of transmission during a steady ionophoretic application of Ca^{++} ions (between the first and second arrows). The open circles show intracellular EPSP's (lower specimen traces of A and B) and the filled circles the extracellularly recorded postsynaptic currents (upper specimen traces of A and B). Evidently extracellular calcium is as necessary for synaptic transmission as it is for neuromuscular transmission (Figure 5).

In Figure 9 a wide range of presynaptic depolarization was employed to give very significant evidence on inward transport of Ca^{++} ions in relation to the liberation of transmitter (Katz and Miledi, 1967d). External tetrodotoxin (TTX) was employet to suppress Na_D in Figure 9 and so eliminate impulse transmission, and tetraethylammonium (TEA) ions were electrophoretically injected into the presynaptic fiber in order to eliminate the potassium carrier mechanism that is activated by depolarization, i.e. K_D in Figure 6 (Armstrong and Binstock, 1965).

In Figures 9A–F there was progressive increase in the depolarizing current pulse (upper trace) applied through a presynaptic intracellular electrode. Because of the elimination of the K_D channels, the currents produced a fairly steady presynaptic potential (lower trace) that in F was displaced about 200 mV from the resting potential. The EPSP recorded in the middle trace gives an indication of the release of synaptic transmitter. There is a brief peak in A, and in B a larger initial peak with a delayed decline. In C the large potential change resulted in a considerable release of transmitter through the whole duration of the pulse. In D a still larger change is associated with less release of transmitter, but there is a brief peak of release at the end of the pulse, and these effects increase in E. Finally in F there is almost no release during the very large potential change of 200 mV, but a very large off-effect. Evidently, as suggested by Katz and Miledi (1967d),

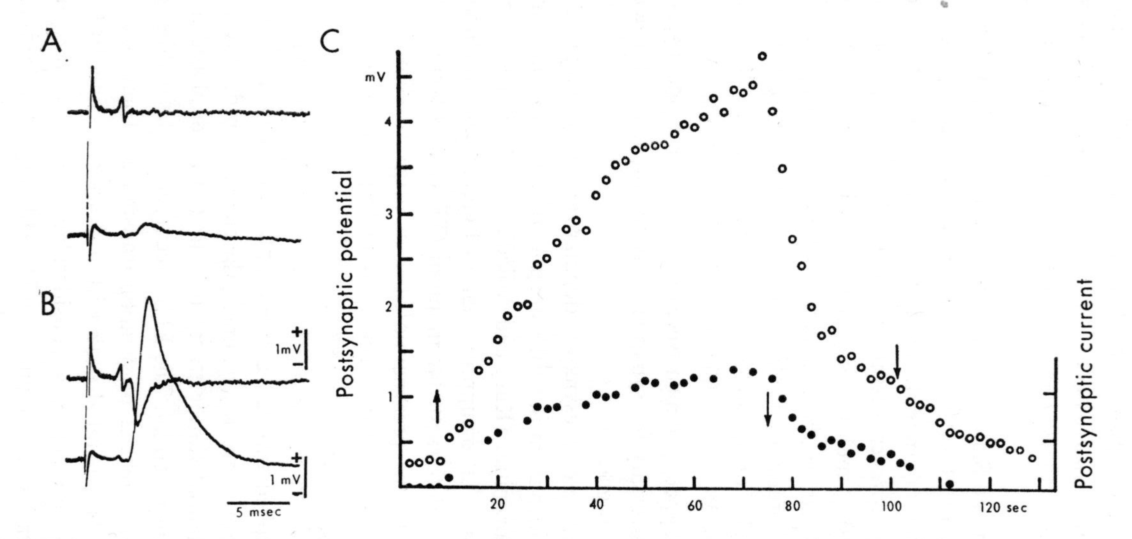

Figure 8 A and B show in upper and lower traces respectively the extracellular and the intracellular postsynaptic records from a squid giant synapse (cf. Figure 7A). In A a nerve impulse gives a diphasic action potential in the upper trace, but almost no EPSP (lower trace) because of blockage by a low Ca solution. In B ionophoretic release of Ca^{++} restores the transmission, as shown in both traces, the nerve impulse in the upper trace being unchanged. C shows time course of this restoration by calcium that was continuously applied ionophoretically between the first arrows. At the third arrow the biassing voltage was reapplied in order to prevent Ca^{++} diffusion. The EPSPs are shown by open circles and the extracellular currents (cf. upper traces in A and B) by the filled circles (Miledi and Slater, 1966)

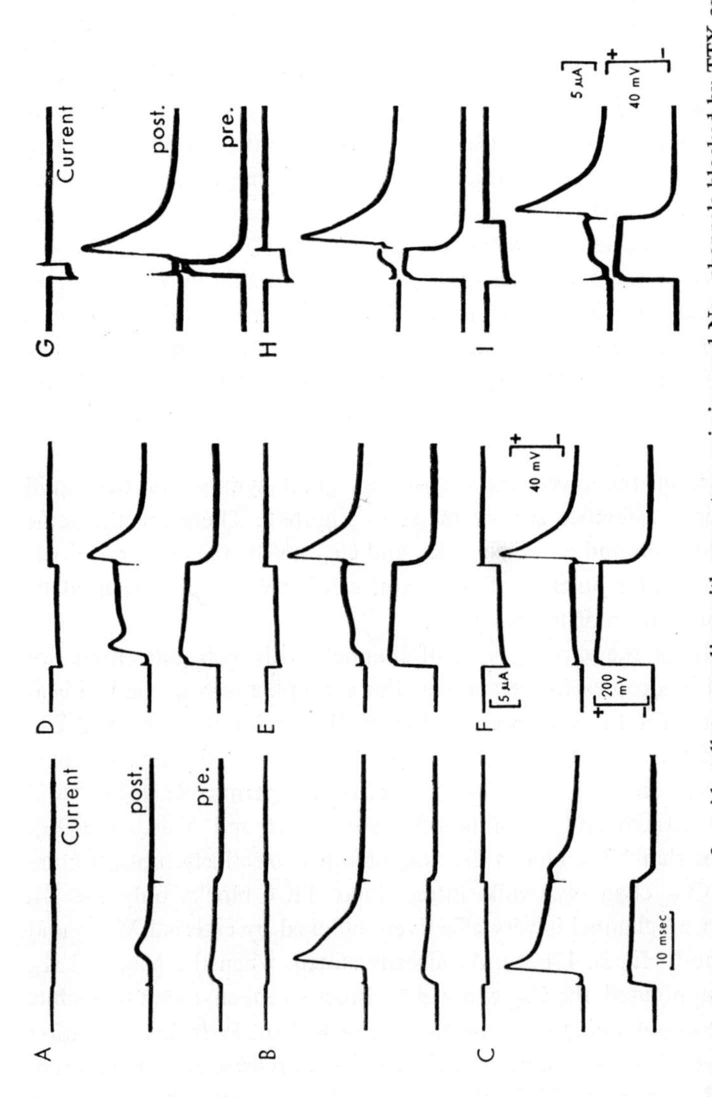

Figure 9 Responses of giant synapse of squid stellate ganglion with nerve transmission and Na_D channels blocked by TTX and K_D channels blocked by intracellular injection of TEA into the presynaptic fibers. In A to F depolarizing current pulses of 18 msec duration were passed through the presynaptic intracellular electrode—uppermost traces of each series—being progressively larger from A to F. Lowermost traces show the presynaptic potentials so induced, and the middle traces are the induced EPSPs recorded from the post-fiber. In G to I the same series was continued with a strong depolarizing pulse, but for briefer intervals as shown. Further description in text (Katz and Miledi, 1967d)

two factors are involved. Membrane depolarization opens the Ca_D gates, just as it does for the Na_D and K_D gates (Hodgkin and Huxley, 1952a, 1952b), and likewise there is a threshold of about 30 mV depolarization. The greater the depolarization, the more effective the opening of the Ca_D gates, but the electrochemical gradient for Ca^{++} ions becomes progressively less, so limiting the Ca^{++} entry and the consequent transmitter release. Finally at about 200 mV, i.e. reversal of membrane potential to about +130 mV, the electrochemical gradient for Ca^{++} is reversed, so virtually no Ca^{++} enters the widely opened gates. However the off-effect in D to F is to be expected if the membrane potential falls more rapidly than the Ca_D gates close, so that momentarily the Ca^{++} ions can move through the gates under a favorable electrochemical gradient.

In Figure 9G–I there was the same large depolarizing current as in F, but it was briefer. A virtually identical off-effect is seen to follow within a millisecond of the current pulse, just as in F. Thus the off-effect is clearly related to the termination of the current pulse. Evidently the explanation of delayed Ca_D gate closure accounts for the off-effects in G to I just as in D to F.

In summary of the investigations on the giant synapse of the squid stellate ganglion, reference can be made to Figure 6. There are the same presynaptic channels and gates, Na_D, K_D and Ca_D, as for the nerve terminals at the neuromuscular junction. Postsynaptically only one change need be made—the elimination of the K_T gates.

The symbols for the various types of channels with their gates must not be interpreted in a too exclusive manner. For example the Na_D channel is at least as effective for Li^+ transport, and even slightly for K^+, Rb^+ and Cs^+ (Chandler and Meeves, 1965). A list has been given above for ions that pass through the Na^+ channel. The Ca_D channel is also permeable to Sr^{++} and Ba^{++} (Dodge, Miledi and Rahamimoff, 1961; Katz and Miledi (1969a). It is important that TTX blocks the Na_D channel selectively, not affecting the K_D and Ca_D channels, while intracellular TEA blocks only the K_D channel. The Ca_D channel is very effectively blocked by external Mn^{++} and Mg^{++} (Katz and Miledi, 1969a). As already stated, when the Na_D and K_D channels are eliminated, the Ca_D channel becomes so effective that it exhibits a regenerative depolarizing response (Katz and Miledi, 1969a). Calculation shows that, even if 4 Ca^{++} ions are required for the release of each quantum (Dodge and Rahamimoff, 1967), the inward calcium current during such a regenerative response gives a Ca^{++} influx 10,000 times greater than that required for transmitter release. Katz and Miledi, (1969a) suggest that this large excess may be necessary in order to give the necessary quadruple action at any one site.

Figure 6 illustrates two features of the five types of ionic channels there displayed. There is firstly the relative specificities of the channels themselves, as symbolized by the various ions principally concerned under normal conditions. Presumably these specificities have chemical bases, and the discovery of these chemical mechanisms would be a great scientific achievement (cf. Pressman, 1968; Eigen, 1969). Secondly, there are the gates controlling these channels, which may be opened by depolarization (Na_D, K_D and Ca_D) or by the action of the synaptic transmitter or some related substance (Na_T and K_T).

It does not seem expedient at this time to consider at length the ionic mechanisms responsible for inhibitory synaptic action. Just as with excitatory synaptic action, the synaptic transmitter increases the permeability of the postsynaptic membrane to certain ions. The permeable species of ions were identified by the changes induced in the inhibitory postsynaptic potential by the intracellular injection of ions. For example, if the inhibitory transmitter increases the permeability to a particular species of anion, then increase in the internal concentration of that anion by electrophoretic injection would cause the inhibitory postsynaptic potential to have an equilibrium at a more depolarized level. In the earlier investigations on spinal motoneurones four species of anions caused this reversible displacement of the E_{IPSP} in the depolarizing direction, whereas five were ineffective (Coombs, Eccles, and Fatt, 1955a). Since in the hydrated state all the permeable anions were much smaller than the impermeable, it was postulated that the inhibitory transmitter merely opened gates on channels that functioned in a sieve-like manner, allowing passage of those hydrated ions below a critical size. This postulate was very thoroughly tested on spinal motoneurones (Araki, Ito, and Oscarsson, 1961; Ito, Kostyuk, and Oshima, 1962), 33 anions being satisfactorily investigated. All eleven species below a critical size (hydrated diameter less than 1.2 times that for potassium) were permeable, whereas all larger (1.24 times potassium and upwards) were impermeable except fo the formate ion (1.35 times potassium). Furthermore, comparable observations were made on inhibitory synapses on motoneurones in fish (Asada, 1963). Except for the formate anomaly, the postulate of permeability according to hydrated ion size was corroborated. This corroboration even extended to inhibitory synaptic action on the ganglion cells of the snail (Kerkut and Thomas, 1964), there being again the formate anomaly, and only the slight difference of a very small permeability to BrO_3 ions.

The experimental testing for permeability to cations provided a more severe challenge. Intracellular injections resulted in complications due to simultaneous anion influx, and K^+ ion permeability was postulated only on the indirect grounds of the E_{IPSP} being at least 10 mV more hyperpolarized than the resting membrane potential and about half way to the E_K value of about -90 to -100 mV (Coombs, Eccles, and Fatt, 1955a; Eccles, 1966). However, preparations that could be examined in the isolated state provided good evidence that at some inhibitory synapses an increased permeability to K^+ ions contributed substantially to the inhibitory action. On the other hand Na^+ ions were excluded very selectively from the inhibitory ionic channels for the good reason that even a small Na^+ permeability would give depolarization and excitation.

In summary Figure 10 shows in A and B the way in which a transmitter could operate to open a gate guarding an ionic channel through the membrane. In C the channel allowed K^+ and Cl^- ions to pass, the electrical model in D showing the ganged switch operating for both K^+ and Cl^- permeabilities. It was postulated that this type of combined cationic and anionic conductance, presumably through the same channels as shown in C, would account for the inhibitory actions in vertebrate nerve cells, and in crustacean stretch receptor cells. By contrast in E and F the ionic conductance is dominated by K^+ ions, and it is postulated that this is due to fixed anionic charges on the walls of the channel. On the other hand, in the preparations listed below G and H the ionic channels are almost exclusively anionic, and fixed cationic charges are shown on the walls.

Since 1965, when that figure was produced (Eccles, 1966), there have been three reports that seem to require a revision or rejection of the hypotheses there modelled. It must first be stated that in the earlier investigations there was only an approximate quantitative evaluation of the relative permeability for the different anions. In a well controlled investigation Takeuchi and Takeuchi (1967) have graded nine anionic species in respect of their permeabilites across the inhibitory postsynaptic membrane of crustacean muscle. On the whole this graded series is in accord with the criterion of the hydrated ion size except for the formate anomaly. But to assess the significance of their new data one would require that impermeable anions also be plotted for such features as the naked ion size and the limiting equivalent conductivity, which is a reciprocal measure of hydrated ion size.

More difficult to evaluate is the report of an extensive investigation of anion permeability for the inhibitory action on cortical neurones (Kelly, Krnjevic, Morris, and Yem, 1969). Unfortunately, in all tests of anion action on IPSPs the anions were applied by diffusion out of the intracellular electrode. Thus it was not possible to test for recovery from the observed changes, as can be done after cessation of the injection, which was an essential feature of the motoneuronal investigations. There was some similarity between these results and the earlier investigations on motoneurones, but the reported differences were extreme, particularly in respect of organic

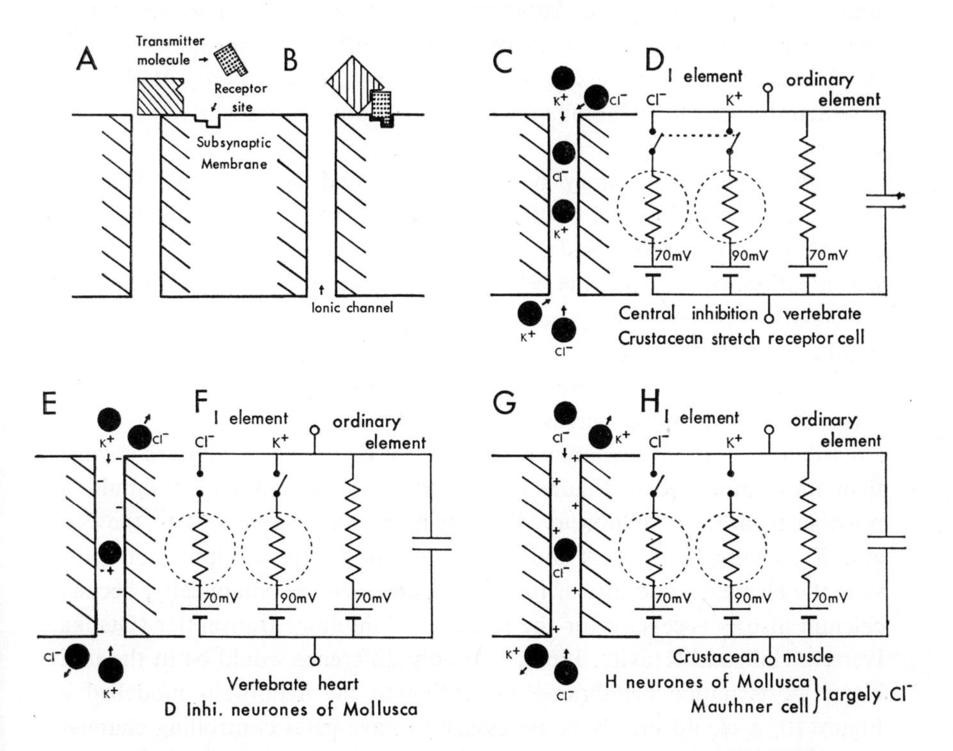

Figure 10 Diagrams summarizing the hypotheses relating to the ionic mechanisms employed by a variety of inhibitory synapses in producing IPSPs. A, B. Schematic representation of the way in which a synaptic transmitter molecule could effect a momentary opening of a channel through the subsynaptic membrane by causing the opening of a gate. In B the transmitter molecule is shown in close steric relationship both to a receptor site and to the gate which has opened. As a consequence ions can move freely through the channels in the subsynaptic membrane for the duration of the transmitter action on it. C is a schematic representation of a channel through an activated inhibitory subsynaptic membrane showing the passage of both chloride and potassium ions that is postulated for IPSP production at central inhibitory synapses, and D is a diagram resembling Figure 1 F, but showing the inhibitory element as being composed of potassium and chloride ion conductances in parallel, each with batteries given by their equilibrium potentials, and being operated by a ganged switch, closure of which symbolizes activation of the inhibitory subsynaptic membrane. E, F and G, H represent the conditions occurring at inhibitory synapses where there is predominantly potassium or chloride ionic conductance as indicated. It is assumed that the channels in E and G are restricted to cation or anion permeability respectively by the fixed changes on their walls as shown (Eccles, 1966)

anions such as propionate, butyrate, benzoate and glutamate which were observed often to reverse the IPSPs. The authors formulate a complex hypothesis, partly of pore size for inorganic anions, and partly of chemical interaction with the membrane for the organic anions. Evidently much more investigation is required.

Another perplexing story has been presented by Lux and Schubert (1969), who have injected electrophoretically a variety of metallic cations and also various amino acids. There were several strange features in this investigation, particularly the very long duration of the changes induced in the IPSPs, recovery being incomplete even after 90 minutes; whereas 3 minutes was sufficient in our earlier investigations. If seems that such prolonged changes may arise because of the induction of cellular changes rather than by a simple increase in a diffusible ion.

Evidently inhibitory synaptic action may be due to more complex changes than those modelled in Figure 10. Nevertheless, in summary it should be possible to draw an inhibitory diagram much like Figure 6, with the Na_D and K_D channels as there indicated both in the presynaptic and postsynaptic elements, and also with the Ca_D channels presynaptically, because calcium also is necessary for the release of inhibitory transmitter (Otsuka, Iversen, Hall, and Kravitz, 1966). The only difference would be in the inhibitory postsynaptic membrane. According to the hypothesis modelled in Figure 10, it would merely be necessary to have gates controlling channels across the postsynaptic membrane, that allowed passage to ions smaller than a hydrated size of about 3 Å, and with or without fixed anionic or cationic changes on their walls. However the recent reports have now called into question this simple model, but any clear alternative models have not yet been proposed. There is much work for the future.

Acknowledgement

This investigation was supported by Grant No. RO 1 NB 0822102 from the National Institute of Neurological Diseases and Stroke, U.S. Public Health Service.

REFERENCES

Araki, T., M. Ito, and O. Oscarsson. Anion permeability of the synaptic and non-synaptic motoneurone membrane, *J. Physiol.*, **159**, 410–435 (1961).

Armstrong, C. M., and L. Binstock. Anomalous rectification in the squid giant axon injected with tetraethylammonium chloride, *J. gen. Physiol.*, **48**, 859–872 (1965).

Asada, Y. Effects of intracellularly injected anions on the Mauthner cells of goldfish, *Jap. J. Physiol.* **13**, 583–598 (1963).

Chandler, W. K., and H. Meeves. Voltage clamp experiments on internally perfused giant axons, *J. Physiol.*, **180**, 788–820 (1965).

Cole, K. S. Dynamic electrical characteristics of the squid axon membrane, *Arch. Sci. Physiol.*, **3**, 253–256 (1949).

Cole, K. S. *Membranes, Ions and Impulses.* Berkeley and Los Angeles, University of California Press, (1968).

Cole, K. S., and H. J. Curtis. Electric impedance of the squid giant axon during activity, *J. gen. Physiol.* **22**, 649–670 (1939).

Coombs, J. S., D. R. Curtis, and J. C. Eccles. The interpretation of spike potentials of motoneurones, *J. Physiol.* **139**, 198–231 (1957).

Coombs, J. S., J. C. Eccles, and P. Fatt. The specific ionic conductances and the ionic movements across the motoneuronal membrane that produce the inhibitory post-synaptic potential, *J. Physiol.* **130**, 326-373 (1955a).

Coombs, J. S., J. C. Eccles, and P. Fatt. Excitatory synaptic action in motoneurones, *J. Physiol.* **130**, 374–395 (1955b).

Dodge, F. A. Jr., R. Miledi, and R. Rahamimoff. Strontium and quantal release of transmitter at the neuromuscular junction, *J. Physiol.* **200**, 267–283 (1969).

Dodge, F. A., and R. Rahamimoff. Co-operative action of calcium ions in transmitter release at the neuromuscular junction, *J. Physiol.* **193**, 419–433 (1967).

Eccles, J. C. *The neurophysiological basis of mind: The principles of neurophysiology.* Oxford: Clarendon Press (1953).

Eccles, J. C. *The Physiology of Nerve Cells.* Baltimore: Johns Hopkins Press (1957).

Eccles, J. C. *The Physiology of Synapses.* Berlin, Göttingen, Heidelberg: Springer-Verlag (1964).

Eccles, J. C. Ionic mechanisms of excitatory and inhibitory synaptic action, *Ann. N.Y. Sci.* **137**, 473–494 (1966).

Eigen, M. Alkali ion carrier; dynamics and selectivity. The Neurosciences Vol. 2, Neurosciences Research Program.

Fatt. P. Sequence of events in synaptic activation of motoneurone, *J. Neurophysiol*, **20**, 61–80 (1957).

Fatt, P. and B. Katz. An analysis of the end-plate potential recorded with an intra-cellular electrode, *J. Physiol.* **115**, 320–370 (1951).

Furukawa, T., and A. Furukawa. Effects of methyl- and ethyl-derivatives of NH_4^+ on the neuromuscular junction, *Jap. J. Physiol.* **9**, 130–142 (1959).

Furukawa, T., T. Takagi, and Sugihara. Depolarization of end-plates by acetylcholine externally applied, *Jap. J. Physiol.* **6**, 98–107 (1956).

Gage, P. W., and C. M. Armstrong. Miniature end-plate currents in voltage-clamped muscle fibres, *Nature*, **218**, 363–365 (1968).

Gage, P. W., and J. W. Moore. Synaptic current at the squid giant synapse, *Science* **160**, 510–512 (1969).

Ginsborg, B. L. Ion movements in junctional transmission, *Pharmacol. Rev.* **19**, 289–316 (1967).

Hodgkin, A. L. The conduction of the nervous impulse, Liverpool: Liverpool University Press, (1964).

Hodgkin, A. L., and A. F. Huxley. Currents carried by sodium and potassium ions through the membrane of the giant axon of Loligo, *J. Physiol.* **116**: 449–472 (1952a).

Hodgkin, A. L., and A. F. Huxley. A quantitative description of membrane current and its application to conduction and excitation in nerve, *J. Physiol.* **117**, 500–544 (1952b).

Hodgkin, A. L., and B. Katz. The effect of sodium ions on the electrical activity of the giant axon of the squid, *L. Physiol.* **108**, 37–77 (1949).

Ito, M., P. G. Kostyuk, and T. Oshima. Further study on anion permeability of inhibitory post-synaptic membrane of cat motoneurones, *J. Physiol.* **164**, 150–156 (1962).

Katz, B. Microphysiology of the neuro-muscular junction. The chemo-receptor function of the motor end-plate, *Bull. Johns Hopk. Hosp.* **102**, 296–312 (1958).

Katz, B. The transmission of impulses from nerve to muscle, and the subcellular unit of synaptic action, *Proc. Roy. Soc. B*, **155**, 455–479 (1962).

Katz, B. *Nerve Muscle, and Synapse*, New York: McGraw-Hill Book Co. (1966).

Katz, B. *The Release of Neural Transmitter Substances.* Liverpool: Liverpool University Press, (1969).

Katz, B., and R. Miledi. Propagation of electric activity in motor nerve terminals, *Proc. Roy. Soc. B*, **161**, 439–468 (1965a).

Katz, B., and R. Miledi. Measurement of synaptic delay and time course of acetylcholine release at neuromuscular junction, *Proc. Roy. Soc. B*, **161**, 469–481 (1965b).

Katz, B., and R. Miledi. Effect of calcium on acetylcholine release from motor nerve terminals, *Proc. Roy. Soc. B*, **161**, 482–489 (1965c).

Katz, B., and R. Miledi. Tetrodotoxin and neuromuscular transmission, *Proc. Roy. Soc. B*, **167**, 2–22 (1967a).

Katz, B., and R. Miledi. Release of acetylcholine from nerve endings by graded electric pulses, *Proc. Roy. Soc. B*, **167**, 23–38 (1967b).

Katz, B., and R. Miledi. The timing of calcium action during neuromuscular transmission *J. Physiol.* **189**, 535–544 (1967c).

Katz, B., and R. Miledi. A study of synaptic transmission in the absence of nerve impulses, *J. Physiol.* **192**, 407–436 (1967d).

Katz, B., and R. Miledi. The effect of local blockage of motor nerve terminals,s *J. Physiol.* **199**, 729–741 (1968).

Katz, B., and R. Miledi. Tetrodotoxin-resistant electric activity in presynaptic terminals, *J. Physiol.* **203**, 459–487 (1969a).

Katz, B., and R. Miledi. Spontaneous and evoked activity of motor nerve endings in calcium Ringer, *J. Physiol.* **203**, 689–706 (1969b).

Kelly, J. S., K. Krnjevic, M. E. Morris,, and G. K. W. Yim. Anionic permeability of cortical neurones, *Exp. Brain Res.* **7**, 11–31 (1969).

Kerkut, G. A., and R. C. Thomas. The effect of anion injection and changes in the external potassium and chloride concentration on the reversal potentials of the IPSP and acetylcholine, *Comp. Biochem. Physiol.* **11**, 199–213 (1964).

Koketsu, K., and S. Nishi. Restoration of neuromuscular transmission in sodium-free hydrazinium solution, *J. Physiol.* **147**, 239–252 (1959).

Lux, H. D., and P. Schubert. Postsynaptic inhibition: intracellular effects of various ions in spinal motoneurons, *Science*, **166**, 625–626 (1969).

Maeno, T. Analysis of sodium and potassium conductances in the procaine end-plate potential, *J. Physiol.* **183**, 592–606 (1966).

Marmont, G. Studies on the axon membrane. I. A new method, *J. cell comp. Physiol.* **34**, 351–382 (1949).

Miledi, R. Spontaneous synaptic potentials and quantal release of transmitter in the stellate ganglion of the squid. *J. Physiol.* **192**, 379–406 (1967).

Miledi, R., and C. R. Slater. The action of calcium on neuronal synapses in the squid, *J. Physiol.* **184**, 473–498 (1966).

Nastuk, W. L. Some ionic factors that influence the action of acetylcholine at the muscle end-plate membrane, *Ann. N.Y. Acad. Sci.* **81**, 317–327 (1959).

Nastuk, W. L., and A. L. Hodgkin. The electrical activity of single muscle fibres, *J. cell comp. Physiol.* **35**, 39–74 (1950).

Otsuka, M., L. L. Iversen, Z. W. Hall, and E. A. Kravitz. Release of gamma-aminobutyric acid from inhibitory nerves of lobster, *Proc. Nat. Acad. Sci.***56**: 1110–1115 (1966).

Pressman, P. C. Ionophorous antibiotics as models for biological transport, *Fed. Proc.* **27**, 1283–1288 (1968).

Takeuchi, A., and N. Takeuchi. Active phase of frog's end plate potential, *J. Neurophysiol.* **22**, 395–411 (1959).

Takeuchi, A., and N. Takeuchi. On the permeability of the end-plate membrane during the action of transmitter, *J. Physiol.* **154**, 52–67 (1960a).

Takeuchi, A., and N. Takeuchi. Further analysis of relationship between end-plate potentials and end-plate current, *J. Neurophysiol.* **23**, 397–402 (1960b).

Takeuchi, A., and N. Takeuchi. Anion permeability of the inhibitory postsynaptic membrane of the crayfish neuromuscular junction, *J. Physiol.* **191**, 575–590 (1967).

Takeuchi, N. Some properties of conductance changes at the end-plate membrane during the action of acetylcholine, *J. Physiol.* **167**, 128–140 (1963a).

Takeuchi, N. Effects of calcium on the conductance change of the end-plate membrane during the action of transmitter, *J. Physiol.* **167**, 141–155 (1963b).

PAPER 14

Pharmacological Characterization of Axonal and End-Plate Membranes

TOSHIO NARAHASHI

Department of Physiology and Pharmacology,
Duke University Medical Center,
Durham, North Carolina

I. INTRODUCTION

Physiologists have not always been reluctant to use pharmacological agents for the study of fundamental properties of excitable tissues. A number of approaches and techniques developed as a result of basic neurophysiological researches have also been utilized for the study of drug action. Such interactions between neurophysiology and neuropharmacology have had a long history since the last century when electricity began to be used as a major signal and tool for examining the excitability of various nerve and muscle tissues. Usefulness of chemicals as tools was in fact pointed out more than one century ago (Bernard, 1857). One of the wellknown examples for the use of chemicals as tools is nicotine. The ability of nicotine to block ganglionic transmission without affecting nerve conduction was utilized to study the organization of the autonomic nervous system (Langley and Dickinson, 1889). Another example is found in curare, which has long been used to block neuromuscular and ganglionic transmissions of cholinergic type without impairing the conduction of nerve and muscle themselves (Bernard, 1857).

However, it does not seem to be fair to make a statement that there has not been even a slight hesitation or reluctance on the part of physiologists to take full advantage of certain chemicals for purely fundamental researches. Or, one might say that physiologists have not necessarily been enthusiastic in searching chemicals which might be useful for their basic researches. Tetrodotoxin (TTX) has brought us a new era. It is not too much to say that, if it had not been for TTX, the progress in certain aspects of modern physiology of nerve and muscle would not have been as remarkable as it is. Similarly important is the fact that TTX triggered an era in which a number of investigators have been exploring the possibility of using a variety of drugs and chemicals as tools for the study of normal functioning of excitable tissues.

In the present article, an attempt is made to characterize and compare the ionic conductances of axonal membranes and end-plate membranes by the use of a variety of pharmacological agents. It should be emphasized that the author does not intend to give a comprehensive review on a wide variety of chemicals used as tools. Interested readers are urged to refer to review articles on the use of pharmacological agents for the basic neurophysiological study (O'Brien, 1969; Narahashi, 1972).

II. METHODS

A. Voltage clamp

Voltage clamp techniques have proved to be the most straightforward approach to the study of membrane ionic conductances. The techniques were first extensively and successfully used by Hodgkin, Huxley, and Katz for squid giant axons (Hodgkin, Huxley, and Katz, 1952; Hodgkin and Huxley, 1952a, b, c, d). Internal axial wire electrodes were used to establish space clamp conditions which were prerequisite for voltage clamping. Since then elaborative analyses and improvements of voltage clamp conditions allowed one to measure the membrane conductances in much higher precision (Moore and Cole, 1963; Cole and Moore, 1960). The voltage clamp techniques were also applied to other excitable tissues. Thanks to the development of the sucrose-gap voltage clamp method (Julian, Moore, and Goldman, 1962a, b), it is now possible to use small axons other than squid giant axons (500 μ diameter) without introducing a longitudinal electrode inside the axon. In addition to lobster giant axons (80 μ diameter) which were employed by Julian *et al.* (1962a, b), crayfish giant axons (150 μ diameter) (Narahashi, Moore and Shapiro, 1969) and even bundles of uterine smooth muscle (Anderson, 1969), taenia coli (Kumamoto and Horn, 1969) and cardiac tissues (Rougier, Vassort, and Stämpfli, 1968) are in use for the sucrose-gap voltage clamp study.

1. Sucrose-gap voltage clamp

In studies described in the present article, the sucrose-gap voltage clamp technique developed originally for lobster giant axons by Julian *et al.* (1962a, b) was used for giant axons from squid, crayfish and lobster.

Details of the techniques will be described elsewhere. The sucrosegap chamber has been slightly modified almost every year and differs somewhat for each type of giant axons, but the principle remains essentially unaltered. A narrow portion (50–150 μ in width) of an isolated giant axon was externally perfused with the physiological saline solution and insulated from both ends of the axon by means of two streams of the isotonic sucrose solution. Each of these two areas covered by the sucrose varied between 1 and 3 mm wide depending on the model of the chamber. One of the side pools contained the physiological saline solution. The other pool contained the isotonic KCl solution to depolarize the membrane completely, and served as the zero potential reference. The isotonic sucrose solution was passed through a column of resin to eliminate contaminated ions, and the final conductance usually ranged between 0.5 and 2.0 μmho/cm. The electric resistance across the sucrose gag without nerve preparation amounted to 20 to 100 MΩ. Therefore, the attenuation of the measured membrane potential caused by shortcircuiting is almost negligible.

2. *Comparison of axial wire and sucrose-gap methods*

The axial wire and sucrose-gap voltage clamp methods for axon preparations have both advantages and disadvantages. In the axial wire method, the area of the nerve membrane on which measurements are made is large and stable. This factor, together with the point control system (Moore and Cole, 1963; Cole and Moore, 1960), allows highly accurate measurements to be made for a reasonably long period of time.

The sucrose-gap method has the following disadvantages:

a) The gap between two sucrose streams must be very narrow to establish space clamp conditions, so that the membrane area on which measurements are made is small thereby making the current signal small. More serious difficulty is an unstability of the gap which not only causes fluctuations in the amplitude of recorded membrane current but also could change the pattern of the membrane current when the gap is unstable in such a way as to move from one spot to another. A number of attempts have been made to improve the stability by modifications of the chamber. In 1969, a version was obtained in the author's laboratory which satisfied this condition reasonably well. Another line of modification by the use of an automatic area compensation system was accomplished by Dr. J. W. Moore. The best practical solution for this problem would probably be the application of the automatic compensation system to the 1969 model chamber.

b) Another difficulty encountered with the sucrose-gap method is a limited survival time. In general, a decrease in the fiber diameter tends to shorten the survival time. This is probably due to the leakage of internal ions across the nerve membrane.

c) Due to the hyperpolarization caused by sucrose-saline interface (Blaustein and Goldman, 1966; Julian, Moore, and Goldman, 1962a), it is not possible to measure the absolute magnitude of resting potential. Moreover, the measured membrane potential changes if the width of gap fluctuates, because the latter affects the intensity of hyperpolarizing current flowing through the nodal membrane.

Despite the serious disadvantages associated with fluctuations of the gap, the sucrose-gap method offers two great advantages:

a) Since only a small portion of the axon preparation is used for measurements at a time, measurements can be made on several areas, which are called artifical nodes, by simply moving the preparation along the longitudinal axis. In fact, at least four or five artificial nodes, or even more, are available from one squid giant axon.

b) Since there is no need to insert any electrode inside the axon, small nerve fibers such as those from crayfish and lobster can be used.

However, with small giant axons (40 μ diameter) from the cockroach, sucrose had to be replaced by oil as an insulator (Pichon and Boistel, 1967). This is because the sucrose solution causes a depletion of internal ions so quickly that the axon membrane does not survive long enough to make any reasonable amount of measurements.

3. *Tissues other than giant axons*

Single nodes of Ranvier are convenient material for voltage clamp, because a small membrane area of the node permits space clamp conditions to be established without introducing internal electrodes. Vaseline gap in conjunction with a special feed-back device gave successful results (Dodge and Frankenhaeuser, 1958). The technique has recently been improved substantially, and it is now possible to achieve fairly good high-frequency responses which are required in voltage clamp experiments for mammalian nodes of Ranvier at body temperatures (Nonner, 1969; Nonner and Stämpfli, 1969).

Skeletal muscle fibers were difficult material for voltage clamp, because no sucrose solution could be used owing to its deteriorating effect. Applications of three microelectrodes made it possible to observe the late steady-state current, but were not completely satisfactory for the peak transient current (Adrian, Chandler, and Hodgkin, 1966)

Cardiac tissues are also important yet difficult material for the voltage clamp study. Several different techniques were successfully developed including intracellular microelectrodes and sucrose gap (Mascher and Peper,

1969; Rougier, Vassort, and Stämpfli, 1968; Deck, Kern, and Trautwein, 1964; Dudel, Peper, Rüdel and Trautwein, 1966).

Voltage clamp techniques were applied to post-junctional membranes also. Takeuchi and Takeuchi (1959, 1960) successfully applied the technique to the end-plate membrane of the frog for the first time. The end-plate membranes have since been subjected to voltage clamp analyses by several investigators (Kordaš, 1968, 1969; Gage and Armstrong, 1968; Oomura and Tomita, 1960). Cell bodies and postsynaptic membranes were also studied by voltage clamp. These include supramedullary nerve cells of the puffer fish (Hagiwara and Saito, 1959 a), *Onchidium* nerve cells (Hagiwara and Saito, 1959 b; Hagiwara and Kusano, 1961), *Aplysia* or *Helix* nerve cells (Frank and Tauc, 1964; Chamberlain and Kerkut, 1969; Alving, 1969) and the motoneurons of the cat (Araki and Terzuolo, 1962; Frank, Fuortes) and Nelson, 1959).

4. *End-plate voltage clamp*

In the present study, voltage clamp of the end-plate membranes of the frog sartorius muscle fiber was performed by the two-microelectrode method originally developed by Takeuchi and Takeuchi (1959). The electric circuit used in the present study is published elsewhere (Deguchi and Narahashi, 1971). Since the end-plate of the frog muscle is confined in an area of about 100 μ long and probably not more than a few microns wide, space clamp conditions can be established by inserting a current microelectrode in the end-plate region.

Unlike axon membranes, the time course of each ionic component of the end-plate current (e.p.c.) is not greatly different. Since the e.p.c. had been found to contain sodium and potassium components (Takeuchi and Takeuchi, 1960), measurements were made of the e.p.c.'s at the equilibrium potential for potassium (E_K) and at the equilibrium potential for sodium (E_{Na}). Only the sodium component of the e.p.c.'s should flow at E_K and only potassium component should flow at E_{Na}. However, normal muscle fibers could not be clamped at E_{Na} which was about +50 mV, because the fibers contracted at that membrane potential. Decoupling between electrical excitation and contraction was achieved by the glycerol treatment developed for skeletal muscle fibers by Eisenberg and Gage (1967) and Gage and Eisenberg (1967). The muscle preparation was first soaked in a Ringer's solution to which 400 mM glycerol was added. After about 1 hour, the preparation was brought back to normal Ringer's solution. This caused decoupling between electrical excitability and contraction, and the muscle fiber was capable of producing normal-sized action potentials without initiating contraction. The only change in the action potential by the glycerol treatment was a disappearance of the negative after-potential.

B. Internal perfusion

Internal perfusion techniques offer a chance to freely control internal as well as external phase in terms of ionic compositions, drug application, and pH value. Two methods were developed independently of each other. Baker, Hodgkin, and Shaw (1961) squeezed out the axoplasm of a squid giant axon by means of a small roller and inflated the crashed axon by perfusion with an internal medium. Oikawa, Spyropoulos, Tasaki, and Teorell (1961) inserted two glass capillaries from each end of a squid giant axon until they met at the middle of the preparation, and pulled them apart while the axoplasm was washed out by means of an internal medium. Most investigators working on internal perfusion, except for the Cambridge group and ourselves, are using the Tasaki's tunnel method or its modification.

The internal perfusate has been much improved thanks to elaborative examinations by Tasaki and his group (Tasaki, Singer, and Takenaka, 1965). The best internal anion in terms of survival time was fluoride, and phosphate, aspartate, and glutamate were also reasonably good. Bromide, iodide and thiocyanate were among the worst. This anion sequence closely follows the lyotropic series. The internal solutions which are currently in use by many investigators including ourselves contain 300 to 500 mM KF or K-glutamate and a small amount of phosphate or tris buffer. Sometimes a low concentration (10–50 mM) of sodium is added for voltage clamp experiments to define the sodium equilibrium potential.

III. EXPERIMENTAL RESULTS ON AXONAL MEMBRANES

It has been well established that there are at least three components in membrane ionic conductances of squid giant axons, i.e. a) the mechanism whereby the sodium conductance is increased upon depolarization, b) the mechanism whereby the increased sodium conductance is decreased upon sustained depolarization, and c) the mechanism whereby the potassium conductance is increased upon depolarization. Mechanism (a) is directly responsible for the rising phase of the action potential, and mechanisms (b) and (c) are responsible for the falling phase.

Exactly speaking, the terms "sodium conductance" and "potassium conductance" should be avoided. This is because not only sodium ions but also other ions such as lithium, potassium, rubidium and cesium could flow to varying extents through the "channels" which are normally utilized by sodium (Chandler and Meves, 1965; Moore, Blaustein, Anderson, and Narahashi, 1967, Rojas and Atwater, 1967). Similarly, the "channels" through which potassium ions normally flow could also be utilized by ribidium ions (Moore *et al.*, 1967). Therefore, the terms "peak transient conductance" and "late steady-state conductance "can describe the situation more accurately. It should be noted that the "channels" mentioned above do not necessarily mean anatomical pores or holes or charrier mechanisms but merely refer to the conceptual pathways through which ions flow according to the electro-chemical potential gradient.

These three components of membrane ionic conductances have been demonstrated to have different affinities for a variety of chemical agents. Some of the agents block one of the conductance components selectively, whereas some others block two of them rather indiscriminately, at least phenomenologically. These different actions on membrane conductances lead to various types of changes in action potential; the action potential may be completely blocked, may be prolonged in its falling phase sometimes resembling cardiac action potentials, or may be initiated repetitively by a single stimulus.

A. Tetrodotoxin

1. *History*

Existence of a poison in puffer fish has long been known since the early Egyptian and Chinese era. A number of records of death caused by poisoning with the puffer fish can be found in the literature. The poison has drawn much attention in Japan where the puffer fish is served as one of the most delicious fish. This is why pharmacological studies of the puffer fish poison had been carried out mostly in Japan until very recently.

The first extraction of the puffer fish poison, tetrodotoxin (TTX), was done by Tahara (1910). TTX is contained in a variety of organs of the puffer fish, especially those in the family Tetraodontidae, but is concentrated mostly in the ovary and liver, and in some species in the skin and intestines. It is possible to extract 10 g of purified TTX from 1 ton of the ovary. Detailed review of history and studies of TTX until mid-1965 is given by Kao (1966).

Several pharmacological investigations were in fact undertaken using multifiber preparations of nerve and muscle (Kuga, 1958; Kuriaki and Wada, 1957; Kurose, 1943; Matsumura and Yamamoto, 1954; Wada, 1957; Yano, 1937). However, it was not until 1960 that TTX was subjected to modern electrophysiological analyses (Narahashi, Deguchi, Urakawa, and Ohkubo, 1960). This study was by no means an accidental one. In the summer of 1959 at the University of Tokyo, we were studying the mode of action of another toxin called maltoxin obtained from the malt rootlet. Maltoxin turned out to be a neuromuscular blocking agent which suppressed the sensitivity of the end-plate membrane to acetylcholine (ACh) (Urakawa, Narahashi, Deguchi, and Ohkubo, 1960). During the course of the experiments, we thought that TTX might have a very similar effect in view of the data in the literature available at that time. However, microelectrode experiments with the TTX-treated muscle fibers revealed quite a different and unique action; the action potential was blocked without any appreciable change in resting potential or in resting membrane resistance, and delayed rectification was still observable in TTX (Narahashi *et al.*, 1960). These observations lent support to the concept that TTX blocks the sodium conductance increase selectively without affecting the potassium conductance increase. The observation was later confirmed by Nakajima, Iwasaki, and

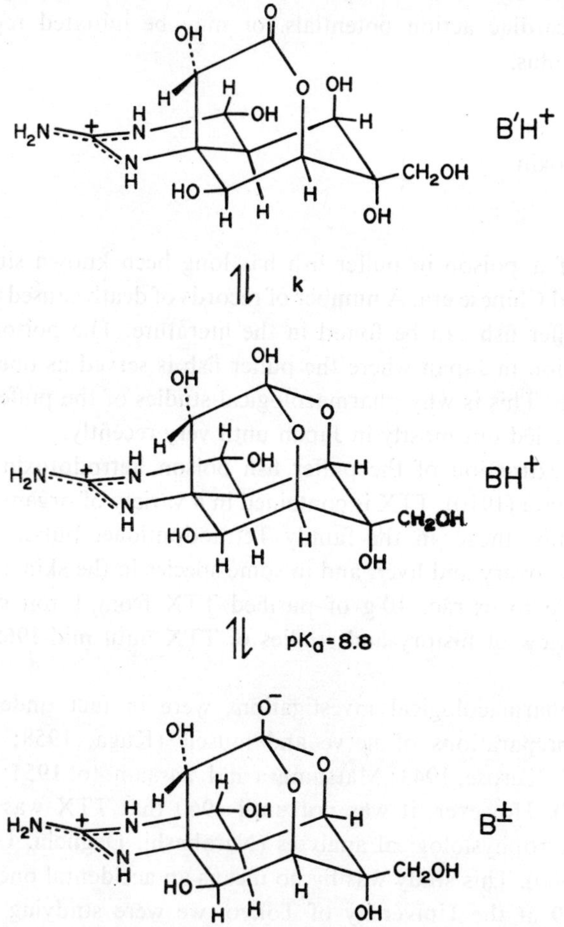

Figure 1 Chemical structures and dissociation of tetrodotoxin (Narahashi, Moore, and Frazier, 1969)

Obata (1962). However, it was obviously necessary to conduct voltage clamp experiments to demonstrate the validity of this hypothesis.

The voltage clamp experiments on TTX action were performed with lobster giant axons in December 1962 through January 1963 at Duke University. The results unequivocally demonstrated that TTX selectively inhibits the sodium conductance increase (Narahashi, Moore, and Scott, 1964). It was really fortunate and timely that the studies of the chemical structure of TTX conducted by two Japanese and two American groups resulted in the general acceptance of the formulae given in Figure 1 at the International Conference on Chemistry of Natural Products held in Kyoto, Japan, in 1964. Without accurate information on the chemical structure, TTX would not have received much attention as a tool in neurophysiology. Detailed accounts of the chemistry of TTX have been published (Goto, Kishi, Takahashi, and Hirata, 1965; Tsuda, Ikuma, Kawamura, Tachikawa, Sakai, Tamura and Amakasu, 1964; Woodward, 1964; Mosher, Fuhrman, Buchwald, and Fisher, 1964).

The specific blocking action of TTX was then confirmed with squid giant axons (Nakamura, Nakajima, and Grundfest, 1965b), with lobster giant axons (Takata, Moore, Kao, and Fuhrman, 1966), with frog nodes of Ranvier (Hille, 1968), with cockroach giant axons (Pichon, 1969), and with electroplaques (Nakamura, Nakajima, and Grundfest, 1965a). Furthermore, the characteristics of TTX blockage were explored in more detail as described in the following section.

2. *Characteristics of conductance blockage*

In our earlier study, it was found that TTX blocked the peak transient inward (sodium) current without any effect on the late steady-state (potassium) current (Narahashi *et al.*, 1964). In the experiments by Takata *et al.* (1966), neither the kinetics of these currents nor the steady-state sodium inactivation was affected by TTX. However, the observation by Nakamura *et al.* (1965b) indicated that the outward peak transient current beyond about +40 mV membrane potential was much less inhibited than the inward peak transient current by the action of TTX. This point was not clear in our earlier work (Narahashi *et al.*, 1964). In this connection, a question arises as to whether TTX has any affinity for the direction of the transient current. Another question, which is not totally unrelated to the above question, is whether TTX blocks the movement of sodium ions per se or TTX blocks the channel for the peak transient current regardless of the kind of ions flowing through that channel. Similarly, one might ask whether the late steady-state channel is unaffected by TTX regardless of the kind of ions and regardless of the direction of current.

These questions were answered by appropriately designed experimentation with squid giant axons (Moore *et al.*, 1967). In the first place, observations of membrane currents beyond the E_{Na} value demonstrated that TTX did indeed block both inward and outward transient currents equally. More confirmative observations were made in a solution in which cesium was replaced for sodium. Since cesium hardly crosses the nerve membrane through the peak transient channel (Pickard, Lettvin, Moore, Takata, Pooler, and Bernstein, 1964; Chandler and Meves, 1965; Moore, Anderson, Blaustein, Takata, Lettvin, Pickard, Bernstein, and Pooler, 1966), the sodium equilibrium potential is shifted toward the inside negative membrane potential thereby causing large outward transient currents upon depolarization. The outward transient current was in fact effectively inhibited by TTX.

Lithium can substitute for sodium in producing normal-sized action potentials (Overton, 1902; Hodgkin and Katz, 1949; Narahashi, 1963). Under voltage clamp conditions, the inward transient current is carried mostly by lithium, while the outward transient current is carried mostly by sodium. Both of these currents were equally blocked by TTX (Figure 2).

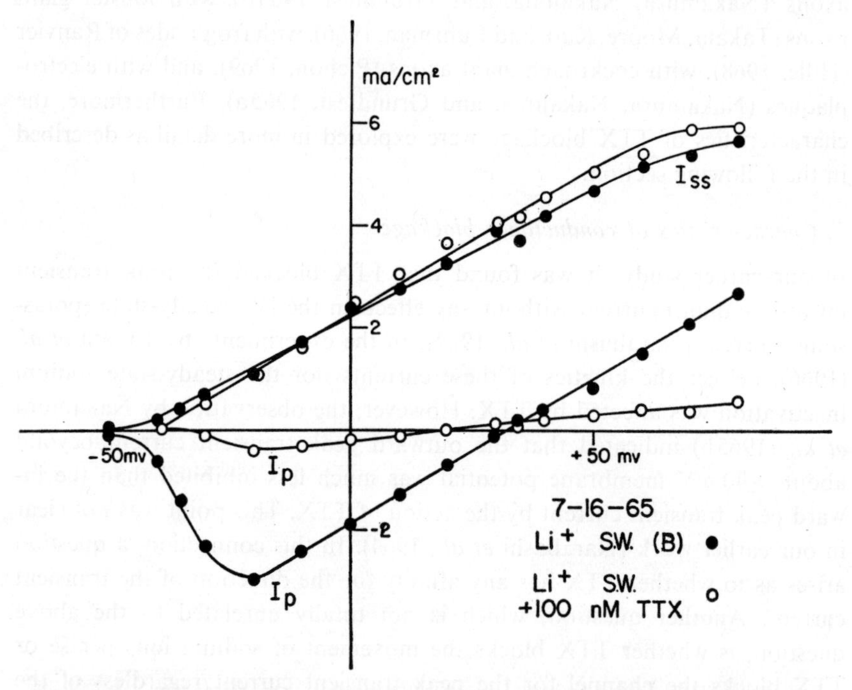

Figure 2 Current-voltage relationships for peak transient lithium current (I_p) and for steady-state potassium current (I_{ss}) in a squid giant axon bathed in a sea water in which lithium is replaced for sodium (Li⁺S.W.) and in Li⁺S.W. added with 100 nM tetrodotoxin (TTX) (Moore, Blaustein, Anderson, and Narahashi, 1967)

These observations demonstrate that TTX blocks the peak transient current regardless of the direction of current and regardless of the ion that carries the current.

Not only sodium and lithium but also potassium, rubidium and even cesium can flow through the peak transient channel to a limited extent (Chandler and Meves, 1965). The potassium component of the peak transient current has been shown to be blocked by TTX (Rojas and Atwater, 1967). Hydrazine and guanidine ions could also be carried through the transient channel, and the current was blocked by TTX (Tasaki and Singer, 1966).

In a solution in which potassium is substituted for sodium, the steady-state currents flow in inward direction at certain membrane potentials. Similar steady-state currents are observed in a solution in which rubidium is replaced for sodium. Both inward and outward steady-state currents, either in the high potassium or in the high rubidium medium, were not affected by TTX. Thus it is clear that the late steady-state conductance increase is insensitive to TTX regardless of the direction of current flow or regardless of the kind of ions involved.

3. *Site of action in nerve membranes*

Tetrodotoxin is hardly soluble in most organic solvents (Mosher *et al.*, 1964). It is therefore reasonable to assume that TTX cannot penetrate the lipid nerve membrane. This property of TTX gives us an excellent opportunity to study its site of action in the nerve membrane.

In spite of the highly potent blocking action of TTX when applied outside the nerve membrane, TTX was practically inert when applied inside

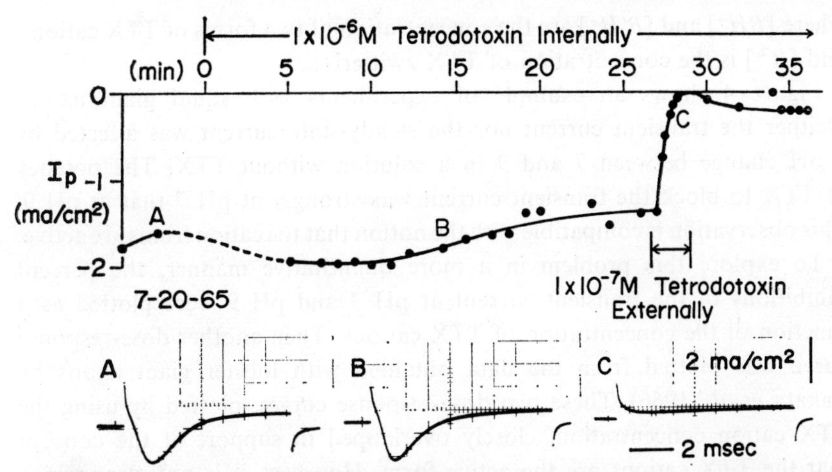

Figure 3 Changes in peak transient sodium current (I_p) during application of 1×10^{-6} M tetrodotoxin (TTX) internally and 1×10^{-7} M TTX externally in an internally perfused squid giant axon (Narahashi, Anderson, and Moore, 1966)

of the squid axon (Narahashi, Anderson, and Moore, 1966, 1967). An example of such experiments is illustrated in Figure 3. The peak transient current remained unchanged while TTX was internally perfused at a high concentration of 10^{-6} M for a period of 27 minutes of observation. Introduction of TTX in the external phase at a lower concentration of 10^{-7} M quickly abolished the transient current. In separate experiments, the action potential was not affected while TTX was internally perfused at a very high concentration of 10^{-5} M for a period of 17 minutes of observation. Since externally applied TTX is able to block the transient current or action potential at a concentration of 3×10^{-8} M within a few minutes, it can be concluded that the inner surface of the nerve membrane is less sensitive to TTX than the outer surface by a factor of 300 or more.

4. *Active form*

Tetrodotoxin exists in two cation forms and a zwitterion form as is shown in Figure 1. Since the pK_a value for the dissociation is estimated as 8.7 (Goto *et al.*, 1965) or 8.84 (Tsuda *et al.*, 1964), it is possible to determine the active form of TTX by comparison of the blocking potency at different pH levels (Narahashi, Moore, and Frazier, 1969). The ratio of the total concentration of TTX cations to the concentration of TTX zwitterion is given by the following equation

$$\log \frac{[BH^+] + [B'H^+]}{[B^\pm]} = pK_a - pH \qquad (1)$$

where $[BH^+]$ and $[B'H^+]$ are the concentration of two forms of TTX cations, and $[B^\pm]$ is the concentration of TTX zwitterion.

Figure 4 shows an example of experiments with squid giant axons. Neither the transient current nor the steady-state current was affected by a pH change between 7 and 9 in a solution without TTX. The potency of TTX to block the transient current was stronger at pH 7 that at pH 9. This observation is compatible with the notion that the cation froms are active.

To explore this problem in a more quantitative manner, the percent inhibitions of the transient current at pH 7 and pH 9 were plotted as a function of the concentration of TTX cations. Then another dose-response curve was plotted from the data obtained with lobster giant axons by Takata *et al.* (1966). These two dose-response curves, plotted by using the TTX cation concentration, closely overlapped in support of the concept that the TTX cations are the active form. However, it is not clear which cation form is active. The value for the equilibrium constant between the two TTX cations is necessary for this determination.

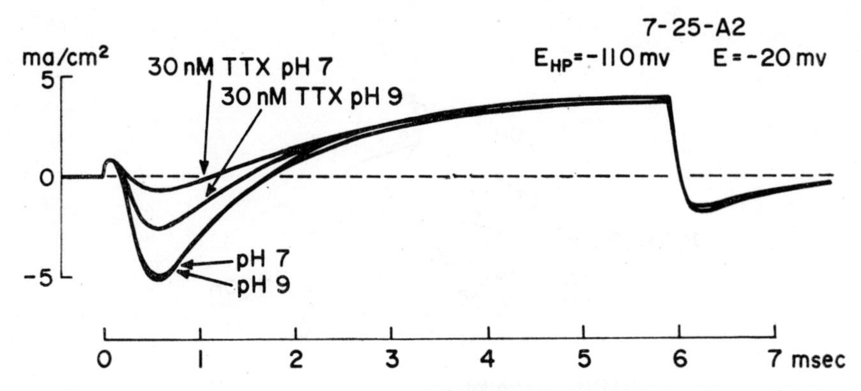

Figure 4 Membrane currents associated with step depolarizations of a constant amplitude in a squid giant axon bathed in artificial sea water at pH 7 or pH 9 with or without 30 nM tetrodotoxin (TTX) (Narahashi, Moore, and Frazier, 1969)

Similar results were obtained by Camougis, Takman, and Tasse (1967) and by Hille (1968) with frog nerves, and by Ogura and Mori (1968) with crayfish abdominal nerve cords.

5. *Structure-activity relationship*

A variety of TTX derivatives have been tested for their potency to block the nerve conduction and for their toxicity to mice (Deguchi, 1967; Tsuda *et al.*, 1964; Narahashi, Moore, and Poston, 1967). The chemical structures of these derivatives are shown in Figure 5, and the summarized data on the nerve blocking concentration are given in Table 1.

Table 1 Comparison of tetrodotoxin and its derivatives for their nerve blocking concentration

Derivative	Blockage* of the sciatic nerve (nM)	Blockage† of lobster nerve (nM)
Tetrodotoxin	16	30
Deoxytetrodotoxin	120	200
Methoxytetrodotoxin	450	
Ethoxytetrodotoxin	960	
Tetrodaminotoxin	1,700	2,000
11-Monoformylanhydrotetrodotoxin	1,800	
Anhydrotetrodotoxin	2,800	3,000
6, 11-Diacetylanhydrotetrodotoxin	36,000	
Tetrodonic acid	>590,000	>100,000

* From Deguchi (1967).
† From Narahashi, Moore, and Poston (1967).

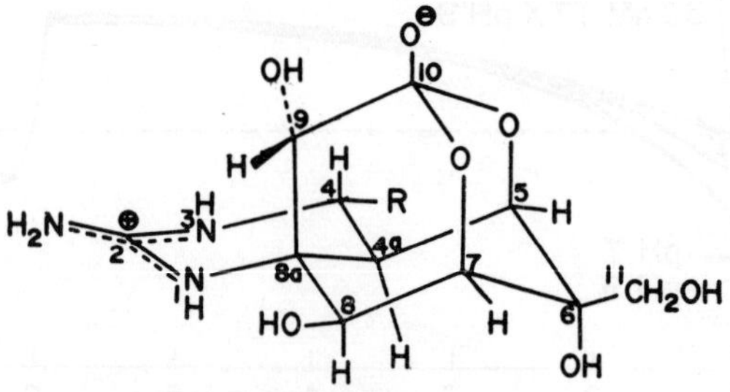

R=OH, Tetrodotoxin
R=H, Desoxytetrodotoxin
R=OMe, Methoxytetrodotoxin
R=OEt, Ethoxytetrodotoxin
R=NH_2, Tetrodaminotoxin

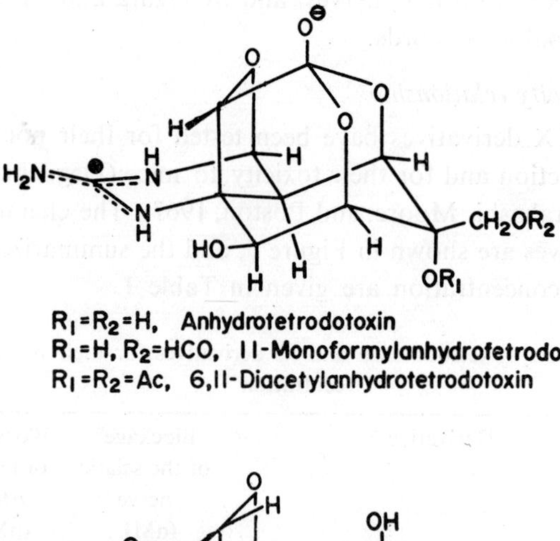

R_1=R_2=H, Anhydrotetrodotoxin
R_1=H, R_2=HCO, 11-Monoformylanhydrotetrodotoxin
R_1=R_2=Ac, 6,11-Diacetylanhydrotetrodotoxin

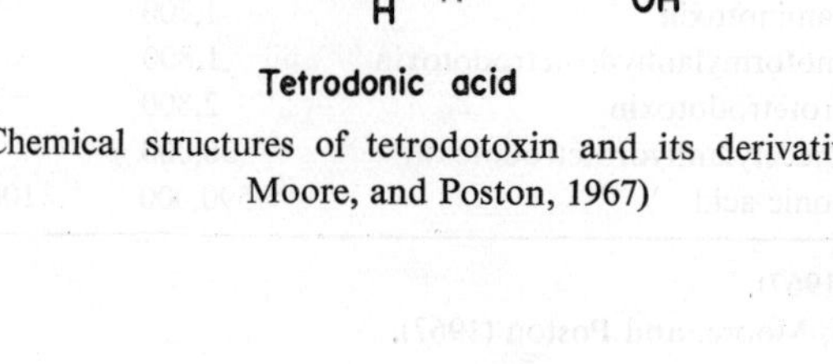

Tetrodonic acid

Figure 5 Chemical structures of tetrodotoxin and its derivatives (Narahashi, Moore, and Poston, 1967)

All of the derivatives except for tetrodonic acid were somewhat effective in blocking the nerve conduction when appropriate concentrations were used. In addition, deoxytetrodotoxin, anhydrotetrodotoxin and tetrodaminotoxin were found to block the peak transient current selectively in the same manner as TTX itself. However, a question arose as to whether the derivative samples were contaminated by TTX itself. This problem would become serious in view of the highly potent action of TTX. For this reason, the derivative samples used in our experiments were checked for their contaminations by means of nuclear magnetic resonance by Drs. H. S. Mosher and L. J. Durham in Stanford University. The deoxytetrodotoxin sample was found to contain as much as 15 per cent TTX which was more than enough to account for the apparent blocking action of this sample in terms of the TTX contamination alone. With regard to tetrodaminotoxin and anhydrotetrodotoxin which exhibited the blocking action at a concentration about 100 times higher than the blocking concentration of TTX itself, it was not possible to detect one per cent contamination in any reliable manner. Thus, it was concluded that deoxytetrodotoxin and tetrodonic acid were almost completely inert. It remains to be seen whether tetrodaminotoxin and anhydrotetrodotoxin have any blocking potency. It is of great importance that a reduction of TTX at carbon-4 position to yield deoxytetrodotoxin completely abolishes the potency. This is in sharp contrast with saxitoxin which exhibits almost the same blocking potency despite its somewhat different chemical structure (see Section III B).

6. *Number of receptors on nerve membranes*

The number of TTX molecules adsorbed on the nerve membrane was measured on the lobster leg nerve (Moore, Narahashi, and Shaw, 1967). This nerve preparation offers several advantages for certain types of experiments:

a) The nerve contains a large number of medium and small nerve fibers, so that the total membrane surface area is fairly large.

b) The nerve can be isolated very easily. It is isolated by chopping off a walking leg, breaking it at a joint, and pulling both sides apart. It takes only 10 seconds or so to isolate a nerve.

c) A long stretch of about 4 cm is obtained making the handling very easy.

A nerve preparation was mounted in a chamber and a small amount (50 μl) of 300 nM TTX was applied. The magnitude of action potential was then plotted against time to obtain a TTX blockage curve. The TTX solution was sucked up and reapplied to the second fresh nerve preparation, and another TTX blockage curve was obtained. This process was repeated several times using the same TTX solution. Thus a family of blockage curves was obtained. In separate control experiments, the blockage curves for calibration of TTX concentrations were obtained using different TTX concentrations, and were used to calculate the amount of TTX adsorbed to all of the test preparations except for the last. The total membrane area of the nerve fibers was measured under light and electron microscopes, and the extracellular sodium space was also measured. From these measurements, the amount of TTX adsorbed on the unit area of the nerve membrane was computed. It turned out that only 13 molecules of TTX were adsorbed on 1 μ^2 area of the nerve membrane. Because of uncertainty about the possible adsorption of TTX on membranes other than those of the nerve, this figure should be regarded as the upper limit.

By comparing the observed dose-response curve for TTX with the calculation, Hille (1968) concluded that TTX interacts with the receptors on one-for-one basis. It then follows that interactions of 13 or less receptors per 1 μ^2 membrane with TTX molecules are enough to cause complete blockage.

Using the figure of 200 mmho/cm^2 for the maximum sodium conductance of the lobster giant axon membrane (Narahashi *et al.*, 1964; Pooler, 1968), the conductance of each sodium channel is calculated as a maximum of 0.15 nmho. This value is of the same order of magnitude as that calculated by Hille (1968).

B. Saxitoxin

Saxitoxin (STX) is the toxic substance obtained from toxic Alaska butter clams, *Saxidomas giganteus.* There is evidence that STX originally comes from the dinoflagellate, *Gonyaulax catanella* (Schantz, Lynch, Vayvada, Matsumoto, and Rapoport, 1966; Kao, 1966). The chemical structure of STX has recently been identified (Wong, Oesterlin, and Rapoport, 1971), and is shown in Figure 6.

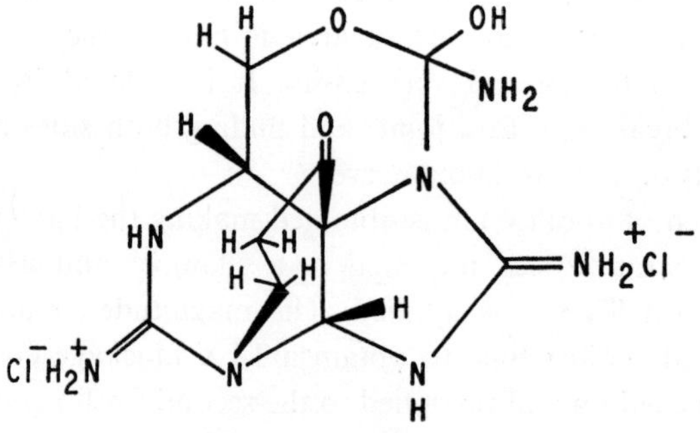

SAXITOXIN

Figure 6 Structure of saxitoxin

In view of the fact that the chemical structure of STX is somewhat different from that of TTX, and that the blocking potency of TTX is extremely sensitive to a small change in structure as has been observed with deoxytetrodotoxin, it is of great interest that STX exerts essentially the same effect as that of TTX on the membrane conductance (Narahashi, Haas, and Therrien, 1967; Nakamura *et al.*, 1965a; Hille, 1968). An example of current-voltage curves before and during STX treatment is illustrated in Figure 7. Saxitoxin is also inert from inside the squid nerve membrane (Narahashi, 1971). The only difference between STX and TTX in their action at the cellular level is that the recovery by washing is faster after STX than after TTX (Narahashi, Haas, and Therrien, 1967).

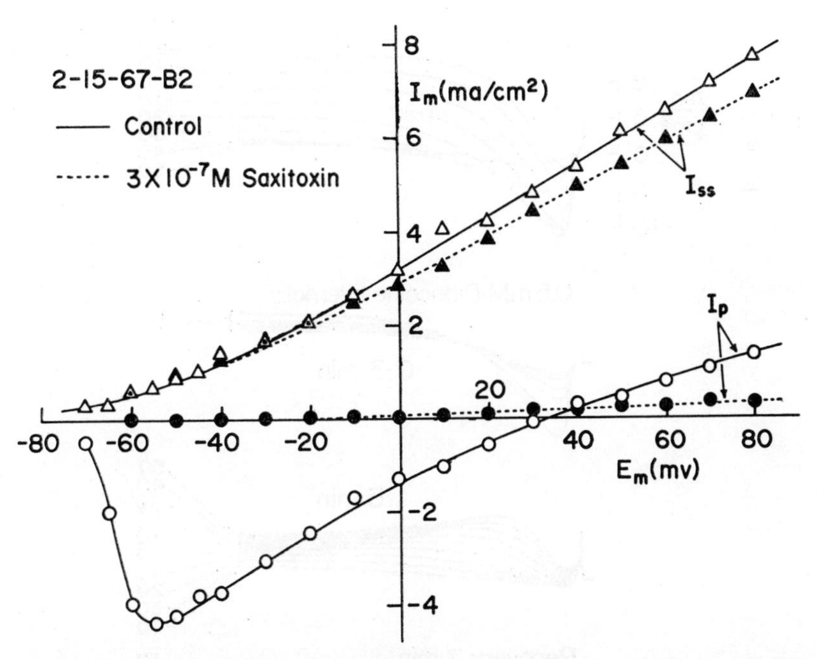

Figure 7 Current-voltage relationships for peak transient sodium current (I_p) and for steady-state potassium current (I_{ss}) in a lobster giant axon before and during application of saxitoxin

C. Local anesthetics

1. *Characteristics of conductance blockage*

It has long been known that local anesthetics block nerve conduction without greatly changing the resting potential (see Shanes (1958) and Ritchie and Greengard (1966) for detailed references). This action of local anesthetics is sometimes called "stabilization". However, it was not until

1959 that local anesthetics were subjected to voltage clamp analyses. Using squid giant axons, Taylor (1959) and Shanes, Freygang, Grundfest, and Amatniek (1959) found that cocaine and procaine inhibited both peak transient and late steady-state conductance increases.

Since then similar results were obtained with other local anesthetics using squid or lobster giant axons as material (Blaustein and Goldman, 1966; Blaustein, 1968b; Narahashi, Anderson, and Moore, 1967; Narahashi, Moore, and Poston, 1969).

Unlike TTX, most local anesthetics are able to penetrate the lipid nerve membrane in the uncharged molecular form. Thus the blocking action is expected to be exerted from either side of the nerve membrane. This has

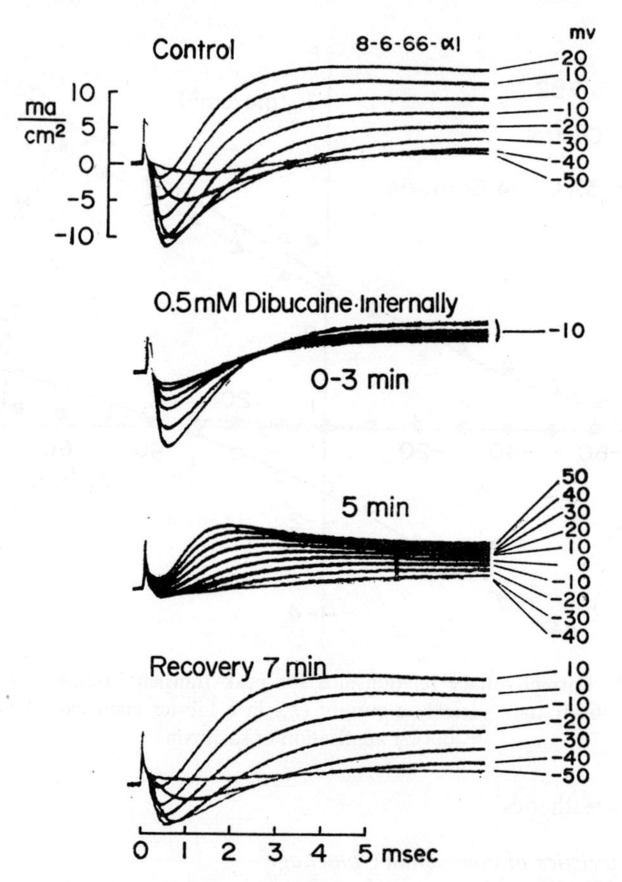

Figure 8 Families of membrane currents associated with step depolarizations in an internally perfused squid giant axon before and during application of dibucaine internally, and after washing with normal internal medium. The second set of records shows changes in membrane current associated with a constant potential step during the course of blockage (Narahashi, Moore, and Poston, 1969)

actually been the case (Narahashi, Anderson, and Moore, 1967; Narahashi, Moore, and Poston, 1969).

An example of the effect of local anesthetics on the membrane currents is shown in Figure 8. The family of membrane currents at the top was obtained before application of dibucaine. The second group of records was obtained immediately before (0 minute) and after starting internal perfusion of dibucaine at a concentration of 0.5 mM. It is seen that both the transient current and the steady-state current are suppressed during a 3-min period of dibucaine perfusion. The third family of membrane currents was recorded 5 minutes after beginning of dibucaine perfusion. Three remarkable effects of dibucaine are observable:

a) The transient current is suppressed.
b) The steady-state current is also suppressed.
c) There appears a hump during the course of the late current.

The bottom records show the recovery after 7-min washing.

The current-voltage relations plotted from Figure 8 are illustrated by Figure 9. The late current is plotted both at the hump and at the steady state. A linear current-voltage relation is obtained at the hump, whereas

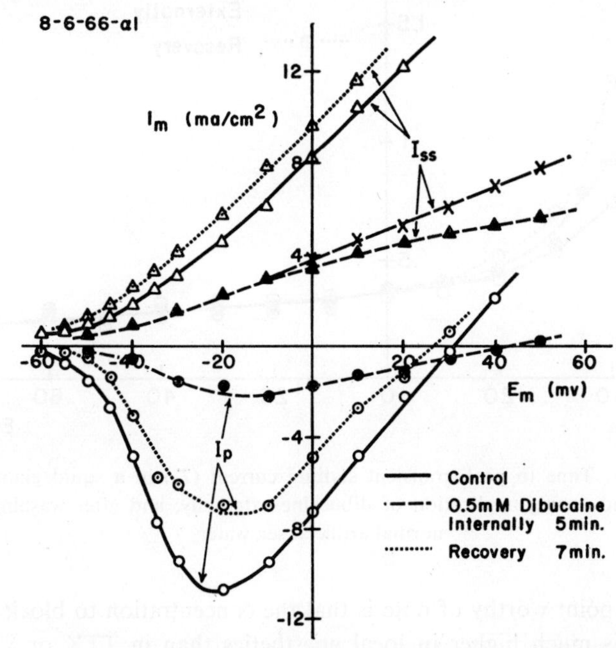

Figure 9 Current-voltage relations for peak transient sodium current (I_p) and for steady-state potassium current (I_{ss}) in an internally perfused squid giant axon before and during application of dibucaine internally, and after washing with normal internal medium (Narahashi, Moore, and Poston, 1969)

the relation of the steady state shows a curvature. This suggests that the plot at the hump represents the conductance for the late current, and the decline of the current after the hump is indicative of an inactivation process such as normally observed in association with the peak current. In fact, Armstrong (1969) observed a more remarkable hump after injection of n-pentyltriethylammonium into the squid giant axon.

However, such a hump was not always observed with other local anesthetics. Dibucaine, tropine p-tolyl acetate hydrochloride (tertiary tropine), and tropine p-tolyl acetate methiodide (quaternary tropine) caused the hump to appear, whereas procaine never produced the hump.

The kinetics of the peak transient current were also affected by some of the local anesthetics. Procaine, tertiary tropine, and quaternary tropine slowed the time for the transient current to reach its peak, whereas dibucaine had no remarkable effect on the kinetics (Figure 10).

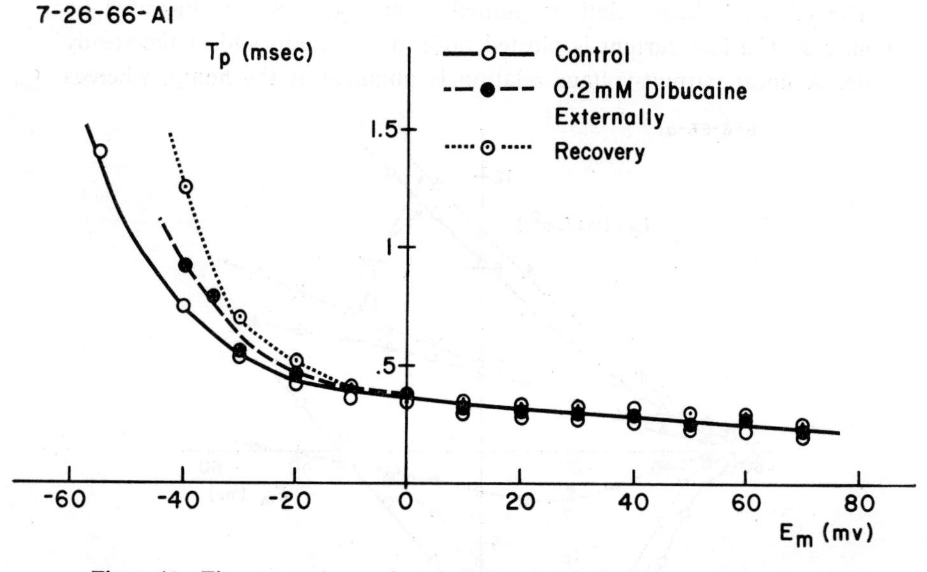

Figure 10 Time to peak transient sodium current (T_p) in a squid giant axon before and during application of dibucaine externally, and after washing with normal artificial sea water

Another point worthy of note is that the concentration to block the conductances is much higher in local anesthetics than in TTX or STX. For example, the effect of 10^{-3} M procaine is comparable to that of 10^{-8} M TTX or STX. Thus the difference in the effective concentration is of the order of 10^5.

2. *Site of action and active form*

Most of the local anesthetics which are in clinical use are basic tertiary amines, so that they exist in the uncharged molecular form and in the charged cationic form. The ratio of the charged to the uncharged form can be calculated by the Henderson-Hasselbach equation

$$\log \frac{[BH^+]}{[B]} = pK_a - \text{pH} \tag{2}$$

where $[BH^+]$ and $[B]$ are the concentrations of charged and uncharged forms, respectively. Since most local anesthetics have a pK_a value near the physiological pH, both the charged and the uncharged forms exist under normal conditions. The question as to which form is responsible for blocking action has received much attention since the early part of this century, and a number of studies have been conducted. Details of the history are reviewed elsewhere (Narahashi and Frazier, 1971; Ritchie and Greengard, 1966; Ariëns, Simonis, and van Rossum, 1964). In short, these studies were based on the observations of the effect of pH changes on the local anesthetic potency.

However, the situation is not so simple as one might expect. Some investigators obtained results in which the blocking potency was increased upon raising the pH, whereas others obtained the opposite results. Generally speaking, earlier studies agreed in that the uncharged form was responsible for blockage because raising the pH increased the potency. However, the data in support of the charged form being active began to accumulate since 1960, and seem to have generally been accepted (Ritchie and Greengard, 1966). On the contrary, there are still many observations which cannot easily be accounted for by that notion.

We decided to reinvestigate this problem by more straightforward approaches. One of the important factors which have generally been neglected in the past is the site of action in the nerve fiber. Although the nerve sheath surrounding the nerve fibers received much attention for its role as a diffusion barrier (Ritchie and Greengard, 1966), the possibility of the nerve excitable membrane as another diffusion barrier was not taken into consideration. For example, if the site of action were located on the inner surface of the nerve membrane, the local anesthetic would have to penetrate two diffusion barriers before exerting its blocking action.

The experiments were performed during the summers of 1967 and 1968 at the Marine Biological Laboratory, Woods Hole, Massachusetts. The detailed experimental results and analyses have been published elsewhere (Narahashi, Yamada, and Frazier, 1969; Narahashi, Frazier, and Yamada,

1970; Frazier, Narahashi, and Yamada, 1970). Three major approaches were used:

a) Using internally perfused squid giant axons, tertiary derivatives of lidocaine were applied either externally or internally at different pH values while keeping the pH in the opposite phase constant, and the blocking potency was compared between high and low pH values. On the assumption that the uncharged form was freely permeable to the nerve membrane, the relationship between the effective dose fifty (ED 50) to block the action potential 50 per cent and the pH was calculated for three possible active forms, i.e. 1) the internally and externally present uncharged forms, 2) the internally present charged form, and 3) the externally present charged form. If one of these forms is really active, the observations should fit the calculations in which that form is assumed as active. Another method of analyses was also used. The data were expressed as dose-response curves in which the dose was calculated as the concentration of each of the three possible active forms. The measurements should fall into a smooth curve if the assumption of the active form is correct.

b) Tertiary derivatives of lidocaine were applied either inside or outside at a constant pH value and the pH in the opposite phase was altered. The pH change in the phase where the site of action is located should affect the blocking potency of the anesthetic applied to the opposite phase, whereas the pH change in the phase opposite to the site of action should not affect the potency of the anesthetic applied to the site of action.

c) Permanently charged quaternary compounds were compared for their blocking potency from outside and inside the nerve membrane. Because

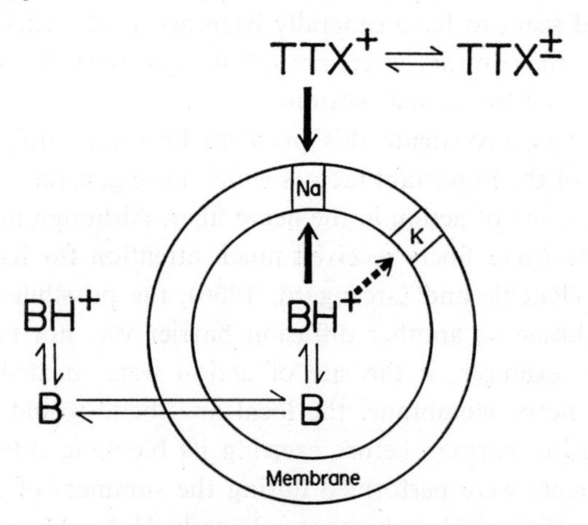

Figure 11 Diagram for the site of action and active form of tetrodotoxin (TTX) and of a tertiary amine local anesthetic. BH^+ and B refer to the charged and uncharged forms of the local anesthetic

they can hardly penetrate the nerve membrane owing to the positive charge, the blocking potency should be much higher from the phase where the site of action is located.

The experimental results will not be described here repeatedly. In summary, all of the three kinds of analyses unequivocally support the notion that local anesthetics penetrate the nerve membrane in the uncharged form and block the action potential from inside the nerve membrane in the charged form. This is in sharp contrast with TTX whose cation forms block the action potential only from the outside of the nerve membrane (Section III A 3 and 4). Figure 11 diagrammatically shows the sites of action and active forms of local anesthetics and TTX.

It should be noted that all of the tertiary and quaternary compounds used in this study are able to block both the peak transient and the late steady-state conductance increases (Narahashi, Frazier, and Moore, 1968; Frazier, Narahashi, and Moore, 1968, 1969). Thus it can be said that each of the transient and steady-state conductance channels is blocked by these local anesthetics from inside the nerve membrane.

D. Barbiturates

1. *Characteristics of conductance blockage*

Although barbiturates are not used as local anesthetics, they can block the conduction of the isolated nerve when applied at a relatively high concen-

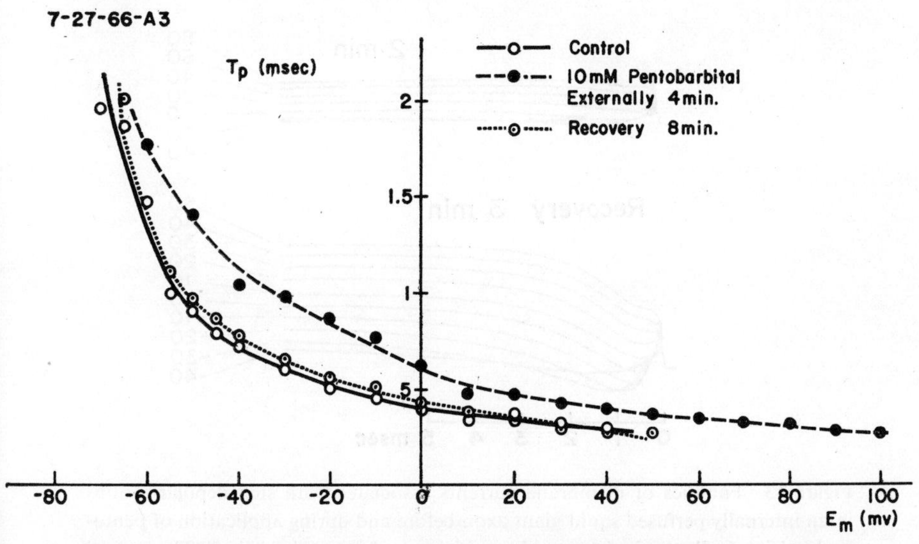

Figure 12 Time to peak transient sodium current (T_p) in a squid giant axon before and during application of pentobarbital externally, and after washing with normal artificial sea water

tration. Voltage clamp experiments with squid and lobster giant axons have revealed that pentobarbital and thiopental block both the peak transient and the steady-state conductance increases (Blaustein, 1968a; Narahashi, Moore and Poston, 1969). The kinetics for the transient conductance increase were slowed (Figure 12), but no hump was observed during the late current (Figure 13). Pentobarbital exhibited its blocking action from either side of the nerve membrane.

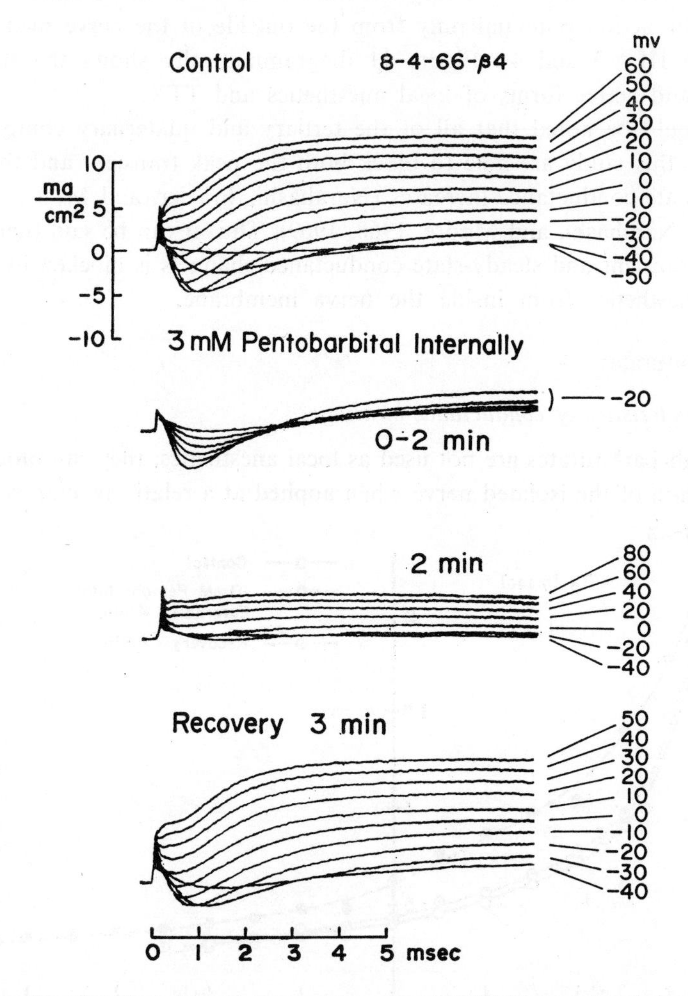

Figure 13 Families of membrane currents associated with step depolarizations in an internally perfused squid giant axon before and during application of pentobarbital internally, and after washing with normal internal medium. The second set of records shows changes in membrane current at a constant potential step during the course of blockage (Narahashi, Moore, and Poston, 1969)

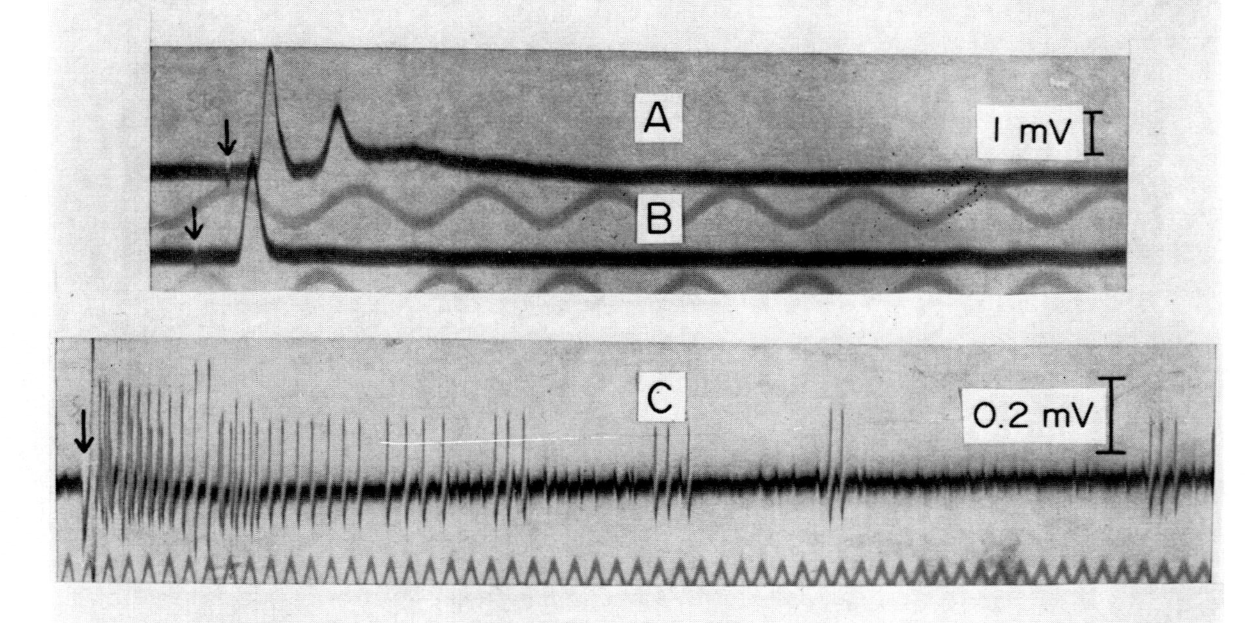

Figure 14 The effect of DDT on the action potential recorded externally from the nerve bundle of the larva of the beetle, *Allomyrina dichotomus* L. A and B, normal responses produced by a supramaximum and a submaximum stimulus, respectively. C, repetitive discharges produced by a supramaximum stimulus 74 minutes after treatment with 3×10^{-6} M DDT. The arrows indicate stimuli, and the time marker 33 c.p.s. (Yamasaki and Ishii (Former name of Narahashi, 1952a)

2. *Site of action and active form*

Barbiturates are dissociated into anions according to the following equation:

$$\log \frac{[A]}{[A^-]} = pK_a - \text{pH} \tag{3}$$

where $[A]$ and $[A^-]$ are the concentrations of the uncharged molecular form and the charged anion form, respectively.

Unlike local anesthetics, only a few studies have been undertaken concerning the active form and site of action of barbiturates. The results of these studies are controversial. By observations of cell division of sea urchin eggs and movement of sea urchin larvae, Clowes, Keltch, and Krahl (1940) suggested that barbiturates penetrated the membrane in the uncharged form and exerted the inhibitory action also in the uncharged form. On the contrary, Blaustein (1968a) concluded that the charged anionic form was active on the basis of the observation that pentobarbital was able to block the action potential of the lobster giant axon more effectively at a high pH than at a low pH. However, Krupp, Bianchi, and Suarez-Kurtz (1969) recently found that pentobarbital and phenobarbital were more effective at a low pH than at a high pH in blocking the action potential when applied to the desheathed frog nerve.

This problem has been reinvestigated by more straightforward approaches similar to those used for local anesthetics (Narahashi, Frazier, Deguchi, Cleaves, and Ernau, 1970, 1971). In short, it is concluded that pentobarbital penetrates the nerve membrane in the uncharged form and blocks the action potential also in the uncharged form. When compared at the same pH, pentobarbital was much more potent from inside than from outside of the nerve membrane. This can be accounted for by the assumption that the pentobarbital molecules are unevenly distributed in the nerve membrane and are more concentrated at the external surface than at the internal surface. It is, however, not possible to determine the site of action of the uncharged form without further experimentation. Thus, pentobarbital is different either from TTX or from local anesthetics in its uncharged active form. This difference implies that both conductance channels could be blocked by entirely different mechanisms.

E. DDT

1. *Effects on action potential*

The insecticide DDT has long been known to produce repetitive discharges in the nervous system. Some of the sensory nerves were especially sensitive,

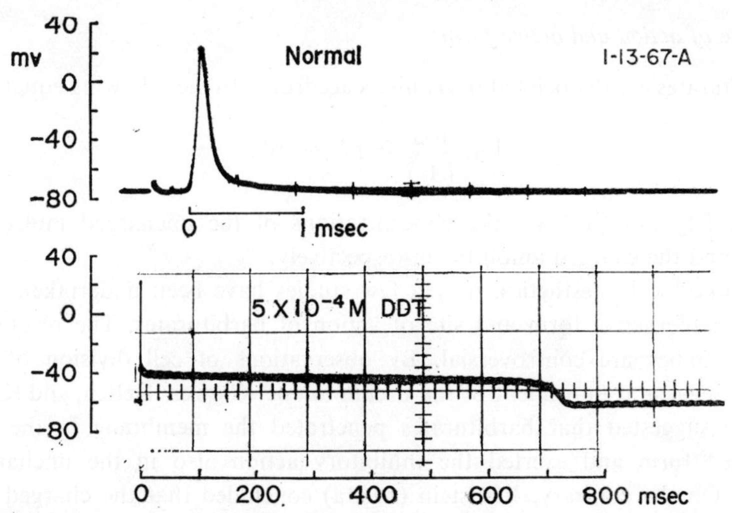

Figure 15 The effect of DDT on the internally recorded action potential from a lobster giant axon. Note a marked prolongation of the falling phase of the action potential

and those from the cockroach leg initiated trains of impulses in response to a weak concentration of DDT (10^{-8} M) (Roeder and Weiant, 1948; Yamasaki and Ishii, 1954a, b; Lalonde and Brown, 1954). Repetitive discharges were also induced in the peripheral motor nerve and the central nerve cord of insects (Yamasaki and Ishii, 1952a, b, 1954a; Yamasaki and Narahashi, 1958, 1962; Narahashi, 1964; Welsh and Gordon, 1947; Gordon and Welsh, 1948). An example of such records is illustrated in Figure 14.

DDT has an additional effect on the nerve. In 1949, Shanes observed a prolongation of the falling phase of the externally recorded action potential in the DDT-poisoned crab nerve. This effect of DDT was confirmed with the cockroach nerve (Yamasaki and Ishii, 1952b), and further studies were made of its mechanism by means of external and internal electrodes (Yamasaki and Narahashi, 1957a, b; Narahashi and Yamasaki, 1960b, c). An example of the prolonged action potential is shown in Figure 15. It was suggested that the mechanism whereby the transient conductance is decreased upon sustained depolarization or sodium inactivation, the mechanism whereby the steady-state conductance is increased upon depolarization or potassium activation, or both, are inhibited by DDT thereby increasing and prolonging the negative after-potential.

2. *Effects on membrane conductances*

Voltage clamp experiments with lobster giant axons have demonstrated the validity of the previous hypothesis (Narahashi and Haas, 1967, 1968).

Figure 16 shows families of membrane currents associated with step depolarizations. The top family of currents represents the normal control. The second set was recorded from another axon poisoned with DDT. At least one remarkable change is easily recognized, i.e. the peak transient current, though rises almost normally, falls very slowly, and is followed by an *inward* steady-state current at certain membrane potentials. It is unlikely that the inward steady-state current is entirely due to potassium flow, because the potassium concentration gradient does not seem to undergo

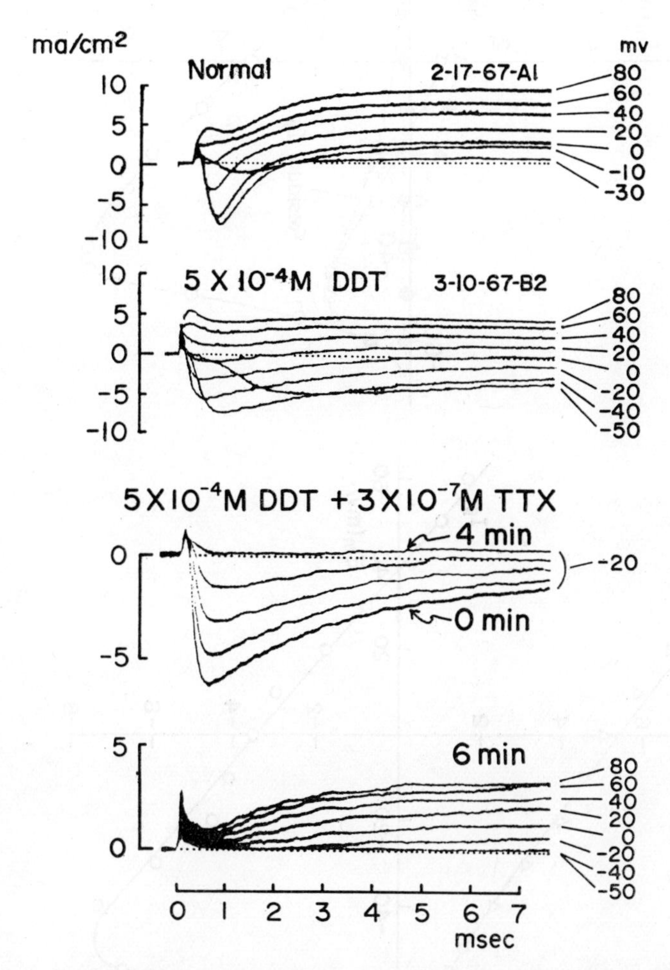

Figure 16 Families of membrane currents associated with step depolarizations in a normal lobster giant axon and in another axon treated with DDT and with DDT and tetrodotoxin (TTX). The third set of records represents changes in membrane current during the course of TTX action. The dotted line in each set refers to the zero base line (Narahashi and Hass, 1967)

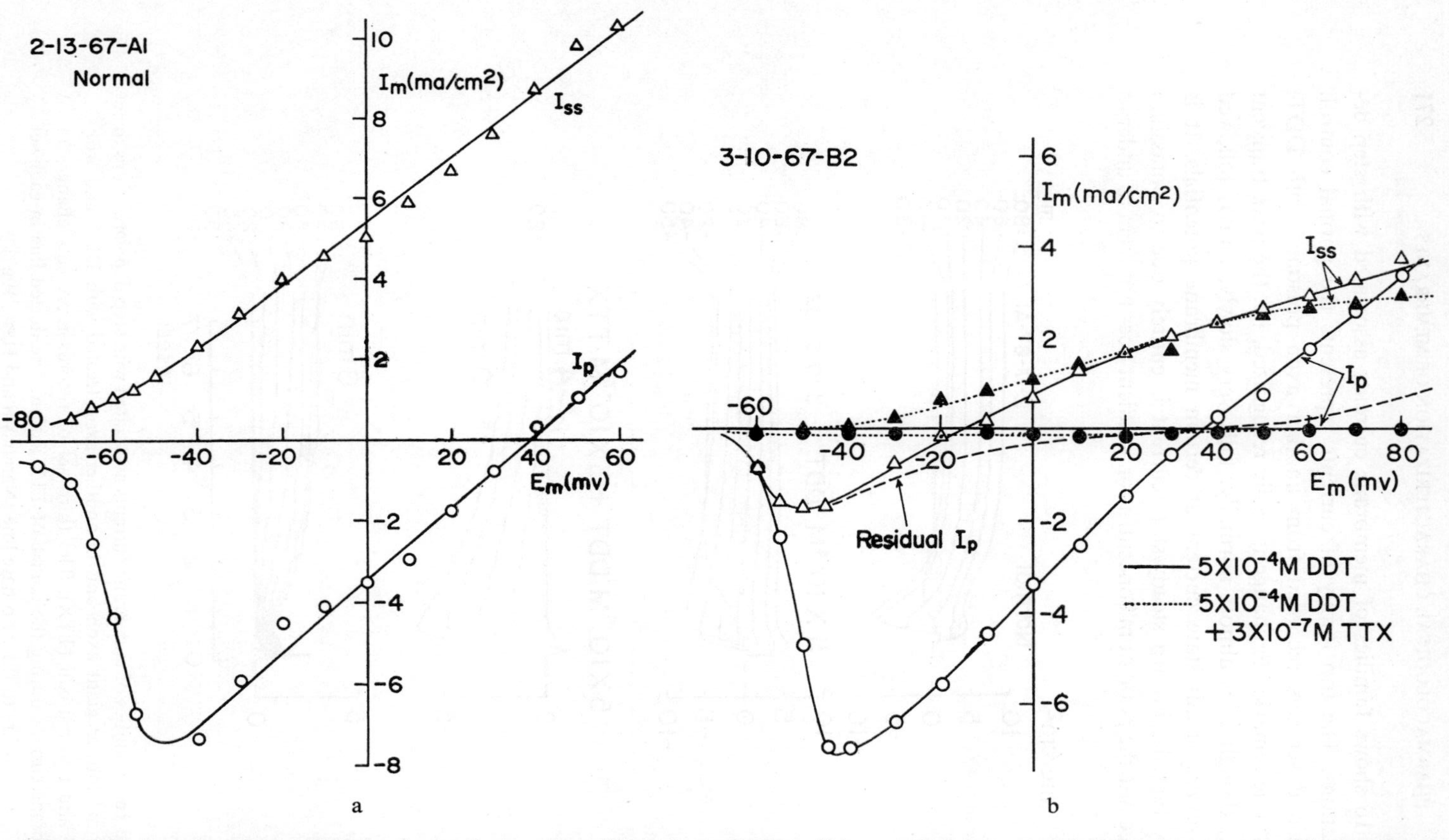

Figure 17 Current-voltage relations for peak transient (I_p) and steady-state (I_{ss}) currents in a normal lobster giant axon, and in another axon treated with DDT and with DDT and tetrodotoxin (TTX). The broken line shows the residual component of I_p and was obtained by subtracting I_{ss} in DDT plus TTX from I_{ss} in DDT (Narahashi and Haas, 1967)

a remarkable change in view of the fact that the resting potential remains unaltered. Therefore, it is possible for that current to be carried by sodium ions.

Tetrodotoxin proved to be highly useful to dissociate the sodium component from the potassium component in the membrane current. The third set of Figure 16 shows the membrane currents associated with a constant magnitude of step depolarization immediately before (0 minute) and after application of TTX to the DDT-poisoned axon. The peak transient (sodium) current is seen to be blocked completely during a 4-minute perfusion of TTX, while the inward steady-state current is now converted into a small outward current. The bottom set of Figure 16 shows a family of membrane currents associated with various step depolarizations 6 minutes after starting perfusion, and represents the potassium currents in the DDT poisoned axon.

Current-voltage relationships in a normal and a DDT-poisoned axon are illustrated in Figure 17. The current-voltage curve measured at the peak of the transient current (open circles) was not appreciably affected by DDT.

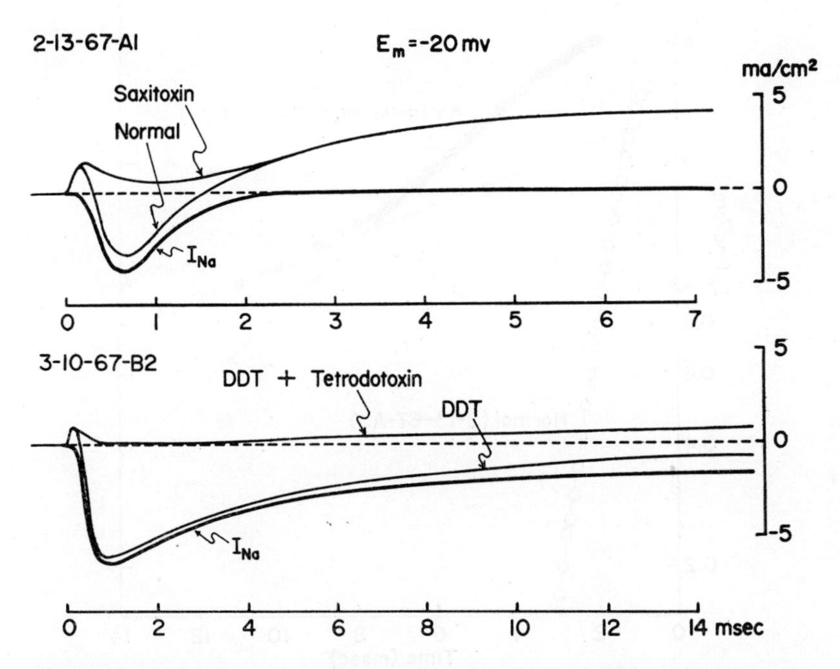

Figure 18 Separation of membrane current into peak transient sodium current (I_{Na}) and steady-state potassium current by the use of saxitoxin or tetrodotoxin at a concentration of 3×10^{-7} M in a normal and a DDT-treated lobster giant axon. I_{Na} was obtained by subtraction of the steady-state current recorded in saxitoxin or tetrodotoxin from the total current (Narahashi and Haas, 1968)

The curve measured at the steady state (open triangles) shows the so-called "negative conductance" in the membrane potential ranging from −60 mV to −50 mV, and the current flows in inward direction between −60 mV and −15 mV. The application of TTX abolished the peak current (filled circles), and converted the steady-state current into an outward current in the entire range of membrane potentials (filled triangles). Because TTX blocks the transient sodium current selectively, the difference between the steady-state current before application of TTX (open triangles) and that after application of TTX (filled triangles) represents the sodium component in the steady-state current. This residual sodium current is drawn by a broken line in Figure 17. It should be noted that both the peak sodium current and the residual sodium current reverse their sign at the same membrane potential of +38 mV. It is also clear that the steady-state potassium current is suppressed by DDT (filled triangles).

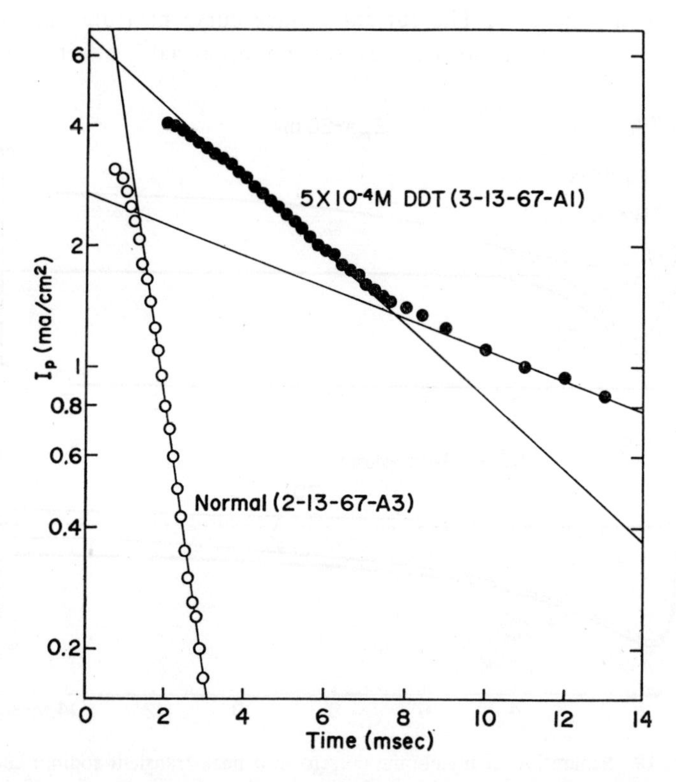

Figure 19 Semilogarithmic plot of the time course of the falling phase of the peak transient sodium current (I_p) in a normal and in a DDT-poisoned lobster giant axon. I_p was obtained by the method shown in Figure 18. The straight lines were fitted by eye (Narahashi and Moore, 1968a)

The time courses of the transient sodium and steady-state potassium currents are depicted in Figure 18. The sodium component is obtained by subtracting the current in STX or TTX, which represents the potassium component, from that in normal solution or in DDT solution. The time course of the falling phase of the sodium current is then plotted in Figure 19. The sodium current in the DDT-poisoned axon is not only turned off very slowly but also broken down into at least two components, the terminal phase being extremely slow. The initial phase in DDT is slowed by a factor of 4.5 on an average. Hille (1968) also observed a remarkable slowing of the sodium inactivation by DDT in nodes of Ranvier.

The kinetics other than the falling phase of the sodium current were also slowed, but to a lesser extent. The time for the transient sodium current to reach its peak was slowed by a factor of 1.3 to 1.9 depending on the membrane potential at which the measurement was made. The onset of the steady-state potassium current was slowed by a factor of 1.4 to 1.5.

F. Allethrin

1. *Effects on action potential*

The insecticide allethrin is a derivative of pyrethrin I which is one of the active ingredients of the classical insecticide pyrethrum. The chemical structure of allethrin is shown in Figure 20. It is the allethrolone ester of chrysanthemummonocarboxylic acid.

It is well known that pyrethrum stimulates the nerve to produce repetitive discharges and then paralyzes it (Hayashi, 1939; Lowenstein, 1942; Welsh and Gordon, 1947; Yamasaki and Ishii, 1952a; Lalonde and Brown, 1954). In 1962, allethrin was subjected to detailed microelectrode analyses for its mechanism of action at the cellular level (Narahashi, 1962a, b). At a relatively weak concentration of 10^{-6} M, allethrin increased the negative after-

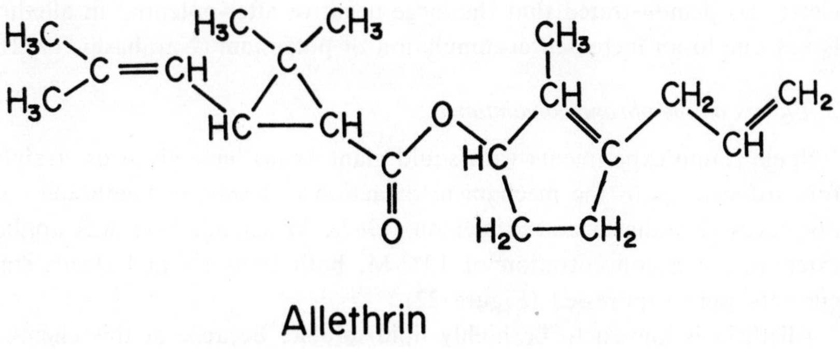

Allethrin

Figure 20 Chemical structure of allethrin

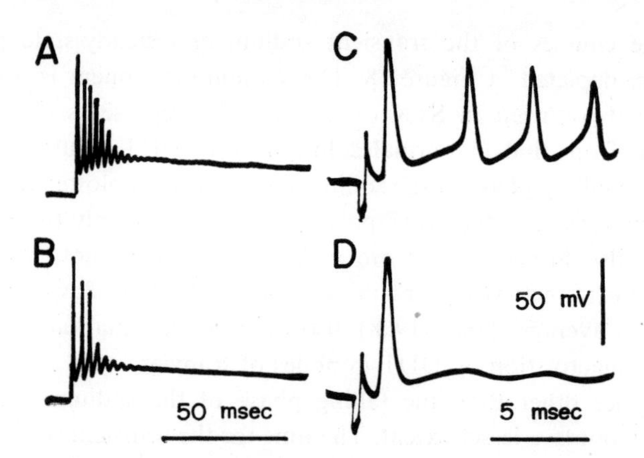

Figure 21 Internally recorded action potentials from a cockroach giant axon poisoned with 1×10^{-6} M allethrin. Temperature 33°C (A), 28°C (B), 26.5°C (C), and 26°C (D) (Narahashi, 1962a)

potential in the cockroach giant axon. At a higher concentration of 3×10^{-6} M, the increase in negative after-potential was followed by a conduction block. The resting potential slightly declined, but the depolarization was not large enough to account for the blockage by itself. Repetitive after-discharges were induced by a single stimulus during the course of allethrin poisoning. The production of after-discharges was highly temperature-dependent, the critical temperature beyond which after-discharges were initiated being estimated to be around 26°C (Figure 21).

Under normal conditions, the action potential of the cockroach giant axon is followed by an undershoot of a few millivolt amplitude which is in turn followed by a small (about 1 mV) negative after-potential. The negative after-potential was shown to be due to the accumulation of potassium in the immediate vicinity of the nerve membrane (Narahashi and Yamasaki, 1960a). Experimental analyses by means of intracellular microelectrodes demonstrated that the large negative after-potential in allethrin is not due to an increased accumulation of potassium (Narahashi, 1962b).

2. *Effects on membrane conductance*

Voltage clamp experiments with squid giant axons have given us straightforward answers to the mechanism of action in terms of membrane conductances (Narahashi and Anderson, 1967). When allethrin was applied externally at a concentration of 10^{-5} M, both transient and steady-state currents were suppressed (Figure 22).

Allethrin is known to be highly lipid-soluble. Because of this chemical property, it is surprising to see an additional effect of allethrin when applied

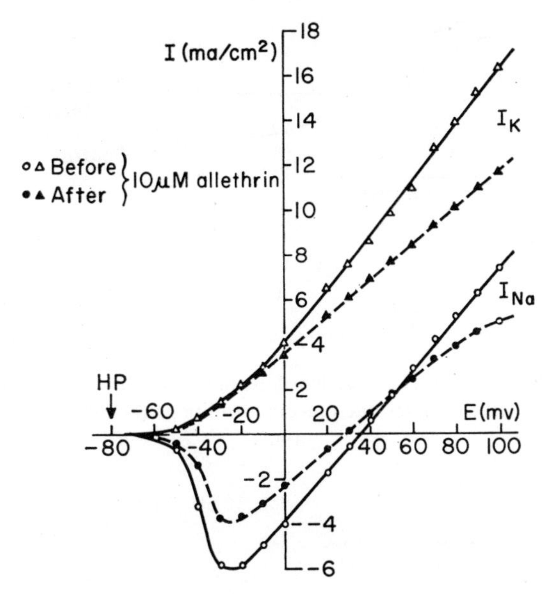

Figure 22 Current-voltage relations for peak transient sodium current (I_{Na}) and for steady-state potassium current (I_K) in a squid giant axon before and after treatment with allethrin externally

inside. Unlike cockroach giant axons, no significant increase in negative after-potential was observed in squid giant axons when allethrin was applied outside. However, when applied inside, allethrin caused a profound augmentation and prolongation of the negative after-potential which lasted as long as 30 seconds (Figure 23). As might be expected from such effect on the action potential, the membrane currents in the squid axon perfused internally with allethrin were somewhat different from those in the axon perfused externally with allethrin. An example of the records is shown in Figure 24. Three effects are apparent:

a) The transient current is suppressed.
b) The outward steady-state current is suppressed.
c) At certain membrane potentials, the steady-state current flows in inward direction.

Current-voltage relationships are illustrated in Figure 25. The transient currents measured at the peak were suppressed by allethrin (filled circles). The steady-state current flowed in inward direction in the membrane potential ranging from -45 mV to -5 mV (filled triangles). The inward steady-state current could not be attributed to potassium, because a potassium concentration gradient, with high inside and low outside, was maintained at a constant value by continuous perfusions in both phases. Therefore, it

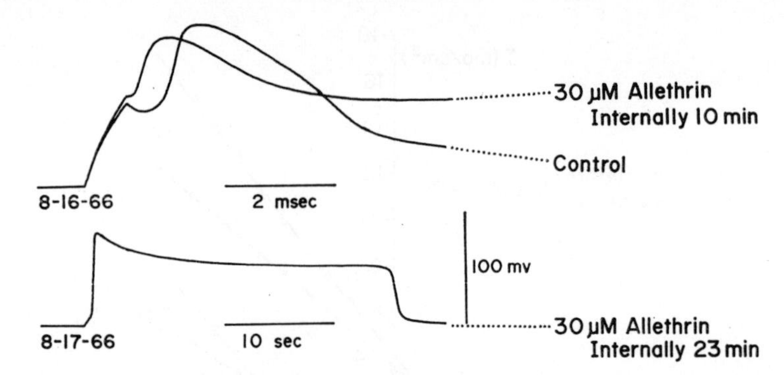

Figure 23 Prolongation of the action potential of squid giant axons by internal perfusion of allethrin under sucrose-gap conditions (Narahashi and Anderson, 1967)

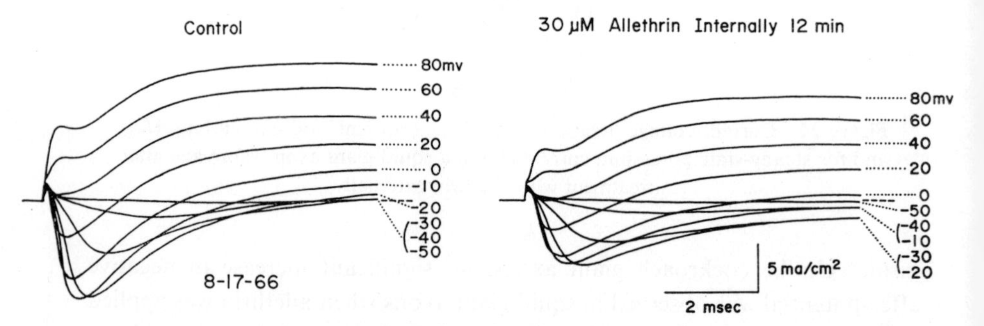

Figure 24 Families of membrane currents associated with step depolarizations in a squid giant axon before and during internal perfusion with allethrin (Narahashi and Anderson, 1967)

is reasonable to assume that the steady-state current in the allethrin perfused axon contains a sodium component. The residual sodium current was obtained in the following way:

Before application of allethrin, potassium current starts flowing outwardly when the membrane is depolarized beyond -25 mV (open triangles, Figure 25). This membrane potential coincides with that where the inward steady-state current in allethrin attains a maximum (filled triangles). Therefore, it is assumed that the potassium current starts flowing outwardly in the allethrin-poisoned axon when the membrane is depolarized beyond this value. At the reversal potential for the peak transient current in the allethrin-poisoned axon ($+40$ mV), there should be no transient sodium current flowing, so that no sodium component is contained in the steady-state current. Hence, the potassium component in the steady-state current of the allethrin-poisoned axon can be obtained by connecting the zero

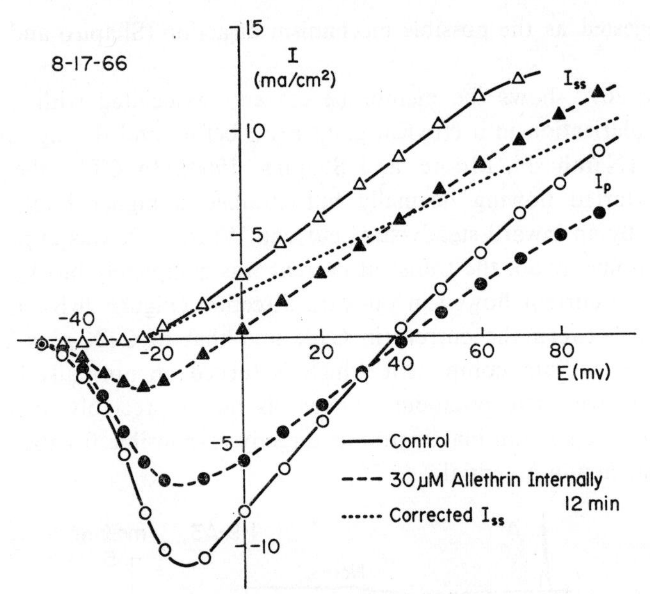

Figure 25 Current-voltage relations for peak transient (I_p) and steady-state (I_{ss}) currents in a squid giant axon before and during internal perfusion with allethrin. The dotted line represents the potassium current corrected for the residual sodium current as described in the text (Narahashi and Anderson, 1967)

current at -25 mV membrane potential with the apparent steady-state current at $+40$ mV (dotted line in Figure 25). It now becomes clear that the steady-state potassium current is suppressed by internally applied allethrin. The difference between the apparent steady-state current (filled triangles) and the steady-state potassium current (dotted line) represents the residual sodium current.

It is concluded that allethrin, when applied internally exerts an additional effect, i.e. a slowing of the sodium inactivation. This effect and the suppression of the potassium activation are responsible for the prolongation of the action potential.

The kinetics for the turn-on of the transient current were slightly slowed by either external or internal application of allethrin. However, the kinetics for the turn-on of the steady-state potassium current remained unaffected.

G. *Condylactis* toxin

Condylactis toxin (CTX) is the poison from the Bermuda amemone, *Condylactis gigantea*. The molecular weight has been estimated as 10,000–15,000 (Shapiro, 1968). CTX greatly prolonged the action potential and initiated repetitive discharges in the giant axon of the lobster or the slow-adapting stretch receptor of the crayfish, and the slowing of the sodium inactivation was suggested as the possible mechanism of action (Shapiro and Lilleheil, 1969).

Figure 26A shows the membrane currents associated with a constant step depolarization in a crayfish giant axon before and during application of CTX (Narahashi, Moore, and Shapiro, 1969). In CTX, the transient current started flowing normally but attained a higher level, and was followed by an inward steady-state current. When TTX was applied to the CTX-poisoned axon, the transient current was completely blocked and the steady-state current flowed in outward direction (Figure 26B). Hence, the difference between the current in CTX and that in CTX plus TTX represents the sodium component which is turned on normally but turned off very slowly. The potassium current is not appreciably affected. The slowing of the sodium inactivation is directly responsible for the prolongation of the action potential.

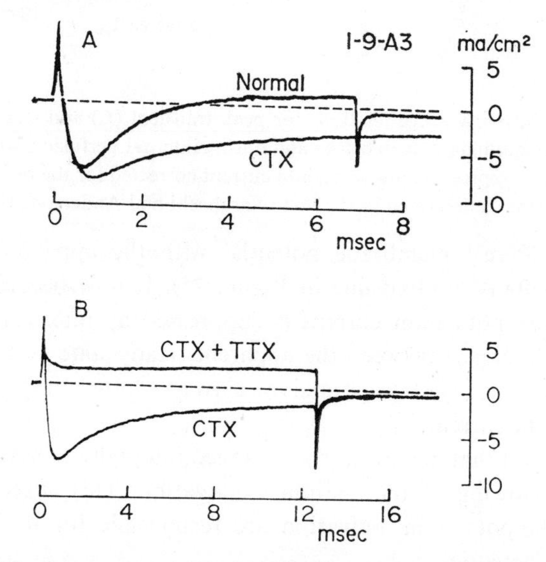

Figure 26 Membrane currents associated with a constant step depolarization in a crayfish giant axon. A, Before and during application of *Condylactis* toxin (CTX) at a concentration of 0.2 mg/ml. B, During application of CTX and after application of CTX plus 3×10^{-7} M tetrodotoxin (TTX) (Narahashi, Moore, and Shapiro, 1969)

IV. EXPERIMENTAL RESULTS ON END-PLATE MEMBRANES

When a microelectrode is inserted in the end-plate region of a muscle fiber immersed in a physiological saline which contains d-tubocurarine, a small, local transient depolarization is observed upon nerve stimulation. This is the end-plate potential (e.p.p.). However, when another microelectrode is inserted in the same end-plate area, it is possible to voltage clamp the end-plate membrane, because the area occupied by the end-plate is much smaller than the space constant of the muscle fiber. Under these conditions, a nerve stimulus elicits an end-plate current (e.p.c.). To observe the e.p.c., curare or equivalent neuromuscular blocking agents are not absolutely necessary, because the voltage clamp by means of intracellular micro-electrodes is sufficient to prevent action potentials from being produced from the end-plate in question, although no space clamp conditions are established with respect to the entire muscle fiber. However, later part of the e.p.c. recorded may be disturbed by the action potential and contraction originated in another end-plate of the same muscle fiber.

Takeuchi and Takeuchi (1959, 1960) performed extensive experimental analyses on the ionic mechanism of end-plate membranes. The peak amplitudes of e.p.c. measured at various membrane potentials were plotted against the membrane potential. The measurements fell on a straight line and were extrapolated to obtain the reversal or equilibrium potential for e.p.c. (Ee.p.c.). Thus a value of -15 mV was obtained. Changes in external sodium or potassium concentration or in internal sodium concentration by iontophoretic injection of sodium resulted in shifts of the equilibrium potential for e.p.c., and the result was compatible with the notion that the conductance change at the end-plate involved both sodium and potassium components. Replacement of external chloride with glutamate did not cause the shift in the equilibrium potential for e.p.c., suggesting that chloride was not involved in the end-plate conductance change.

In view of these considerations, it is possible to measure sodium and potassium components of the e.p.c. separately, if measurements are made of the e.p.c. at the equilibrium potentials for potassium (E_K) and sodium (E_{Na}) respectively. E_K and E_{Na} are estimated as -100 mV and $+50$ mV in the frog muscle, respectively. Clamping the membrane potential at $+50$ mV will cause severe contraction, but this can be avoided by means of the glycerol treatment method described before (Section II A 4).

A. Normal end-plate membrane

Normal end-plate membrane here refers to the end-plate preparation treated with glycerol to block the excitation-contraction coupling. There is some uncertainty about the internal concentrations of sodium and potassium in the glycerol-treated muscle. This problem is discussed in detail elsewhere (Deguchi and Narahashi, 1971). In short, errors in the measurement of e.p.c. caused by possible changes in internal ionic concentrations are relatively small.

The records of the e.p.p.'s and the e.p.c.'s measured at E_K and E_{Na} from a glycerol-treated muscle fiber are shown in the left-hand column of Figure 27. The sodium component of e.p.c. (I_{Na}) attained a larger amplitude and decayed more slowly than the potassium component (I_K) (Table 2). Close examination has also revealed that I_{Na} was slightly slower in its rising phase than I_K (Table 2). The peak amplitude of e.p.c. was measured at various membrane potentials between -110 mV and $+50$ mV, and the resultant current-voltage curve is shown in Figure 28 (open circles). The

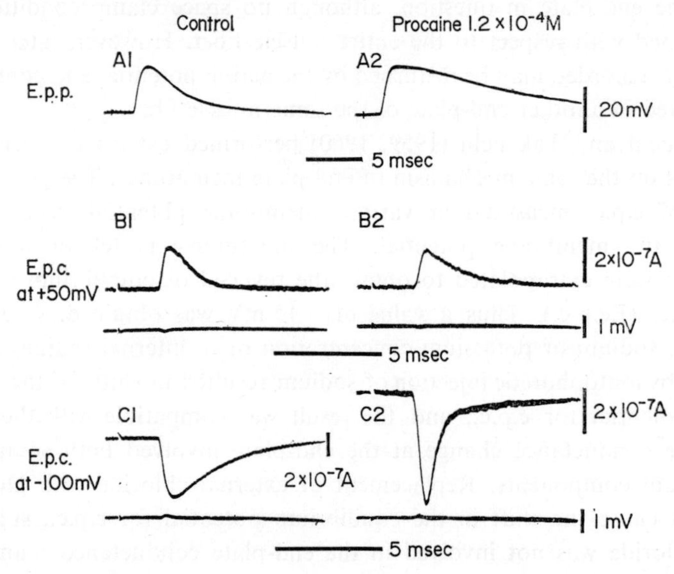

Figure 27 End-plate potential (e.p.p.) under current clamp conditions and end-plate current (e.p.c.) under voltage clamp conditions in a frog sartorius muscle fiber before and during application of procaine 1.2×10^{-4} M. Upward potential deflection shows depolarization, and upward current deflection shows outward current. The e.p.c. recorded at $+50$ mV membrane potential (sodium equilibrium potential) represents potassium current, and the e.p.c. at -100 mV (potassium equilibrium potential) represents sodium current

Table 2 Peak amplitude and time course of sodium (I_{Na}) and potassium (I_K) components of end-plate current recorded from the control glycerol-treated muscle fibers of the frog. Data are given by mean ± S.E. with the number of measurements in parentheses

	Measurement at (mV)	Peak amplitude ($\times 10^{-7}$ A)	Time to peak (m sec)	Half decay time (m sec)
I_{Na}	-100	6.26 ± 0.712 (11)	0.67 ± 0.035 (9)	1.59 ± 0.052 (9)
I_K	$+50$	3.62 ± 0.647 (9)	0.54 ± 0.019 (19)	0.76 ± 0.042 (19)
I_{Na}/I_K		1.72	1.24	2.09

over-all conductance is seen to be constant over the entire range of membrane potentials studied and the e.p.c. reverses its polarity at a membrane potential of +1.5 mV.

B. Effect of procaine on membrane conductances

Procaine has been shown to be able to block the neuromuscular transmission at a concentration insufficient to impair the conduction of nerve or muscle (Feng, 1940, 1942; Harvey, 1939). The e.p.p. observed in the procainized preparation was greatly prolonged in its falling phase (Furukawa, 1957, Maeno, 1966). It has also been demonstrated that procaine, when applied at low concentrations enough to block the neuromuscular transmission, was without effect on the release of the transmitter substance acetylcholine from the nerve terminals (Maeno, 1966). Therefore, the change in the shape of e.p.p. by procaine is entirely due to the post-junctional action.

Maeno (1966) made an attempt to interpret the mechanism underlying the prolongation of the e.p.p. in the procainized muscle fiber. Based on the observations of the extracellulary recorded e.p.p., which was a measure of e.p.c., it was suggested that the sodium component in the e.p.c. was selectively affected by procaine thereby causing a prolongation of the e.p.p. This problem would best be studied by means of voltage clamp techniques whereby the sodium and potassium components can be measured separately at E_K and E_{Na}, respectively. Such an attempt was indeed made by Gage and Armstrong (1968) who observed spontaneous miniature e.p.c.'s before and during application of procaine. It was found that the initial falling phase of the sodium miniature e.p.c. was accelerated and was followed by a slow terminal phase by the action of procaine.

Another attempt was also made by a somewhat different method to interpret the effect of lidocaine derivatives on the e.p.p. (Steinbach, 1968a, b). Many of the lidocaine derivatives prolonged the e.p.p. A computer was programmed according to his kinetic equations so as to provide a current which was able to cancel the recorded e.p.p. when applied to the end-plate region through an internal microelectrode. Thus, the current from the output of the computer should have been equal and opposite in sign to the real e.p.c. The change in the shape of the artificial e.p.c thus measured at normal resting potentials after application of the lidocaine derivatives was very similar to that observed by Gage and Armstrong (1968) in the real sodium e.p.c. after application of procaine. Since the normal resting potential is about −90 mV and close to E_K, the artificial e.p.c. observed represents mostly the sodium e.p.c.

Voltage clamp analyses have been performed with the sartorius muscle of the frog and the detailed account of the study is given elsewhere (Deguchi and Narahashi, 1971). The glycerol treatment was used throughout to avoid mechanical movements. A typical example of the effect of 1.2×10^{-4} M procaine on the e.p.p. and e.p.c. is illustrated in the right-hand column of Figure 27. The e.p.p. was greatly prolonged in its falling phase. The most conspicuous effect on the e.p.c. was observed in the falling phase of I_{Na}. Its initial falling phase was markedly accelerated and was followed by a very slow terminal phase. On the contrary, I_K remained almost unchanged. The amplitudes of I_{Na} and I_K were only slightly suppressed by a low concentration ($1.2 \times 1o^{-4}$ M) of procaine. However, they both were suppressed when a high concentration (3.6×10^{-4} M) of procaine was applied.

Average values for the effect of procaine on the kinetics of I_{Na} and I_K are given in Table 3. In addition to the marked acceleration of the initial falling phase of I_{Na}, procaine slightly accelerated the rising phase of I_{Na} and slowed the falling phase of I_K. The rising phase of I_K remained essentially unchanged. Thus the observations by Gage and Armstrong (1968), by Maeno (1966), and by Steinbach (1968a, b) on the effects of procaine and lidocaine derivatives on the falling phase of I_{Na} have been confirmed and extended. Another important feature of the present study is that I_{Na} and I_K are affected by procaine quite differently. None of the effects of procaine on the kinetics is the same between I_{Na} and I_K.

Table 3 Effects of procaine on the kinetics of the sodium (I_{Na}) and potassium (I_K) components of end-plate current. The data are given as the mean values relative to the control before application of procaine with the number of measurements in parentheses

Component of end-plate current	Procaine concentration (M)	Time to peak	Half decay time
I_{Na}	1.2×10^{-4}	0.85 (7)	0.35 (7)
	3.6×10^{-4}	0.60 (3)	0.16 (3)
I_K	1.2×10^{-4}	1.10 (7)	1.25 (7)
	3.6×10^{-4}	0.94 (3)	1.51 (3)

The amplitudes of the e.p.c. at the peak and 8 milliseconds after the onset of the e.p.c. in the procainized muscle fiber are plotted as a function of membrane potential in Figure 28. The equilibrium potential for e.p.c. remained unchanged by application of procaine at a concentration of 1.2×10^{-4} M. Unlike the peak e.p.c., the residual current in procaine was not completely potential-independent and became zero at zero membrane potential. No outward residual current was observed in the inside positive membrane potential. Since I_K is over at 8 msec after the e.p.c. onset, the current voltage curve for the residual current does not contain the potassium component. However, it is also unlikely that the residual current is entirely due to sodium, because it becomes zero at zero membrane potential.

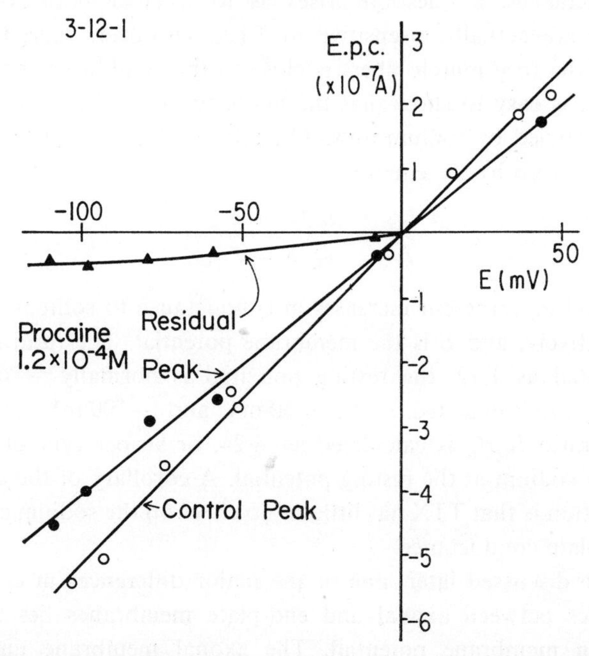

Figure 28 Current-voltage relations at the end-plate of a frog sartorius muscle fiber. The peak amplitude of end-plate current (e.p.c.) before and during application of procaine 1.2 × 10^{-4} M, and the amplitude of the e.p.c. 8 msec after its onset in procaine are plotted as a function of the holding membrane potential

V. DISCUSSION

A. Comparison of axonal and end-plate membranes for pharmacological characteristics

1. *Tetrodotoxin and saxitoxin*

In spite of the potent blocking action on the peak transient sodium conductance, TTX has been demonstrated to exert no effect on the sensitivity of the end-plate membrane to ACh (Furukawa, Sasaoka, and Hosoya, 1959; Katz and Miledi, 1967). Similar observations were made of STX (Kao and Nishiyama, 1965). This is of particular interest because the depolarization of the end-plate membrane in response to ACh involves increases in conductances to both sodium and potassium. Therefore, it is concluded that the conductance mechanism or channel for the sodium current in the end-plate is qualitatively different from that in the axonal membrane.

In the absence of voltage clamp data on the effect of TTX on the end-plate conductances, a question arises as to whether both conductance components are actually insensitive to TTX. However, since the resting potential of the frog muscle fibers is close to the equilibrium potential for potassium, it is easy to show that the majority of the e.p.c. at the resting potential is carried by sodium ions. The ratio of I_{Na} to I_K at the end-plate membrane is given by the equation

$$\frac{I_{Na}}{I_K} = \frac{g_{Na}(E - E_{Na})}{g_K(E - E_K)} \tag{4}$$

where g_{Na} and g_K represent increases in conductance to sodium and potassium, respectively, and E is the membrane potential. The ratio g_{Na}/g_K has been estimated as 1.72, the resting potential is normally -90 mV, and E_{Na} and E_K are calculated to be $+50$ mV and -100 mV, respectively. Hence, the ratio I_{Na}/I_K is calculated as -24, or 96 per cent of the e.p.c. is carried by sodium at the resting potential. A corollary of the calculation and observation is that TTX has little or no effect on the sodium component of the end-plate conductance.

As will be discussed later, one of the major differences in conductance characteristics between axonal and end-plate membranes lies in the dependency on membrane potential. The axonal membrane undergoes a highly potential-dependent conductance change upon depolarization, whereas the end-plate membrane exhibits a linear current-voltage relation or a constant over-all conductance. However, it is not clear from the presently available data whether the potential dependency of conductance is related to the sensitivity to TTX. It should be borne in mind that TTX inhibits the sodium component in the conductance of the resting membrane as well (Freeman, 1969). A question remains unanswered as to whether the sodium conductances at rest and during activity involve a common membrane channel.

2. *Local anesthetics*

Another striking difference between the axonal membrane and the end-plate membrane is found for their response to local anesthetics. Although procaine and certain lidocaine derivatives block both the electrical excitability of the axonal membrane and the chemical excitability of the end-plate membrane, the ionic mechanisms involved in the blockage are not the same. In the squid axon membrane, the increases in sodium and potassium conductances are inhibited while the kinetics are only slightly affected by procaine; the time to peak sodium current is slightly prolonged. Dibucaine inhibits both conductance components without affecting the kinetics.

On the contrary, procaine has a remarkable influence on the kinetics of the sodium component of the e.p.c.; its initial falling phase is greatly accelerated and is followed by a very slow terminal phase which is not present before treatment with procaine. The kinetics of the potassium component are affected only slightly, its falling phase being slowed. The amplitudes of both components of the e.p.c. are suppressed. The change in the kinetics of I_{Na} is the major factor responsible for the slowing of the falling phase of the e.p.p. Similar changes in e.p.p. have been observed with several lidocaine derivatives (Steinbach, 1968a, b). In view of these considerations, it appears that the effects of local anesthetics on the membrane conductances are qualitatively different between the axonal and end-plate membranes.

The site of action of local anesthetics appears to be different between these two membranes. In the axonal membrane, lidocaine derivatives have been demonstrated to block the action potential from inside the membrane in the charged form. Steinbach (1968a) studied the effect of pH on the potency of lidocaine, two tertiary lidocaine derivatives and procaine in augmenting the terminal slow phase of the e.p.p. The increase in the potency of lidocaine (pK_a 7.85) and a lidocaine derivative 14,465 (pK_a 7.35) by a decrease in external pH from 7.8 to 6.5 runs parallel with the increase in the respective anesthetic cation concentration. However, the cation concentration of another lidocaine derivative L-30 (pK_a 8.96) and procaine (pK_a 8.92) does not increase with the increase in pH, and their potency is not appreciably affected by the pH change.

These results obtained by Steinbach (1968a) seem to be, at first sight, compatible with the notion that the cationic charged form of these anesthetics acts on the end-plate membrane from outside. However, other possibilities must be excluded before drawing a conclusion. The following analysis clearly shows that this is actually the case.

The analysis is essentially the same as that adopted for the nerve blocking action of tertiary amine local anesthetics (Narahashi, Frazier, and Yamada, 1970). Two major assumptions are made: a) Only the uncharged form of anesthetics is freely permeable through the membrane so that it is distributed in both external and internal phases in the same concentration. b) The charged and uncharged forms of anesthetics and the hydrogen ions are distributed uniformly in each phase. The justification for these assumptions are given in the earlier papers (Narahashi, Frazier, and Yamada, 1970; Narahashi and Frazier 1971).

On these assumptions, the concentrations of the charged and uncharged forms of anesthetics in each phase can be calculated by the Henderson-Hasselbach equation (2). The result of such calculations for externally applied lidocaine is shown in Figure 29. The ordinate represents the logarithm of the effective dose fifty (ED 50) to exhibit 50 percent maximum effect, unity concentration being assigned to ED 50 of the active form. The abscissa represents the pH value.

There are three possible active forms, i.e. the externally or internally present uncharged form, the externally present charged form, and the internally present charged form. Each curve in Fig. 29 represents the following situations:

a) Curve A represents the ED 50-pH relationship if the externally or internally present uncharged form is assumed as active. The internal pH of muscle fibers actually changes slightly when the external pH is altered (Bianchi and Strobel, 1968). However, the change in internal pH does not affect the potency of anesthetics in this case, because the uncharged form is in the same concentration in both external and internal phases. Therefore, the potency of lidocaine should become lower as the external pH is lowered from 7.8 to 6.5 contrary to the observation by Steinbach (1968a).

b) If the internally present charged form is active, curve B represents the case where the internal pH is kept constant regardless of external pH changes. Curve B' represents the case where the internal pH half follows the external pH change; for example, when the external pH is altered by unity, the internal pH changes by 0.5. Since the change in the internal pH following the external pH change is smaller than that (Bianchi and Strobel, 1968), the actual situation may be represented by a curve somewhere between curve B and curve B′. In either case, however, the potency of lidocaine should decrease as the external pH is lowered from 7.8 to 6.5. This is again opposite to the observation by Steinbach (1968a).

c) If, however, the externally present charged form is active, the calculation fits the observation. In this case, curve C represents the ED 50-pH relationship. Changes in internal pH in response to external pH changes do not affect the potency. Thus the potency increases as the external pH is lowered from 7.8 to 6.5 in agreement with the observation (Steinbach, 1968a).

The ED 50-pH curves shown in Figure 29 can easily be modified to other tertiary amine anesthetics having different values of pK_a. Curves A and B may be shifted along the 45-degree line so that the inflection occurs at the pK_a. Curve C may also be shifted along the abscissa so that the inflection occurs at the pK_a. The data on 14,465, L-30 and procaine (Steinbach, 1968a) have been analyzed in the same manner, and the observations agree with the calculation when the externally present charged form is assumed as active. Thus, it is concluded that at the end-plate membrane tertiary amine local anesthetics act on the external surface of the membrane in the charged form in support of the concept proposed by Steinbach (1968a).

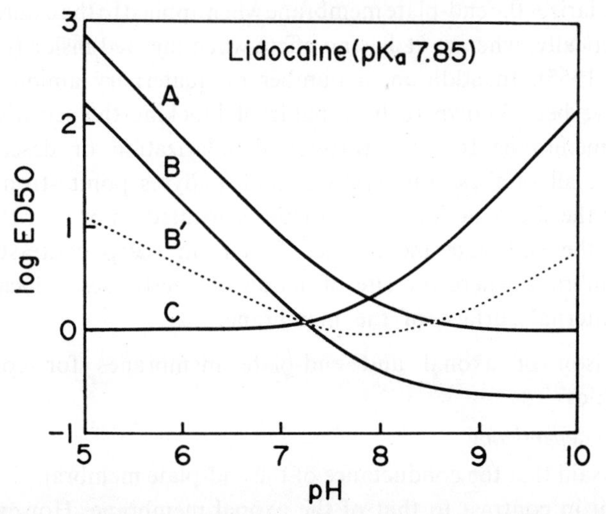

Figure 29 Calculated relationships between the logarithm of the number of unit concentrations of lidocaine applied externally and the external pH. Unity concentration is assigned to the concentration of the active form that blocks the end-plate potential by 50% (ED 50). Curve A is for the case where the uncharged form of lidocaine, distributed uniformly between the external and internal phases of the muscle fiber, is assumed as the active form. Curve B is for the case where the internally present charged form is assumed as active. Curve *B'* is the same as curve *B* except that the internal pH half follows the external pH change. Curve *C* is for the case where the externally present charged form is assumed as active. See text for further explanation

It is well documented by experiments that the majority of quaternary ammonium compounds hardly penetrates the membrane because of their permanent positive charges whereas tertiary compounds can do so in the uncharged form (e.g. Krahl, Keltch, and Clowes, 1940; Ritchie, Ritchie, and Greengard, 1965a, b; Eldefrawi and O'Brien, 1967; O'Brien, 1967; Bianchi and Bolton, 1967; Rothenberg, Sprinson, and Nachmansohn, 1948; Schanker, Nafpliotis, and Johnson, 1961; Whitcomb, Friess, and Moore, 1958). In fact, two quaternary lidocaine derivatives, QX-314 and QX-572, and hemicholinium-3 which is also in quaternary form, exhibit highly asymmetrical effect on the nerve membrane, blocking the conduction much more strongly from inside than from outside the membrane. This also supports the concept of impermeability of quaternary compounds. Therefore, the fact that quaternary lidocaine derivatives, when applied externally, are effective in changing the shape of the e.p.p. (Steinbach, 1968a) is in keeping with the aforementioned conclusion that anesthetics affect the end-plate membrane from outside.

In support of this concept, ACh which is a quaternary ammonium compound depolarizes the end-plate membrane when applied to the external surface iontophoretically, whereas it has no effect when injected inside (del Castillo and Katz, 1955). In addition, a number of quaternary ammonium compounds have been known to be capable of blocking the sensitivity of the end-plate membrane to ACh through depolarization or desensitization. Collectively, all of these observations and analyses point strongly to the notion that the site of action of anesthetics is located on or near the external surface of the end-plate membrane. This is in sharp contrast with the axonal membrane where the site of action of anesthetics is located on or near the internal surface of the membrane.

B. Comparison of axonal and end-plate membranes for conductance characteristics

1. *Potential dependency*

It has been said that the conductance of the end-plate membrane is potential-independent in contrast to that of the axonal membrane. However, unlike the axonal membrane where the total conductance can easily be separated into the peak transient sodium component and the late steady-state potassium component, no such separation of the end-plate conductance into sodium and potassium components has been achieved over the entire range of membrane potentials. As described before, the e.p.c. can be separated into the two components only at E_{Na} and E_K, where only one of the components can be observed. It follows that no statement can be made concerning the potential dependency of *each* component. One of the straightforward approaches to this problem would be to use agents which block one of the components selectively without affecting the other. Unfortunately, no such agent has been discovered yet. In order to find such agents, voltage clamp experiments would be necessary because it is not easy to distinguish between a selective blocker of I_{Na} and I_K from the observation of the e.p.p. alone. In both cases, the e.p.p. measured at the resting potential which is close to E_K will be suppressed.

2. *Ionic channels*

Net transfer of charge during the transmitter action at an end-plate can be calculated from the record of the e.p.c. From record Cl of Figure 27, the transfer of charge associated with the e.p.c. is measured as 1.6×10^{-9} coulombs. This corresponds to a net inward transport of sodium ions in an amount of 1.6×10^{-14} moles. Since record Cl of Figure 27 is made at -100 mV (E_K) which is close to the normal resting potential of -90 mV, the net inward transport of univalent cations of the resting potential should be very close to 1.6×10^{-14} moles. This amount is of the same order of magnitude as $2 \sim 4 \times 10^{-14}$ moles calculated by Fatt and Katz (1951) from the e.p.p. record. On the assumption that the nerve terminal at the end-plate has a diameter of 1 μ and extends for a distance of 100 μ (Katz, 1966), the area of the end-plate is calculated as 3×10^{-6} cm^2. Hence, the net inward transport of univalent cations (sodium) per unit area of the end-plate is 5×10^{-9} mole/cm^2. The sodium entry has been measured sa 3.5×10^{-12} mole/cm^2/impulse in *Loligo* axons and $3.7 \sim 3.8 \times 10^{-12}$ mole/cm^2/impulse in *Sepia* axons (Keynes, 1951; Keynes and Lewis, 1951). Therefore, the amount of cations transported at an end-plate during a transmitter action is about 1000 times larger than that in an axon during an action potential.

The peak values for the sodium current and sodium conductance during the transmitter action can be calculated from the e.p.c. measurement. The peak sodium e.p.c. measured at E_K amounts to 6×10^{-7} A (Record Cl of Figure 27). This corresponds to the current density of 200 mA/cm^2. The ionic current (I_i) flowing in inward direction across the nerve membrane at the moment when the rate of rise of the action potential is at a maximum can be calculated from the relation

$$I_i = -C_m \frac{\partial V}{\partial t} \tag{5}$$

where C_m represents the specific membrane capacity, and V and t are the membrane potential change and time, respectively. In squid giant axons, I_i is calculated to be 1 mA/cm^2. This is mostly carried by sodium ions. Thus, the sodium current density is much larger in the end-plate than in the axon membrane by a factor of 200.

The conductance at the peak sodium e.p.c. is measured as 1.3 mho/cm^2 from Figure 28. In the voltage clamped squid axon, the peak transient sodium conductance reaches a maximum value of 0.1 mho/cm^2. Therefore, the end-plate membrane undergoes a much higher sodium conductance increase than the squid axon membrane, the difference being 13-fold. No data are available for the resting conductance of the end-plate membrane. The value for the muscle membrane is estimated as 1/4000 mho/cm^2 (Fatt and Katz, 1951). If the same value holds for the end-plate membrane, then the transmitter action causes the conductance to increase by a factor of 5000. This is much higher than the value obtained from the squid axon membrane where the total conductance increases only 100-fold under voltage clamp conditions.

By comparison of the maximum sodium conductances of the frog end-plate membrane and the lobster or squid axon membrane, it is tempting to estimate the number of sodium channels in the end-plate. This may be done by simply multiplying the upper limit of sodium channels measured in lobster nerve by the difference in the maximum sodium conductance, i.e. $13 \times 20 = 260/\mu^2$. However, this calculation is somewhat questionable, because the sodium channels of the end-plate have been shown to be qualitatively different from those of the axonal membrane as described before (Section V A 1).

The leaky status of the end-plate membrane is understandable in view of the large membrane area surrounding the end-plate. This area is to be depolarized by as much as 40 mV by a local circuit current provided from the strongly activated end-plate membrane. In the axon membrane, however, the local circuit current for impulse propagation is provided from a relatively large membrane area, so that the activation or conductance increase in any particular area need not be as large an intensity as in the end-plate membrane. In this connection, it seems to be somewhat puzzling for the end-plate membrane to undergo conductance increase to both sodium and potassium with almost the same time course, because the intensity of the e.p.c. would be greater and therefore more efficient in producing the e.p.p. if there were only the sodium component in the e.p.c.

Somewhat analogous situation in relation to the large maximum conductance is known in the nodal membrane of myelinated nerve fibers. In the nodes of Ranvier of the frog, *Xenopus*, and rat, the maximum value for the peak sodium current under voltage clamp conditions was estimated as large as 50 to 70 mA/cm^2 (Dodge and Frankenheuser, 1958; Bergmann, Nonner and Stämpfli, 1968; Nonner and Stämpfli, 1969). The maximum peak sodium conductance was then calculated as 0.8 to 1.4 mho/cm^2 which is about 10 times as large as the value in the squid axon. Hille (1968) estimated the maximum sodium conductance as 750 nmho in the frog node of Ranvier. If one assumes a nodal area of 3×10^{-6} cm^2 as calculated before, the maximum sodium conductance is 0.25 mho/cm^2 which is about 2.5 times as large as that of the squid axon membrane. In the impulse conduction along the myelinated nerve fiber, such a high density of current facilitates the depolarization of the adjacent node of Ranvier thereby speeding up the conduction velocity.

C. Single or separate channels?

Since sodium and potassium ions are the major ions transported across the nerve membrane during excitation, one of the most fundamental questions would be whether these two ions are carried through a common channel or through separate channels. Mullins (1959) proposed a model in which a single channel undergoes a modulation to adapt first to sodium and then potassium during excitation. However, recent studies on TTX, TEA and DDT have been interpreted as supporting the concept of separate channels (Narahashi and Haas, 1968; Hille, 1967). This proposal triggered discussions, and factors and experimental evidence for or against the separate channel concept were much argued (Mullins, 1968; Narahashi and Moore, 1968b). The discussions will not be repeated here in detail, but in short, the following points were considered in favor of the separate channel concept:

a) TTX and TEA specifically block the peak transient current and the late steady-state current, respectively. It is at least easier to visualize the existence of separate channels for these two currents, although these observations do not demonstrate it.

b) DDT slows the sodium inactivation and suppresses the potassium current in the lobster giant axon, so that both currents are flowing simultaneously (Narahashi and Haas, 1968). In the node of Ravier of the frog, the sodium inactivation is slowed without accompanying decrease in the potassium conductance (Hille, 1968). Another important feature of DDT action is that the effect on kinetics is different between the sodium inactivation and the potassium activation. The slowing of the latter is only 1.4 to 1.6-fold, whereas that of the former is as much as 4.5-fold (Narahashi and Hass, 1968).

c) Hyperpolarizations of the squid axon membrane slow the onset of the potassium current without much affecting the kinetics of the sodium current (Cole and Moore, 1960; Frankenhaeuser and Hodgkin, 1957).

In addition to these experimental observations, the following data have recently been accumulated in support of the separate channel concept:

d) Ammonium ions can be transported across the squid axon membrane in two distinct time courses comparable to those sodium and potassium currents. Moreover, the peak transient ammonium current is selectively blocked by TTX, whereas the late steady-state ammonium current is selectively blocked by TEA (Binstock and Lecar, 1969). The two distinctly different ammonium currents are difficult to visualize on the basis of the single channel concept.

e) *Condylactis* toxin greatly slows the sodium inactivation without much affecting the increases in the peak transient conductance and in the late steady-state conductance (Narahashi, Moore, and Shapiro, 1969). Therefore, the total conductance, including the residual sodium and the steady-state potassium conductances, exceeds the maximum value for the peak transient sodium or steady-state potassium conductance. This phenomenon is again difficult to explain in terms of the single channel hypothesis.

In view of these considerations, it is generally much easier to account for a number of observations on axon membranes in terms of the separate channel hypothesis, although there is as yet no experimental observation that can definitely demonstrate either of those hypotheses.

In the end-plate membrane, a question also arises as to whether single or separate channels exist. The experimental observations and arguments described above for axon membranes do not directly apply, because the characteristics of end-plate conductance changes are almost entirely different from those of axon membranes. Although data are too scarce to fully discuss this problem, there are some indications that support the concept of separate sodium and potassium channels in the end-plate membrane.

As described before, the time course of sodium current and that of potassium current at the end-plate membrane are not greatly different (Table 2). They reach their peaks with almost the same time course. Potassium current declines approximately twice as fast as sodium current. When treated with procaine, however, the sodium current undergoes remarkable changes in its time course, especially in its falling phase, while the potassium current is affected to a much lesser extent (Table 3). It should be noted that the changes in sodium and potassium currents qualitatively different. These observations are easier to visualize from the viewpoint of separate channels. As in the situation of the axon membrane, however, there is no experimental evidence to prove or disprove either of those hypotheses.

Acknowledgements

The studies described in the present article were supported by grants from The National Institutes of Health (NS 06855 and NS 03437) and from the Grass Foundation, and by contract with The National Institute of Environmental Health Sciences (PH-43-68-73). The author wishes to thank Mrs. C. A. Munday and Mrs. R. M. Crutchfield for their secretarial assistance.

REFERENCES

Adrian, R. H., W. K. Chandler, and A. L. Hodgkin (1966) *J. Physiol.* (London) **186**, 51P.

Alving, B. O. (1969) *J. Gen. Physiol.* **54**, 512.

Anderson, N. C. (1969) *J. Gen. Physiol.* **54**, 145.

Araki, T., and C. A. Terzuolo (1962) *J. Neurophysiol.* **25**, 772.

Ariëns, E. J., A. M. Simonis, and J. M. van Rossum (1964) In: Molecular Pharmacology, Vol. 1, ed. by E. J. Ariëns, Academic Press, New York, p. 287.

Armstrong, C. M. (1969) *J. Gen. Physiol.* **54**, 553.

Baker, P. F., A. L. Hodgkin, and T. I. Shaw (1961) *Nature* (London) **190**, 885.

Bergmann, C., W. Nonner, and R. Stämpfli (1968) *Pflüg. Arch.* **302**, 24.

Bernard, C. (1857) Lecons sur les Effets des Substances Toxiques et Médicamenteuses. J. B. Bailliere et Fils, Paris.

Bianchi, C. P., and T. C. Bolton (1967) *J. Pharmacol. Exp. Ther.* **157**, 388.
Bianchi, C. P., and G. E. Strobel (1968) *Trans. N.Y. Acad. Sci.* (Series II) **30**, 1082.
Binstock, L., and H. Lecar (1969) *J. Gen. Physiol.* **53**, 342.
Blaustein, M. P. (1968a) *J. Gen. Physiol.* **51**, 293.
Blaustein, M. P. (1968b) *J. Geb. Physiol.* **51**, 309.
Blaustein, M. P., and D. E. Goldman (1966) *Biophys. J.* **6**, 453.
Blaustein, M. P., and D. E. Goldman (1966) *J. Gen. Physiol.* **49**, 1043.
Camougis, G., B. H. Takman, and J. R. P. Tasse (1967) *Science* **156**, 1625.
Chamberlain, S. G., and G. A. Kerkut (1969) *Comp. Biochem. Physiol.* **28**, 787.
Chandler, W. K., and H. Meves (1965) *J. Physiol.* (London) **180**, 788.
Clowes, G. H. A., A. K. Keltch, and M. E. Krahl (1940) *J. Pharmacol. Exp. Ther.* **68**, 312.
Cole, K. S., and J. W. Moore (1960) *Biophys. J.* **1**, 1.
Cole, K. S., and J. W. Moore (1960) *J. Gen. Physiol.* **44**, 123.
Deck, K. A., R. Kern, and W. Trautwein (1964) *Pflüg. Arch.* **280**, 50.
Deguchi, T. (1967) *Japan. J. Pharmacol.* **17**, 267.
Deguchi, T., and T. Narahashi (1971) *J. Pharmacol. Exp. Ther.* **176**, 423.
del Castillo, J., and B. Katz (1955) *J. Physiol.* (London) **128**, 157.
Dodge, F. A., and B. Frankenhaeuser (1958) *J. Physiol.* (London) **143**, 76.
Dudel, J., K. Peper, R. Rüdel, and W. Trautwein (1966) *Pflüg Arch.* **292**, 255.
Eisenberg, R. S., and P. W. Gage (1967) *Science* **158**, 1700.
Eldefrawi, M. E., and R. D. O'Brien (1967) *J. Exp. Biol.* **46**, 1.
Fatt, P., and B. Katz (1951) *J. Physiol.* (London) **115**, 320.
Feng, T. P. (1940) *Chinese J. Physiol.* **15**, 367.
Feng, T. P. (1941) *Biol. Symp.* **3**, 121.
Frank, F., and L. Tauc (1964) In: The Cellular Functions of Membrane Transport, ed. by J. F. Hoffman, Prentice-Hall, Inc., Englewood Cliffs, N. J., p. 113.
Frank, K., M. G. F. Fuortes, and P. G. Nelson (1959) *Science* **130**, 38.
Frankenhaeuser, B., and A. L. Hodgkin (1957) *J. Physiol.* (London) **137**, 217.
Frazier, D. T., T. Narahashi, and J. W. Moore (1968) *Proc. Int. Union Physiol. Sci.*, Vol. 7, XXIV Int. Congr. p. 143.
Frazier, D. T., T. Narahashi, and J. W. Moore (1969) *Science* **163**, 820.
Frazier, D. T., T. Narahashi, and M. Yamada (1970) *J. Pharmacol. Exp. Ther.* **171**, 45.
Freeman, A. R. (1969) *Fed. Proc.* **28**, 333.
Furukawa, T. (1957) *Japan. J. Physiol.* **7**, 199.
Furukawa, T., T. Sasaoka, and Y. Hosoya (1959) *Japan. J. Physiol.* **9**, 143.
Gage, P. W., and R. S. Eisenberg (1967) *Science* **158**, 1702.
Gordon, H. T., and J. H. Welsh (1948) *J. Cell. Comp. Physiol.* **31**, 395.
Goto, T., Y. Kishi, S. Takahashi, and Y. Hirata (1965) *Tetrahedron* **21**: 2059.
Hagiwara, S., and N. Saito (1959a) *J. Neurophysiol.* **22**, 204.
Hagiwara, S., and N. Saito (1959b) *J. Physiol.* (London) **148**, 161.
Hagiwara, S., and K. Kusano (1961) *J. Neurophysiol.* **24**, 167.
Harvey, A. M. (1939) *Bull. Johns Hopkins Hosp.* **65**, 223.
Hayashi, I. (1939) *Shokubutsu Oyobi Dobutsu* **7**, 2001.
Hille, B. (1967) *J. Gen. Physiol.* **50**, 1287.
Hille, B. (1968) *J. Gen. Physiol.* **51**, 199.
Hodgkin, A. L., and A. F. Huxley (1952a) *J. Physiol.* (London) **116**, 449.
Hodgkin, A. L., and A. F. Huxley (1952b) *J. Physiol.* (London) **116**, 473.

Hodgkin, A. L., and A. F. Huxley (1952c) *J. Physiol.* (London) **116**, 497.
Hodgkin, A. L., and A. F. Huxley (1952d) *J. Physiol.* (London) **117**, 500.
Hodgkin, A. L., A. F. Huxley, and B. Katz (1952) *J. Physiol.* (London) **116**, 424.
Hodgkin, A. L., and B. Katz (1949) *J. Physiol.* (London) **108**, 37.
Julian, F. J., J. W. Moore, and D. E. Goldman (1962a) *J. Gen. Physiol.* **45**, 1195.
Julian, F. J., J. W. Moore, and D. E. Goldman (1962b) *J. Gen. Physiol.* **45**, 1217.
Kao, C. Y. (1966) *Pharmacol. Rev.* **18**, 997.
Kao, C. Y., and A. Nishiyama (1965) *J. Physiol.* (London) **180**, 50.
Katz, B. (1966) Nerve, Muscle, and Synapse. McGraw-Hill, New York, 193 pp.
Katz, B. and R. Miledi (1967) *Proc. Roy. Soc.* B **167**, 8.
Keynes, R. D. (1951) *J. Physiol* (London) **114**, 119.
Keynes, R. D., and P. R. Lewis (1951) *J. Physiol.* (London) **114**, 151.
Kordaš, M. (1968) *Int. J. Neuropharmacol.* **7**, 523.
Kordaš, M. (1969) *J. Physiol.* (London) **204**, 493.
Krahl, M. E., A. K. Keltch, and G. H. A. Clowes (1940) *J. Pharmacol. Exp. Ther.* **68**, 330.
Krupp, P., C. P. Bianchi, and G. Suarez-Kurtz (1969) *J. Pharm. Pharmacol.* **21**, 763.
Kuga, T. (1958) *Folia Pharmacol. Jap.* **55**, 1257.
Kumamoto, M., and L. Horn (1969) *Physiologist* **12**, 278.
Kuriaki, K., and I. Wada (1957) *Japan. J. Pharmacol.* **7**, 35.
Kurose, T. (1943) *Folia Pharmacol. Jap.* **38**, 441.
Lalonde, D. I. V., and A. W. A. Brown (1954) *Can. J. Zool.* **32**, 74.
Langley, J. N., and W. L. Dickinson (1889) *Proc. Roy. Soc.* (London) **46**, 423.
Lowenstein, O. (1942) *Nature* (London) **150**, 760.
Maeno, T. (1966) *J. Physiol.* (London) **183**, 592.
Mascher, D., and K. Peper (1969) *Pflüg. Arch.* **307**, 190.
Matsumura, M., and S. Yamamoto (1954) *Japan. J. Pharmacol.* **4**, 62.
Moore, J. W., N. Anderson, M. Blaustein, M. Takata, J. Y. Lettvin, W. F. Pickard, T. Bernstein, and J. Pooler (1966) *Ann. N.Y. Acad. Sci.* **137**, 818.
Moore, J. W., M. P. Blaustein, N. C. Anderson, and T. Narahashi (1967) *J. Gen. Physiol.* **50**, 1401.
Moore, J. W., and K. S. Cole (1963) In: Physical Techniques in Biological Research, Vol. 6, ed. by W. L. Nastuk, Academic Press, New York and London, p. 263.
Moore, J. W., T. Narahashi, and T. I. Shaw (1967) *J. Physiol.* (London) **188**, 90.
Mosher, H. S., F. A. Fuhrman, H. D. Buchwald, and H. G. Fischer (1964) *Science* **144**, 1100.
Mullins, L. J. (1959) *J. Gen. Physiol.* **42**, 1013.
Mullins, L. J. (1968) *J. Gen. Physiol.* **52**, 550.
Nakajima, S., S. Iwasaki, and K. Obata (1962) *J. Gen. Physiol.* **46**, 97.
Nakamura, Y., S. Nakajima, and H. Grundfest (1965a) *J. Gen. Physiol.* **49**, 321.
Nakamura, Y., S. Nakajima, and H. Grundfest (1965b) *J. Gen. Physiol.* **48**, 985.
Narahashi, T. (1962a) *J. Cell. Comp. Physiol.* **59**, 61.
Narahashi, T. (1962b) *J. Cell. Comp. Physiol.* **59**, 67.
Narahashi, T. (1963) In: Advances in Insect Physiology, Vol. 1, ed. by J. W. L. Beament, J. E. Treherne, and V. B. Wigglesworth, Academic Press, London and New York, p. 175.
Narahashi, T. (1964) *Japan. J. Med. Sci. Biol.* **17**, 46.
Narahashi, T. (1971) In: Biophysics and Physiology of Excitable Membranes, ed. by W. J. Adelman, Jr., Van Nostrand Reinhold Co., New York, p. 423.

Narahashi, T. (1972) In: Pharmacology of the Cells, ed. by S. Dickstein, C. C Thomas, London, in press.
Narahashi, T., and N. C. Anderson (1967) *Toxicol. Appl. Pharmacol.* **10**, 529.
Narahashi, T., N. C. Anderson, and J. W. Moore (1966) *Science* **153**, 765.
Narahashi, T., N. C. Anderson, and J. W. Moore (1967) *J. Gen. Physiol.* **50**, 1413.
Narahashi, T., T. Deguchi, N. Urakawa, and Y. Ohkubo (1960) *Am. J. Physiol.* **198**, 934.
Narahashi, T., and D. T. Frazier (1971) In: Neurosciences Research, ed. by S. Ehrenpreis and O. C. Solnitzky, Academic Press, New York and London, p. 65.
Narahashi, T., D. T. Frazier, T. Deguchi, C. A. Cleaves, and M. C. Ernau (1970) *Fed. Proc.* **29**, 483 Abs.
Narahashi, T., D. T. Frazier, T. Deguchi, C. A. Cleaves, and M. C. Ernau (1971) *J. Pharmacol. Exp. Ther.* **177**, 25.
Narahashi, T., D. T. Frazier, and J. W. Moore (1968) *Proc. Int. Union Physiol. Sci.*, Vol. 7, XXIV Int. Congr. p. 313.
Narahashi, T., D. T. Frazier, and M. Yamada (1970) *J. Pharmacol. Exp. Ther.* **171**, 32.
Narahashi, T., and H. G. Haas (1967) *Science* **157**: 1438.
Narahashi, T. and H. G. Haas (1968) *J. Gen. Physiol.* **51**, 177.
Narahashi, T., H. G. Haas, and E. F. Therrien (1967) *Science* **157**: 1441.
Narahashi, T., and J. W. Moore (1968a) *J. Gen. Physiol.* **51**, 93s.
Narahashi. T., and J. W. Moore (1968b) *J. Gen. Physiol.* **52**, 553.
Narahashi, T., J. W. Moore, and D. T. Frazier (1969) *J. Pharmacol. Exp. Ther.* **169**, 224.
Narahashi, T., J. W. Moore, and R. N. Poston (1967) *Science* **156**, 976.
Narahashi, T., J. W. Moore, and R. N. Poston (1969) *J. Neurobiol.* **1**, 3.
Narahashi, T., J. W. Moore, and W. R. Scott (1964) *J. Gen. Physiol.* **47**: 965.
Narahashi, T., J. W. Moore, and B. I. Shapiro (1969) *Science* **163**, 680.
Narahashi, T., M. Yamada, and D. T. Frazier (1969) *Nature* **223**, 748.
Narahashi, T., and T. Yamasaki (1960a) *J. Physiol.* (London) **151**, 75.
Narahashi, T., and T. Yamasaki (1960b) *J. Physiol.* (London) **152**, 122.
Narahashi, T., and T. Yamasaki (1960c) *J. Cell. Comp. Physiol.* **55**, 131.
Nonner, W. (1969) *Pflüg. Arch.* **309**, 176.
Nonner, W., and R. Stämpfli (1969) In: Laboratory Techniques in Membrane Biophysics, ed. by H. Passow and R. Stämpfli, Springer-Verlag, Berlin, p. 171.
O'Brien, R. D. (1967) *Fed. Proc.* **26**, 1056.
O'Brien, R. D. (1969) In: Essays in Toxicology, ed. by F. R. Blood, Academic Press, New York and London, p. 1.
Ogura, Y., and Y. Mori (1968) *Europ. J. Pharmacol.* **3**, 58.
Oikawa, T., C. S. Spyropoulos, I. Tasaki, and T. Teorell (1961) *Acta Physiol. Scand.* **52**: 195.
Oomura, Y. and T. Tomita (1960) In: Electrical Activity of Single Cells, ed. by Y. Katsuki, Igaku Shoin Ltd., Tokyo, p. 181.
Overton, E. (1902) *Arch. Ges. Physiol.* **92**, 346.
Pichon, Y. (1969) *C. R. Acad. Sci.* (Paris) **268**, 1095.
Pichon, Y., and J. Boistel (1967) *J. Exp. Biol.* **47**: 343.
Pickard, W. F., J. Y. Lettvin, J. W. Moore, M. Takata, J. Pooler, and T. Bernstein (1964) *Proc. Nat. Acad. Sci.* **52**, 1177.
Pooler, J. (1968) *Biophys. J.* **8**, 1009.
Ritchie, J. M., and P. Greengard (1966) *Ann. Rev. Pharmacol.* **6**, 405.
Ritchie, J. M., B. Ritchie, and P. Greengard (1965a) *J. Pharmacol. Exp. Ther.* **150**, 152.

Ritchie, J. M., B. Ritchie, and P. Greengard (1965 b) *J. Pharmacol. Exp. Ther.* **150**, 160.
Roeder, K. D., and E. A. Weiant (1948) *J. Cell. Comp. Physiol.* **32**, 175.
Rojas, E., and I. Atwater (1967) *Proc. Nat. Acad. Sci.* **57**, 1350.
Rothenberg, M. A., D. B. Sprinson, and D. Nachmansohn (1948) *J. Neurophysiol.* **11**, 111.
Rougier, O., G. Vassort, and R. Stämpfli (1968) *Pflüg. Arch.* **301**, 91.
Schanker, L. S., P. A. Nafpliotis, and J. M. Johnson (1961) *J. Pharmacol. Exp. Ther.* **133**, 325.
Schantz, E. J., J. M. Lynch, G. Vayvada, K. Matsumoto, and H. Rapoport (1966) *Biochemistry* **5**, 1191.
Shanes, A. M. (1949) *J. Gen. Physiol.* **33**, 75.
Shanes, A. M. (1958) *Pharmacol. Rev.* **10**, 59.
Shanes, A. M., W. H. Freygang, H. Grundfest, and E. Amatniek (1959) *J. Gen. Physiol.* **42**, 793.
Shapiro, B. I. (1968) *Toxicon* **5**, 253.
Shapiro, B. I., and G. Lilleheil (1969) *Comp. Biochem. Physiol.* **28**, 1225.
Steinbach, A. B. (1968 a) *J. Gen. Physiol.* **52**, 144.
Steinbach, A. B. (1968 b) *J. Gen. Physiol.* **52**, 162.
Tahara, Y. (1910) *Biochem. Zeitsch.* **10**, 255.
Takata, M., J. W. Moore, C. Y. Kao, and F. A. Fuhrman (1966) *J. Gen. Physiol.* **49**, 977.
Takeuchi, A., and N. Takeuchi (1959) *J. Neurophysiol.* **22**, 395.
Takeuchi, A., and N. Takeuchi (1960) *J. Physiol.* (London) **154**, 52
Tasaki, I., and I. Singer (1966) *Ann. N.Y. Acad. Sci.* **137**, 792.
Tasaki, I., I. Singer, and T. Takenaka (1965) *J. Gen. Physiol.* **48**, 1095.
Taylor, R. E. (1959) *Am. J. Physiol.* **196**, 1071.
Tsuda, K., S. Ikuma, M. Kawamura, R. Tachikawa, K. Sakai, C. Tamura, and O. Amakasu (1964) *Chem. Pharmaceut. Bull.* **12**, 1357.
Urakawa, N., T. Narahashi, T. Deguchi, and Y. Ohkubo (1960) *Am. J. Physiol.* **198**, 939.
Wada, I. (1957) *Folia Pharmacol. Jap.* **53**, 429.
Welsh, J. H., and H. T. Gordon (1947) *J. Cell. Comp. Physiol.* **30**, 147.
Whitcomb, E. R., S. L. Friess, and J. W. Moore (1958) *J. Cell. Comp. Physiol.* **52**, 275.
Wong, J. L., R. Oesterlin, and H. Rapoport (1971) *J. Am. Chem. Soc.* **93**, 7344.
Woodward, R. B. (1964) *Pure Appl. Chem.* **9**, 49.
Yamasaki, T., and T. Ishii* (1952 a) Oyo-Kontyu (*J. Nippon Soc. Appl. Entomol.*) **7**, 157.
Yamasaki, T., and T. Ishii* (1952 b) Oyo-Kontyu (*J. Nippon Soc. Appl. Entomol.*) **8**, 111.
Yamasaki, T., and T. Ishii* (1954 a) *Botyu-Kagaku* **19**: 1; English translation in Japanese Contributions to the Study of the Insecticide-Resistance Problem. Publ. by Kyoto University for the W. H. O., p. 140 (1957).
Yamasaki, T., and T. Ishii* (1954 b) *Botyu-Kagaku* **19**, 39; English translation in Japanese Contributions to the Study of the Insecticide-Resistance Problem. Publ. by Kyoto University for the W. H. O., p. 155 (1957).
Yamasaki, T., and T. Narahashi (1957 a) *Botyu-Kagaku* **22**, 296.
Yamasaki, T., and T. Narahashi (1957 b) *Botyu-Kagaku* **22**, 305.
Yamasaki, T., and T. Narahashi (1958) *Botyu-Kagaku* **23**, 146.
Yamasaki, T., and T. Narahashi (1962) *Japan. J. Appl. Entomol. Zool.* **6**, 293.
Yano, I., (1937) *Fukuoka-igaku-Zassi* **30**, 1669.

---

* Former name of T. Narahashi.

An Extension of Cole's Theorem and its Application to Muscle

R. H. ADRIAN, W. K. CHANDLER, and A. L. HODGKIN

From the Physiological Laboratory, University of Cambridge

In 1941 Cole and Curtis described an elegant theoretical method which is sometimes known as Cole's theorem. Suppose that a steady current is applied through a microelectrode to the inside of a nerve or muscle fibre which is long compared to the space constant. When the potential at the point where current is applied is plotted against the electrode current, a curve is obtained if the relation between membrane current and potential is not linear. From this curve, Cole's theorem enables one to calculate how membrane current varies with membrane potential at any point in the fibre. If i_m is the membrane current per unit length in a fibre with an internal resistance r per unit length, which is surrounded by a large volume of conducting fluid, Cole's method gives

$$i_m = \frac{r}{4} I_0 \frac{dI_0}{dV_0}$$

where V_0 is the internal potential at the point where the electrode current I_0 is applied.

Cole (1961, 1968) has given alternative derivations of this relation, but in neither the original paper with Curtis (1941) nor in the later versions is it very clear that the theory applies only if the fibre is of infinite length. It is also not immediately obvious how one can use the theorem in reverse to calculate electrode current from membrane current. The present note deals with these questions, and shows how a treatment similar to Cole's can be used to predict current and voltage thresholds in a muscle fibre. It also shows how the inverse of Cole's theorem can be applied to a situation in which the fibre is not of infinite length. The experimental methods need no description since they were essentially the same as those used by Adrian, Chandler and Hodgkin (1966, 1970).

THEORY

Consider the case of a steady current I_0 applied at x = 0 to an infinitely long fibre immersed in a large volume of conducting fluid. The functions $f(V)$ and $f_i(V)$ are defined by

$$i_m = f(V), \tag{1}$$

$$\int_0^V i_m \, dV = f_i(V) \tag{2}$$

where i_m is the membrane current per unit length in the steady state and V is the displacement of the membrane potential from its resting value.

Since

$$i_m = -\frac{di}{dx} \tag{3}$$

and

$$ri = -\frac{dV}{dx} \tag{4}$$

where i is the longitudinal current in the interior of the fibre, it follows that

$$i_m \frac{dV}{dx} = ri \frac{di}{dx} \tag{5}$$

or

$$f(V) \frac{dV}{dx} = ri \frac{di}{dx} \tag{6}$$

or

$$\frac{df_i(V)}{dx} = \frac{r}{2} \frac{d(i^2)}{dx}. \tag{7}$$

If $f_i(V)$ and i are zero at $x = \infty$, integration between $x = 0$ and ∞ gives

$$f_i(V_0) = \tfrac{1}{2} r(i_0)^2 \tag{8}$$

where V_0 and i_0 are the potential and longitudinal current at $x = 0$. Now i_0 is half the electrode current I_0 so

$$f_i(V_0) = rI_0^2/8 \tag{9}$$

or

$$I_0 = 2\sqrt{2f_i(V_0)/r}. \tag{10}$$

This is the inverse of Cole's theorem and enables I_0 to be calculated as a function of V_0, if the relation between membrane current and membrane potential is known. On differentiating equation (9) with respect to V_0 we obtain

$$f(V_0) = \frac{rI_0}{4}\frac{dI_0}{dV_0} \tag{11}$$

or

$$i_m = \frac{rI_0}{4}\frac{dI_0}{dV_0} \tag{12}$$

which is the usual form of Cole's theorem.

STABILITY OF CABLE IF MEMBRANE POTENTIAL IS CLAMPED AT $x = 0$

Suppose that the relation between membrane current and voltage is time independent and is given by the curve in Figure 1A. Then by using the inverse of Cole's theorem we can calculate the relation between electrode current and the potential at the point where current is applied, provided that $f_i(V)$ is positive; this is shown in Figure 1B.

The voltages V_α, V_β and V_γ are defined in the following way

$$V_\alpha, \qquad \frac{di_m}{dV} = 0.$$

$$V_\beta, \qquad i_m = 0.$$

$$V_\gamma, \qquad \int_0^V i_m\, dV = 0.$$

For an isolated patch of membrane it is well known that V_α is the constant current threshold or rheobase, and that V_β, the displacement threshold, is the potential which must be exceeded in order to make $\frac{dV}{dt}$ positive in the absence of applied current. Under voltage-clamp conditions all currents are stable and there is no threshold.

If the non-linear elements are connected in a cable it can be seen from Figure 1B that V_β is now the constant current threshold, but that with voltage control we can obtain steady electrode currents up to V_γ. Beyond that point $f_i(V)$ becomes negative, its square root is imaginary and there are no steady state solutions; V_γ is evidently the voltage threshold.

To illustrate the points mentioned in the previous paragraphs we shall assume that the relation between ionic current (steady membrane current)

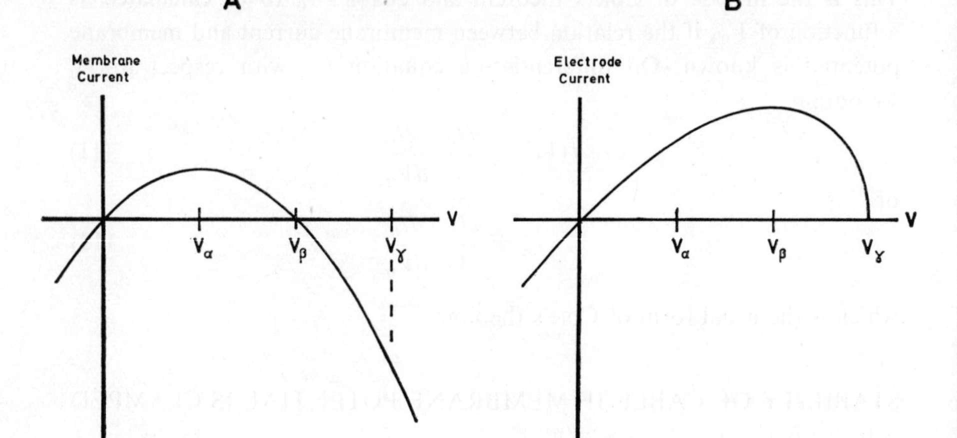

Figure 1 A. Assumed relation between membrane current and membrane potential, V. The curve is

$$f(V) = V - \frac{3V^2}{2}.$$

B. Relation between electrode current and potential at point where current is applied, in an infinite cable with membrane characteristics given in A. The curve is

$$\sqrt{2f_i(V)} = V\sqrt{1-V}$$

where

$$f_i(V) = \int_0^V f(V)\,dV.$$

and potential is the parabola

$$i_m = aV - \frac{3}{2}bV^2 \tag{13}$$

in which case $\left(\dfrac{di_m}{dV}\right)_{V\to 0} = a$, $V_\alpha = \dfrac{a}{3b}$, $V_\beta = \dfrac{2a}{3b}$ and $V_\gamma = \dfrac{a}{b}$ The differential equation for the steady state is

$$\frac{1}{r}\frac{d^2V}{dx^2} = aV - \frac{3bV^2}{2}. \tag{14}$$

After multiplication by $\dfrac{2\,dV}{dx}$, integration between x and infinity gives

$$\frac{dV}{dx} = -r^{\frac{1}{2}}V\sqrt{a - bV}, \tag{15}$$

the integration constant being zero because $\dfrac{dV}{dx}$ and V are zero at $x = \infty$.

Hence the electrode current $I_0 = -\dfrac{2}{r}\left(\dfrac{dV}{dx}\right)_0$ is given by

$$I_0 = 2r^{-\frac{1}{2}}\, V_0 \sqrt{a - bV_0}\,, \tag{16}$$

a result which could be obtained directly by applying the inverse of Cole's theorem [equation (10)] to equation (13). To obtain the distribution of potential we separate the variables in equation (15) and integrate between 0 and x, which gives

$$V = \frac{a}{b}\left\{1 - \tanh^2\left[(ar)^{\frac{1}{2}}\frac{x}{2} + B\right]\right\} \tag{17}$$

where $B = \tanh^{-1}\sqrt{1 - bV_0/a}$.

Curves for different values of V_0 are plotted in Figure 2 and the values of I_0 required to maintain the distribution are given in the legend. The maximum value of V_0 is a/b and if $V_0 > \dfrac{a}{b}$ there are no solutions which satisfy the boundary conditions $V = 0$ at $x = \pm\infty$. Between $V_0 = \dfrac{2a}{3b}$ and $V_0 = \dfrac{a}{b}$, $\dfrac{dI_0}{dV_0}$ is negative and in this region the distribution would be stable if V_0 were held constant, but not if I_0 were held constant.

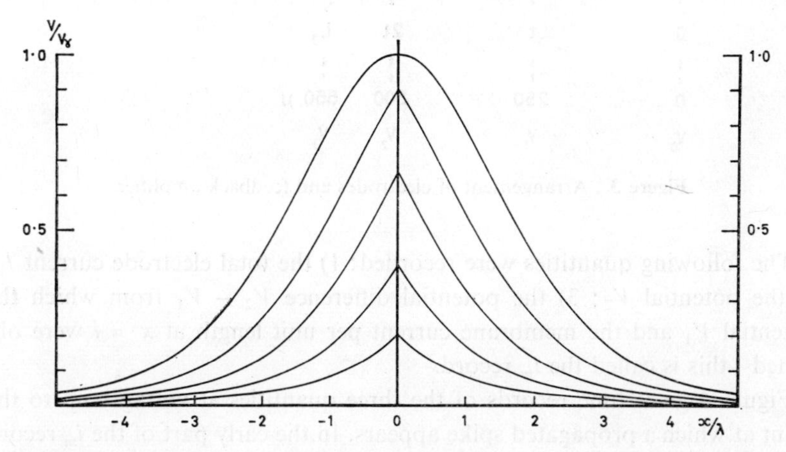

Figure 2 Distribution of potential in infinite cable in which membrane current is related to potential by

$$i_m = aV - \frac{3bV^2}{2}$$

and a current I_0 is applied at $x = 0$.
The curves were plotted from equation (15) with $V_\gamma = \dfrac{a}{b}$ and $\lambda = (ar)^{-\frac{1}{2}}$. The currents required to maintain these distributions are from above down: 0, 0.285, 0.385, 0.310, 0.179, in units of $2V_\gamma(a/r)^{\frac{1}{2}}$.

EXPERIMENTAL RESULTS

Three electrodes were inserted near the end of a muscle fibre in the frog's sartorius muscle with the spacing shown in Figure 3. The feedback amplifier was controlled from the middle electrode and the voltage V_2 underwent a rectangular step. In describing and analysing the results, the origin $x = 0$ is now taken to be at the end of the fibre. The recording electrodes were at l and $2l$ from the end of the fibre where $l = 250\ \mu$ in this experiment.

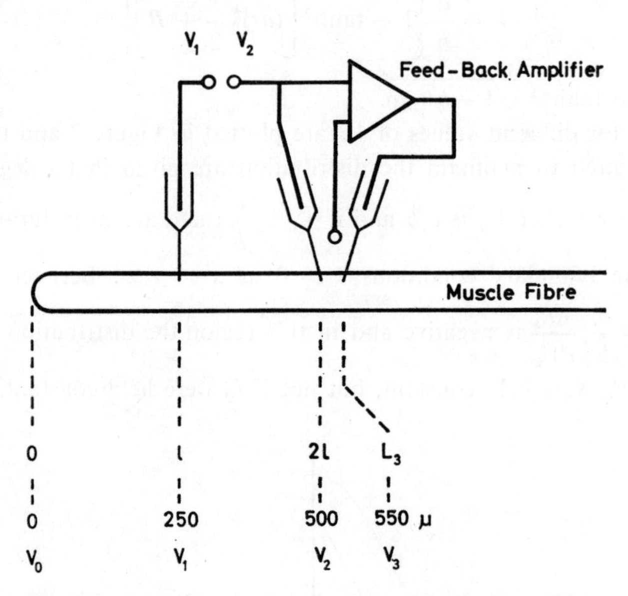

Figure 3 Arrangement of electrodes and feedback amplifier

The following quantities were recorded: 1) the total electrode current I_0; 2) the potential V_2; 3) the potential difference $V_2 - V_1$ from which the potential V_1 and the membrane current per unit length at $x = l$ were obtained—this is called the i_m record.

Figure 4 gives time records of the three quantities at voltages up to the point at which a propagated spike appears. In the early part of the i_m record there is a large capacitative transient. The final time constant of this transient, which is determined by $rc_m l^2$ gave the membrane capacity as about $4\ \mu F/cm^2$. The transient was followed by a phase of ionic current which reversed in sign and became inward as the control voltage V_2 approached threshold. At a fixed control voltage the magnitude of the inward current varied from one record to the next, but there was less scatter when i_m was plotted against V_1.

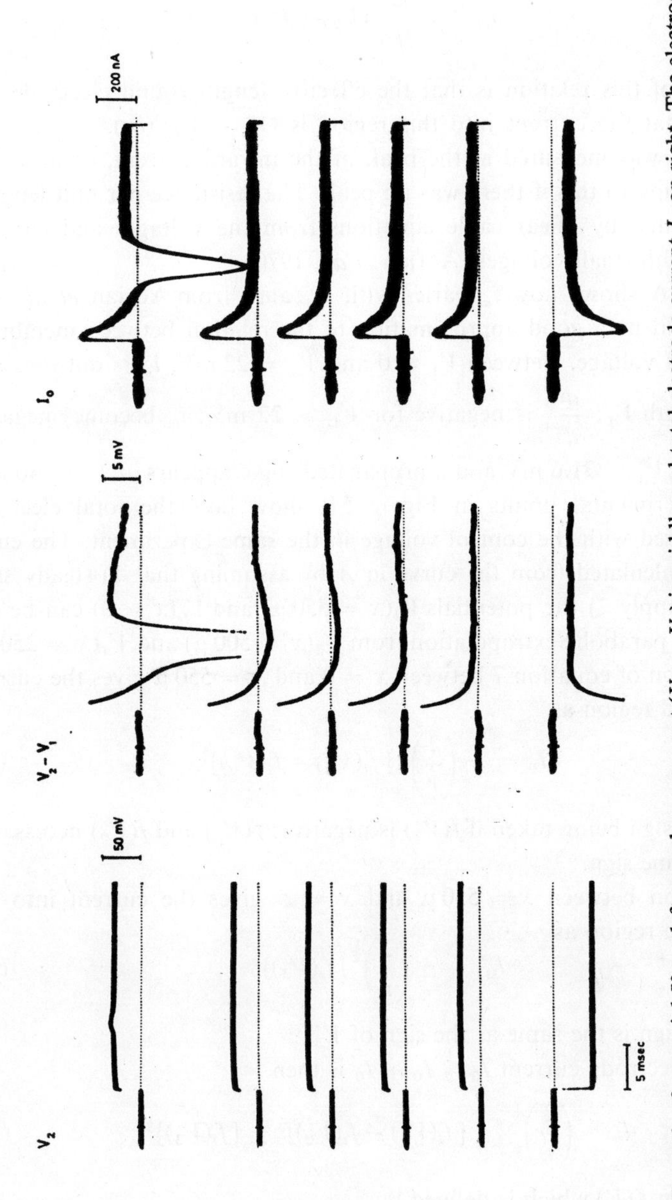

Figure 4 Records of potential V_2 on left, membrane current $(V_2 - V_1)$ in middle and total electrode current I_0 on right. The electrode spacing was as in Figure 3. Resting potential = holding potential = −100 mV; fibre diameter, 75 μ; internal resistance per unit length = 5.8 MΩ/cm; temperature 5.5°C. On the membrane current records 1 mV ≡ 184 nA/cm. The total current was recorded with an exponential delay of 0.6 msec

From Adrian, Chandler, and Hodgkin (1970) the membrane current per unit length at a distance l from the end of the fibre is given by

$$i_m(l) = \frac{2}{3l^2 r}(V_2 - V_1) \tag{18}$$

The basis of this relation is that the effective length round electrode 1 is $3l/2$ and that the current into this region is $(V_2 - V_1)/(lr_i)$.

$V_2 - V_1$ was measured at the peak of the inward current, or at a time corresponding to this if there was no peak. The resistance per unit length r was calculated by linear cable equations from the voltages and currents observed with small voltages (Adrian *et al.*, 1970).

Figure 5A shows how i_m varies with V_1 and, from Adrian *et al.*, this curve should be a good approximation to the relation between membrane current and voltage. Between $V_1 = 0$ and $V_1 = 22$ mV, i_m is outward and increases with V_1; $\frac{di_m}{dV}$ is negative for $V_1 > 22$ mV; i_m becomes negative (inward) at $V_1 > 31.6$ mV and a propagated spike appears at $V_1 \doteqdot 40$ mV.

The experimental points in Figure 5B show how the total electrode current varied with the control voltage in the same experiment. The curve in *B* was calculated from the curve in *A* by assuming that 1) steady state equations apply 2) the potentials $V_3(x = 550\ \mu)$ and $V_0(x = 0)$ can be obtained by a parabolic extrapolation from $V_2(x = 500\ \mu)$ and $V_1(x = 250\ \mu)$.

Integration of equation 7 between $x = 0$ and $x = 550\ \mu$ gives the current into the end region as

$$I_0' = \pm\left(\frac{2}{r}\right)^{\frac{1}{2}}[f_i(V_3) - f_i(V_0)]^{\frac{1}{2}} \tag{19}$$

a negative sign being taken if $f(V_3)$ is negative; $f(V_3)$ and $f(V_0)$ necessarily have the same sign.

Integration between $x = 550\ \mu$ and $x = \infty$ gives the current into the semi-infinite region as

$$I_0'' = \pm\left(\frac{2}{r}\right)^{\frac{1}{2}}[f_i(V_3)]^{\frac{1}{2}} \tag{20}$$

where the sign is the same as the sign of V_3.

The total electrode current $I_0 = I_0' + I_0''$ is then

$$I_0 = \left(\frac{2}{r}\right)^{\frac{1}{2}}\{\pm[f_i(V_3) - f_i(V_0)]^{\frac{1}{2}} \pm [f_i(V_3)]^{\frac{1}{2}}\} \tag{21}$$

The function $f_i(V)$ which is defined by

$$f_i(V) = \int_0^V i_m\, dV$$

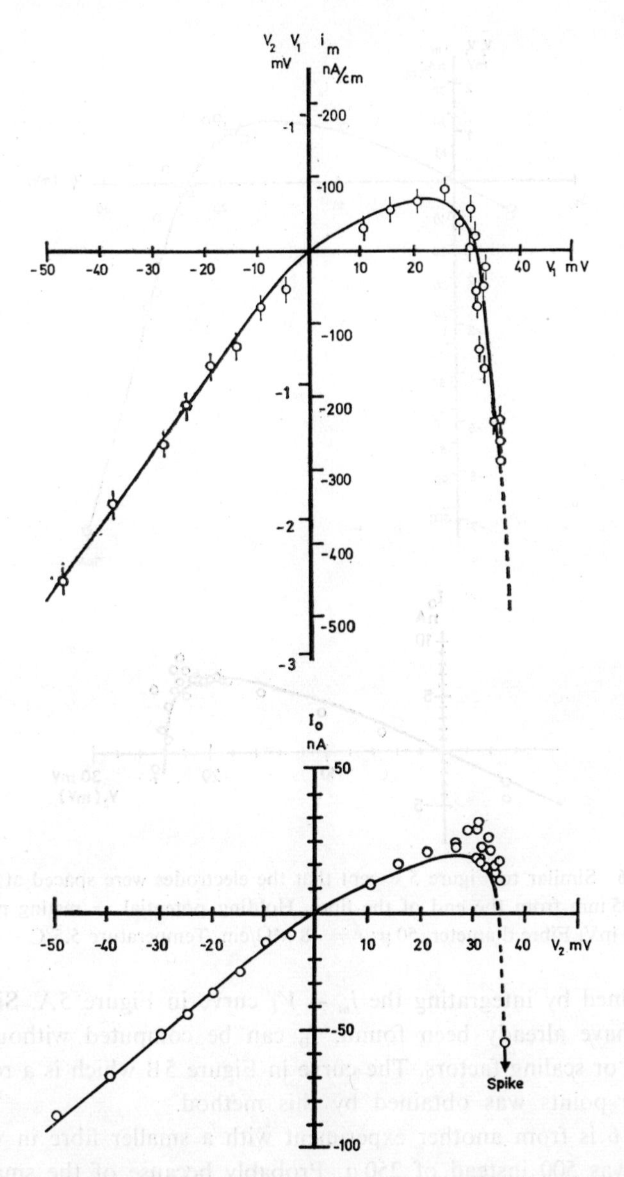

Figure 5 A. Relation between membrane current i_m and displacement of membrane potential from resting value obtained from the experiment of Figure 4. The abscissa is the potential V_1 $(x = l)$ at the time of maximum inward current, measured as a displacement from the holding potential of -100 mV. The ordinate is i_m at $x = l$, obtained from $V_2 - V_1$ by equation 18. B. Relation between electrode current I_0 and control voltage V_2. The circles are values of I_0 measured at the same time as the values of i_m in A. The continuous curve was calculated from the curve in A by equation (21) with r, the internal resistance per unit length $= 5.8$ MΩ/cm

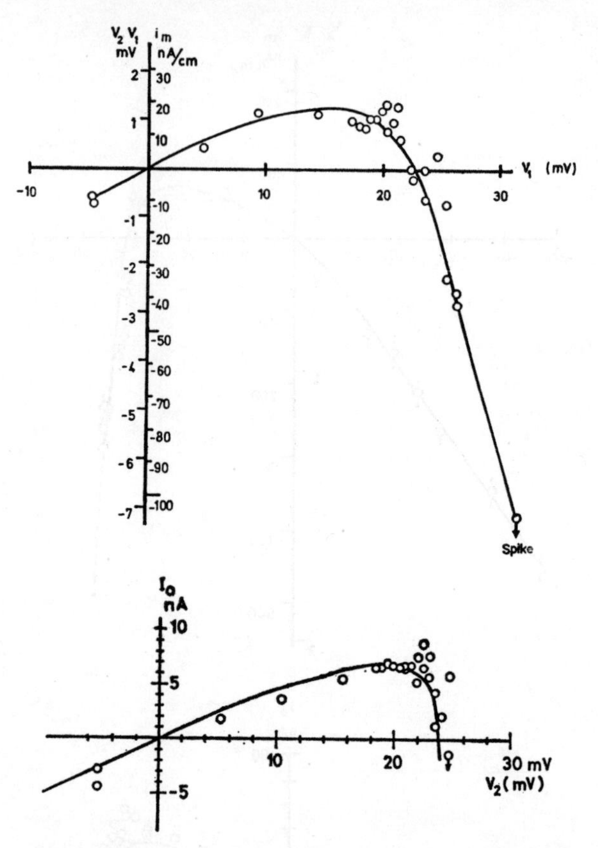

Figure 6 Similar to Figure 5 except that the electrodes were spaced at 0.5, 1.0 and 1.05 mm from the end of the fibre. Holding potential = resting potential = −89 mV. Fibre diameter, 50 μ; r = 18 MΩ/cm. Temperature 5.5°C

was obtained by integrating the $i_m - V_1$ curve in Figure 5A. Since r, V_3 and V_0 have already been found, I_0 can be computed without further constants or scaling factors. The curve in Figure 5B which is a reasonable fit to the points was obtained by this method.

Figure 6 is from another experiment with a smaller fibre in which the length l was 500 instead of 250 μ. Probably because of the smaller fibre diameter the feedback was more effective and the control voltage did not change when a spike appeared. In both experiments, the slope of the curve relating electrode current to control voltage approached $-\infty$ at the voltage at which a spike appeared.

In the limiting case where the length of the end region is infinitessimal the electrode current should be zero at threshold. With a finite length in the end region it is theoretically possible to have a stable distribution even though the electrode current has a sign opposite to that of the control voltage. Thus if $V_3 = V_\gamma$ there is zero current into the semi-infinite region and an inward current into the finite region of the fibre.

The agreement shown in Figures 5 and 6 helps to validate the use of equation 18 to calculate membrane current and shows that non-linear cable theory makes reasonably successful predictions about the threshold under conditions of localised voltage control.

REFERENCES

Adrian, R. H., W. K. Chandler, and A. L. Hodgkin (1966). Voltage clampexperiments in skeletal muscle fibres. *J. Physiol.* **186,** 51–52P.

Adrian, R. H., W. K. Chandler, and A. L. Hodgkin (1970). Voltage clamp experiments in striated muscle fibres. *J. Physiol.* **208,** 607–644.

Cole, K. S., and H. J. Curtis (1941). Membrane potential of the squid giant axon during current flow. *J. gen. Physiol.* **24,** 551–563.

Cole, K. S. (1961). Non-linear current-potential relations in an axon membrane. *J. gen. Physiol.* **44,** 1055–1057.

Cole, K. S. (1968). Membranes, ions and impulses. A chapter of classical biophysics. Univ. California Press.

Research on Nerve and Muscle

A. F. HUXLEY

Royal Society Research Professor
Department of Physiology, University College London

Having changed may field of work, nearly twenty years ago, from nerve conduction to muscle contraction, I found myself unable to write anything very directly appropriate for this volume in honour of K. S. Cole's seventieth birthday. I would have been very sorry, however, to miss this opportunity of contributing something by way of expressing my admiration of his work and my consciousness of its importance in the development of our ideas about excitable membranes, and it was while reflecting in this way on Cole's work that I began to wonder if it would be possible to draw any instructive analogies between the courses that research in these two fields has taken, both in the recent past and during the last seventy to a hundred years.

There are many obvious similarities between the activity of nerve and muscle. Apart from the fact that the first step in most forms of muscular contraction is a membrane potential change, in many cases essentially the same as a nerve action potential, there is the simple feature that the unit response in each case is a short-lasting event, of a physical character. In the nerve impulse, current flows, and there are changes of electric potential; in the muscle twitch, the fibres shorten and the tension rises. In each case the accompanying change of temperature gives a further clue to the nature of the underlying processes. In these respects the training and the general approach required in an investigator are much the same for the two tissues, and perhaps it is chiefly for this reason that many people before me have divided most of their energies between these two fields—notable examples are Julius Bernstein and A. V. Hill.

On the other hand, there is a striking difference between the two fields in that chemical investigations have played a much greater part in relation to contraction than to conduction. In one sense this difference is a matter of opportunity: in muscle there is an almost unlimited quantity of the contractile substance to investigate, while the molecules controlling the permeability changes in a nerve membrane are so scarce that it is only very recently that even their number has been satisfactorily estimated[1] and they have still not been isolated in any useful sense. In view of this, it is lucky that it has turned out that the carriers of charge in nerve conduction are among the simplest of chemical substances, namely sodium and potassium ions, both of which are present in large quantity. If, for example, it had been that the current is carried across the membrane by the movement of charged molecules which never escape altogether from the membrane, then their number could well have been so small that they would probably not yet have been identified, especially if they were of the same order of complexity as, say, ATP, the substance whose role in muscle contraction is in a sense equivalent to that of the sodium ion in nerve conduction.

Apart from the availability of large quantities of material, muscle chemists have an immense advantage over nerve chemists, in that they are able to recognise changes in fragmented or dissolved preparations that correspond to the contraction process. It is true that, just as a potential change in a nerve fibre can only be produced if the membrane is intact, so a muscle can only generate tension if it is continuous and anchored at both ends; but in the case of nerve there is no available method equivalent to watching the shortening of a small piece of a myofibril, or following the change of viscosity or flow birefringence which reflects the shape changes of actomyosin particles in solution. One day, no doubt, someone will find a way of producing and detecting the permeability change in an isolated preparation, but at present the change can only be recognised by detecting a potential change related to the movement of ions, and this in turn requires an intact membrane for the potential change to be developed across.

As far as concerns the nature of the actual contraction process itself, the only directly important result from the tremendous amount of work in the period from 1900 to 1940 on the metabolism of muscle was the emergence of ATP as the substance whose interactions with the contractile proteins produces shortening or the development of tension. Oxidative metabolism, glycolysis and the splitting of phosphagens were shown in turn to be recovery, or backing-up, processes which were important in muscular activity because, and only because, they contributed to the rephosphorylation of ADP to ATP. The central position of ATP was made practically certain in 1939–41[2,3,4] by the direct demonstration of a two-way-interaction: the protein "myosin", long regarded as the contractile material, acts as an ATPase, and ATP causes changes of shape in the myosin molecule. In contrast to the four decades of intensive search that led up to this result, the equivalent point in the investigation of nerve conduction was established by a single series of experiments, namely the demonstration by Hodgkin and Katz[5] that sodium ions are the actual carriers of the current which generates the impulse. Nevertheless, this discovery did not come until a decade after the establishment of the position of ATP in muscular contraction.

So much for the identification of the substances most directly involved in the processes of contraction and conduction. If we turn to ideas about the general nature of the two processes, we find that the boot is on the other foot. It was in 1902 that Bernstein[6] proposed that the change underlying the generation of an action potential is an increase in the ionic permeability of the surface membrane, and this idea was generally accepted although there was no direct proof of it until the beautiful experiment of Cole and Curtis[7]. The equivalent idea in relation to muscle contraction is the sliding-filament theory, which was not put forward[8,9,10] until more than half a century after Bernstein's theory of the action potential, and when it did come it was a direct consequence of new observations that cropped up as soon as there was a revival of serious interest in the structural aspects of muscle.

The sodium theory and the sliding-filament theory thus came rather late in the development of ideas on the mechanism of action of nerve and muscle respectively. Each was accepted quickly after it was published. These are points which could be taken as indicating that the discoveries were in some sense overdue at the time when they were made. Another pointer in this direction is that some of the key pieces of evidence in favour of each of these theories were first obtained long ago. Overton's paper[11] on the "indispensability of sodium (or lithium) ions for the contraction of muscle", published in 1902, has become widely known since Hodgkin and Katz's experiments. I suppose the reason why its conclusions were not generally accepted much earlier is that a *nerve* can be placed in a sodium-free solution for some hours without losing its power of conducting an impulse, although Overton himself[12] gave the correct explanation of this, namely that sodium salts are retained in the spaces between the fibres of a nerve bundle by the perineurial sheath which surrounds it.

As regards the sliding-filament theory, the observation that all or nearly all of the length change of a striated muscle fibre takes place in the I band was made repeatedly during the latter half of the 19th century. It was the basis of a theory of contraction by Krause in 1869[13]; other references will be found in another article that I wrote[14]. The idea that the *A* bands were solid structures was widespread; some but not all of the people who wrote on muscle considered that it consisted of an array of parallel rodlets.

But, as far as I know, no one proposed a second set of filaments passing through the I band and into the adjacent *A* bands; it was generally assumed that the I bands contained a liquid. The rediscovery of constant *A* band width was, of course, one of the chief pieces of evidence for the sliding-filament theory in 1954[9,10], but at that time there was no danger of supposing the I bands to be fluid since filaments passing through the I bands had been demonstrated in several investigations with the electron microscope[15,16,8].

Another aspect of the activity of nerve and muscle where an analogy—perhaps a rather far-fetched one—can be drawn is the mechanism of spatial spread. In the case of nerve, the local-circuit mechanism was proposed by Hermann in the 1870's[17], and was generally accepted long before a quantitative demonstration of its adequacy was finally given by Hodgkin in his 1937 papers[18]. In the case of muscle, the existence of a problem is less obvious. In the older literature there are occasional discussions [19,20] of the question how excitation—an event essentially confined to the surface membrane—can influence the myofibrils in the middle of a muscle fibre so that they begin to contract, but the question did not arise in an acute form until after the second war, when A. V. Hill[21] showed that simple inward diffusion of an activating substance from the surface membrane would not be fast enough to account for the speed of onset of contraction in a twitch. The solution came in 1955–59, partly through experiments on living fibres designed to investigate this question[22], and partly through the discovery of the transverse tubular system with the electron microscope[23,24,25]. Here again the structural side of the evidence had been found with the light microscope at about the turn of the century but was later forgotten: the transverse network was beautifully shown in Golgi preparations of muscle[26,27], and the fact that it consists of tubules open to the external solution was shown (in heart muscle) by injections of Indian ink[26]. But, just as in the matter of the constancy of *A* band width, the old work was quite unknown at the time of the rediscovery. However, at the time of the early observations there was no evidence that the function of this network of tubules was in any way connected with the inward spread of activity. One of the names by which the network was known was "Trophospongium", implying a nutritive function, which, in the absence of direct evidence, is very plausible for a system of tubes by which solutes in the external solution can reach the middle of the fibre.

I mentioned Bernstein as someone who had worked in the fields of both conduction and contraction. In the former, his contributions are universally known, but I think many people are unaware that he was also the author of a long paper on the mechanism of contraction in 1901[28]. This paper consists of a detailed and semi-quantitative discussion of a surface-tension theory of contraction. He shows that forces of the same order of magnitude as the isometric tetanus tension might be produced by surface tension acting within submicroscopic tubes in the *A* bands. The point I want to emphasise is that there is a close parallel between the evidence he brings forward in favour of surface tension as the origin of muscular force, and certain of the evidence which he gave for the ionic theory of bioelectric potentials. In each case thermodynamic arguments were important. For the ionic theory, there was the proportionality between injury potential and absolute temperature, and the cooling of the electric organ when it discharges through an external load; for the surface-tension theory, there was the fact that at least in some muscles the tension developed in a twitch is greater the lower the temperature, and surface tension is about the only familiar physical force which varies in this direction with change of temperature. At the present time none of these arguments is quite as straightforward as they may have seemed sixty or seventy years ago. Changes of permeability contribute to the variation of resting potential with temperature; the cooling of the electric organ is only partly explained in the way Bernstein suggested[29]; while the decrease of twitch height with rise of temperature is clearly due to a decrease in the degree to which the contractile apparatus is turned on, not to a drop in the maximum tension it is capable of producing. If Bernstein was a little lucky in his application of the first two of these arguments, everyone will agree that he deserved it; and he was clearly unlucky in his argument about surface tension.

I have been speaking up to now about ideas and discoveries; there are also parallels between the methods that have been employed in the fields of nerve and muscle, and here I am sure that muscle physiologists have adapted to their own use the mechanical equivalents of electrical methods that had been applied to nerve, notably by Cole. The complex impedance loci so sucessfully used by Cole in his high-frequency analysis of the electrical properties of the surface membranes of nerve fibres and other cells, have been employed since the war on muscle, in relation both to electrical and to mechanical behaviour. In the former case, the complications due to the tubular systems of muscle have been elucidated by Fatt and Falk[30,31], while in the latter, Pringle's analysis[32] of the mechanical oscillatory behaviour of the "asynchronous" flight muscles of certain insects has been carried out almost entirely by sinusoidal analysis, with the results again presented as impedance loci in the way that was made familiar to physiologists by Cole's work. In a similar way, the application of feedback methods to control the length or tension changes in a muscle fibre[33] has followed on from the analogous methods introduced by Cole and by Hodgkin for controlling the membrane potential or current in a giant nerve fibre.

The discovery of these giant nerve fibres in squids, by J. Z. Young in 1936[34], was a key factor in much of the progress in nerve physiology since that date because many of the experimental procedures would have been impracticable without them. A smaller role has been played in muscle physiology by fibres of exceptional size. It is true that several very important experiments have depended on the use of very large muscle fibres from crustaceans, notably the measurement of intracellular pH[35], control of intracellular calcium concentration[36], and the recording of the time course of intracellular calcium concentration after stimulation, by means of the protein aequorin which luminesces in response to calcium ions even at very low concentration[37]. But it would be an exaggeration to claim that giant fibres have been as important in muscle as in nerve physiology. Perhaps the kind of muscle preparations which will take a position equivalent to that of the squid giant axon are those in which direct access to the contractile apparatus has been achieved by eliminating the surface membrane, whether by chemical means, as in Szent-Györgyi's glycerol-extracted muscle, or mechanically, as in Natori's stripped, or "skinned", muscle fibres[38]. Such preparations have already been of great importance and their value is likely to increase in the future.

1. Keynes, R. D., J. M. Ritchie, and E. Rojas (1971) *J. Physiol.* **213,** 235.
2. Engelhardt, W. A., and M. N. Ljubimova (1939) *Nature, Lond.*, **144,** 668.
3. Engelhard, W. A., M. N. Ljubimova, and R. A. Meitina (1941) *C. R. Acad. Sci. URSS*, **30,** 644.
4. Needham, J., S.-C. Shen, D. M. Needham, and A. S. C. Lawrence (1941) *Nature, Lond.*, **147,** 766.
5. Hodgkin, A. L., and B. Katz (1949) *J. Physiol.* **108,** 37.
6. Bernstein, J. (1902) *Pflügers Archiv*, **92,** 521 (p. 560).
7. Cole, K. S., and H. J. Curtis (1939) *J. gen. Physiol.* **22,** 649.
8. Huxley, H. E. (1953) *Biochim. biophys. Acta*, **12,** 387.
9. Huxley, A. F., and R. Niedergerke (1954) *Nature, Lond.*, **173,** 971.
10. Huxley, H. E., and J. Hanson (1954) *Nature, Lond.*, **173,** 973.
11. Overton, E. (1902) *Pflügers Archiv*, **92,** 346.
12. Overton, E. (1904) *Pflügers Archiv*, **105,** 176 (pp. 251–7).
13. Krause, W. (1869) *Die motorischen Endplatten der quergestreiften Muskelfasern.* Hannover: Hahn.
14. Huxley, A. F. (1957) *Progr. Biophys.* **7,** 255.
15. Hall, C. E., M. A. Jakus, and F. O. Schmitt (1946) *Biol. Bull.* Woods Hole, **90,** 32.
16. Draper, M. H., and A. J. Hodge, (1949) *Aust. J. exp. Biol. med. Sci.* **27,** 465.
17. Hermann, L. (1879) *Handbuch der Physiologie*, Bd. 1, Abth. 1, p. 256.
18. Hodgkin, A. L. (1937) *J. Physiol.* **90,** 183 and 211.

19. Retzius, G. (1890) *Biol. Untersuchungen, N. F.* **1,** 51.
20. Tiegs, O. W. (1924) *Aust. J. exp. Biol. med. Sci.* **1,** 11.
21. Hill, A. V. (1949) *Proc. Roy. Soc., B.* **136,** 399.
22. Huxley, A. F., and R. E. Taylor (1955) *Nature, Lond.,* **176,** 1068.
23. Porter, K. R., and G. E. Palade (1957) *J. Biophys. Biochem. Cytol.* **3,** 269.
24. Huxley, A. F. (1959) *Ann. N. Y. Acad. Sci.* **81,** 446.
25. Andersson-Cedergren, E. (1959) *J. Ultrastr. Res.,* Suppl. 1.
26. Nyström, G. (1897) *Arch. Anat. Physiol.* (*Anat. Abt.*) 1897, 361.
27. Veratti, E. (1902) *Mem. R. Ist. Lomb. Sci. Lett.* (*Cl. Sci. math. nat.*), **19,** 87.
28. Bernstein, J. (1901) *Pflügers Archiv,* **85,** 271.
29. Aubert, X., and R. D. Keynes (1968) *Proc. Roy. Soc., B.* **169,** 241.
30. Fatt, P. (1964) *Proc. Roy. Soc., B.* **159,** 606.
31. Falk, G., and P. Fatt (1961) *Proc. Roy. Soc., B.* **160,** 69.
32. Machin, K. E., and J. W. S. Pringle (1960) *Proc. Roy. Soc., B.* **152,** 311.
33. Gordon, A. M., A. F. Huxley, and F. J. Julian (1963) *J. Physiol.* **167,** 42P; (1966) *J. Physiol.* **184,** 143.
34. Young, J. Z. (1936) *Proc. Roy. Soc., B.* **121,** 319.
35. Caldwell, P. C. (1958) *J. Physiol.* **142,** 22.
36. Ashley, C. C., P. C. Caldwell, A. G. Lowe., C. D. Richards, and H. Schirmer (1965) *J. Physiol.* **179,** 32P.
37. Ashley C. C., and E. B. Ridgway (1968) *Nature, Lond.,* **219,** 1168; (1970) *J. Physiol.* **209,** 105.
38. Natori, R. (1954) *Jikei med. J.* **1,** 119.